Education Division

Core Programming Topic at the 2012 AIChE Annual Meeting

Pittsburgh, Pennsylvania, USA
28 October - 2 November 2012

ISBN: 978-1-62276-724-3

Printed from e-media with permission by:

Curran Associates, Inc.
57 Morehouse Lane
Red Hook, NY 12571

Some format issues inherent in the e-media version may also appear in this print version.

Printed by Curran Associates, Inc. (2013)

For permission requests, please contact AIChE
at the address below.

AIChE
3 Park Avenue
New York, NY 10016-5991

Phone: (203) 702-7660
Fax: (203) 775-5177

www.aiche.org

Additional copies of this publication are available from:

Curran Associates, Inc.
57 Morehouse Lane
Red Hook, NY 12571 USA
Phone: 845-758-0400
Fax: 845-758-2634
Email: curran@proceedings.com
Web: www.proceedings.com

TABLE OF CONTENTS

(3a) Nanostructured Materials for Advanced Sensing Platforms

Evan K. Wujcik, Department of Chemical & Biomolecular Engineering, University of Akron, Akron, OH

Research

In past decades, nanostructured materials have shown promise of revolutionizing a wide number of areas of needed development, including theranostic, electronic, and photonic materials. Applications of these fields include yet wider range of specialties, all of which can be incorporated into sensor material design. This shall be the focus of my research laboratory; fundamental material properties and applications of nanomaterials to sensor design. I believe that my background in inorganic nanomaterial synthesis (for photovoltaics and catalysis), cell membrane-nanomaterial interactions, materials & surface science, sensor design, and management have prepared me for a promising career heading a research laboratory. This research laboratory shall span a spectrum of expertise, all of which shall fall under the umbrella of sensor design.

Teaching

In my years of being a graduate student, I have assisted teaching a number of courses and labs, and this fall will be assisting in the development of a chemical & biomolecular engineering elective course. Being the recipient of the University of Akron's Department of Chemical & Biomolecular Engineering **Outstanding Teaching Assistant Award** (Fall 2011), I understand that teaching graduate and undergraduate courses requires a certain level of dedication, which I plan to show through my outreach, elective development, and general teaching philosophy of "talking with" students instead of "talking at" students. I will apply this doctrine to both my course and research instructing, which I shall be presenting in further detail in this poster session.

Evan K. Wujcik, M.S., M.B.A. (LinkedIn Profile: http://www.linkedin.com/in/evankwujcik)

Advisor: Assistant Professor of Chemical & Biomolecular Engineering Chelsea N. Monty

Be on the lookout for my other 2012 AIChE oral/poster presentations:

1. Oral presentation at the 2012 AIChE National Conference (November) in Pittsburgh, PA, U.S.A., Title: Diagnostic Sodium Ion Sensor for the Real-Time Screening and Diagnosis of Cystic Fibrosis
2. Oral presentation at the 2012 AIChE National Conference (November) in Pittsburgh, PA, U.S.A., Title: Membrane Preconcentration Technique for the Colorimetric Detection of Trihalomethanes in Water
3. Poster presentation at the 2012 AIChE National Conference (November) in Pittsburgh, PA, U.S.A., Title: Diagnostic Na+ Sensor for the Real-Time Screening and Diagnosis of Cystic Fibrosis

(3ab) Harnessing Degron-Based Substrates As Tools for the Development and Evaluation of Novel Chemotherapeutics Targeting the Ubiquitin Proteasome System

Adam Melvin, Chemistry, Univeristy of North Carolina, Chapel Hill, NC

Aberrant protein expression can lead to dysregulation of essential cell signaling pathways culminating in a variety of cancers. These abnormal protein levels are the result of improper activity of the ubiquitin proteasome system (UPS), specifically the family of E3 ubiquitin ligases which are responsible for the recognition and degradation of target proteins. E3 ligases, the final component of a cascade of E1, E2, and E3 enzymes, recognize a specific amino acid sequence, or degron, on a target protein which facilitates their binding and, ultimately, the transfer of the small protein ubiquitin to a proximal lysine residue. Ubiquitin polymerization results in target protein recognition by the proteasome, where it is degraded into small peptide fragments. The large number of E3 ligases (>600) confer the specificity necessary for this degradation pathway and, as such, offer an attractive target for potential chemotherapeutics. For example, blocking expression of the E3 ligase Hdm2 has been shown to stabilize p53, a powerful tumor suppressor. Further, in recent years it has been demonstrated that improper E3 ligase signaling results in defects in a variety of cellular responses such as the cell cycle, cell migration, epithelial-mesenchymal transition, intracellular trafficking, and DNA damage repair.

In my research I propose to harness small (~25-30 amino acid) degron-based substrates as novel tools for understanding the intracellular dynamics of E3 ligase/degron binding. While current research mainly focuses on the identification of new E3 ligases and degron targets, I propose to conduct an in-depth study of E3 ligase signaling dynamics across a wide range of cells, protein-protein interactions, and different classes of degrons (i.e. phospho-degrons, oxygen-dependent degrons, and the N-end rule pathway) to identify potential targets for reporters and inhibitors. As reporters, the degron-based substrates could be utilized to evaluate a cellular response to current E3 ligase-based chemotherapeutics (such as MLN-4924). As inhibitors, I believe that degron-based substrates could provide a new method for competitive inhibition of unregulated E3 ligases. As previously mentioned, the binding of the E3 ligase to the target protein is the essential step for the targeted degradation of a protein which can result in improper protein clearance. By competitively targeting the E3 ligase, I aim to abrogate this protein clearance while simultaneously minimizing off-target effects commonly associated with small molecule based therapeutics. Many of the current UPS-targeted chemotherapeutics mainly focus on E3 ligase/degron combinations involved in the cell cycle (i.e. APC/C, p53, and Cdc4); however an increasing amount of evidence suggests that uncontrolled E3 ligase activity plays an important role in the directed migration of cells in response to extracellular gradients of chemoattractants, or chemotaxis. Recent work has identified potential E3 ligases such as c-Cbl, SMURF1/2, and Hdm2 as regulators of cell migration; although many of these studies use bulk, qualitative metrics such as the scratch assay or the Boyden chamber. I intend to expand on this work by taking advantage of the degron-based substrates as both reporters and inhibitors of E3 ligases involved in cell migration. Finally, to evaluate the spatial and temporal dynamics of E3 ligases on chemotaxis I will develop novel microfluidic devices designed to deliver controllable gradients of chemoattractants to both mesenchymal and amoeboid cell migration. Similar to the chemotherapeutics developed against kinases to specifically target cancer, I believe that specifically targeting the E3 ligase/degron interaction will open the door to a new level of chemotherapy drugs targeting an array of cancers.

(3ac) Modeling Complex Structures in Nucleic Acids

Margaret C. Linak, Chemical Engineering and Materials Science, University of Minnesota, Minneapolis, MN

Since the discovery of DNA, researchers have been attempting to decode the detailed structure, properties, and abilities of this molecule. At first approximation, DNA can be thought of as a ling, regular, double-stranded helix encoding the genomic information of life. However, on closer analysis DNA has been found to take on a wide variety of complex shapes and function both *in vitro* and *in vivo*. DNA can be single-, double-, triple-, and even quadruple-stranded in nature and can bind in both the Watson-Crick conformations and also in a wealth of non-canonical configurations that add to its inherent flexibility, structure, and activity. Elucidating the varied structures and behaviors of DNA has historically been an experimental endeavor, due in large part to the difficulties in capturing nucleic acid's complex motions and functions in a tractable computation model. However, as the applications of DNA expand and computational power increases, simulation models are playing an increasingly important role in DNA understanding and engineering.

Throughout the course of my doctoral work, I developed new nucleic acid models able to successfully capture the complex structures and behaviors of DNA and RNA that previously have been left out of such coarse grained models. I also designed specific experimental tests able to validate nucleic acid simulation models. My new models are able to reproduce simple systems such as experimentally gathered single-stranded DNA melting curves and B-form double-stranded DNA. In addition, much more complex systems such as double-stranded DNA twisting behavior (P- and S-DNA transitions), triple-stranded conformations found in H-DNA and during strand invasion, and quadruple-stranded structures such as G-quartets, can be modeled. These complex structures are vital components of cellular recognition, transcription, and regulatory pathways, DNA repair, and telomeres and may play a role in understanding wider, multi-faceted issues in cancer, aging, and disease. Although many of these complex structures have yet to be characterized and there is only indirect experiment evidence for their functions, molecules relying on non-canonical interactions have begun to be used in nanotechnology, genetic engineering, and therapeutic applications. I anticipate that more in depth understanding of nucleic acids will provide a platform upon which I can harness DNA's power and ability in order to engineer these molecules for novel applications.

(3ad) Whole Organ Engineering

Basak Uygun, Center for Engineering in Medicine, Massachusetts General Hospital, Harvard Medical School, Shriners Hospital for Children, Boston, MA

There is a critical shortage of donor organs with the total organ waiting list currently at 110,000 requests and the number is increasing by 5% every year; the total cost of organ failure to the US is estimated at $200B/yr. Tissue engineering offers a solution for the donor organ deficit problem by providing the tools to prepare artificial organ/tissue replacements. However, despite decades of efforts, clinical application has a single success in a relatively simple, sheet-like structure: tissue-engineered skin. The major vertical gap preventing engineering of other, more complex organs is the lack of an ideal transplantable scaffold that has all the necessary microstructure and extracellular cues for cell attachment, differentiation, function and vascularization, which has so far proven difficult to manufacture using a grounds-up approach. Recently, I developed a cadaveric organ decellularization protocol to create a whole-liver scaffold for engineering liver grafts. The decellularized organ matrix scaffold is an ideal platform for preparation of organ grafts because it represents the tissue specific extracellular matrix (ECM) composition, which aids in cell engraftment and function. The perfusion decellularized organ matrix scaffold also has the tissue specific microarchitecture that facilitates perfusion of the graft and subsequently allows delivery of nutrients and removal of waste materials from the cells that are seeded inside. The preserved vascular bed allows connection of the graft directly to the recipient's blood torrent upon transplantation, making it surgically ideal. My *research trajectory* focuses on developing technologies to create tissue engineered organ grafts for transplantation using whole organ scaffolds. Reaching this goal requires parallel research in multiple intertwined projects that are cross cutting across tissue engineering, developmental biology, and surgical research. Therefore, my objective as an independent research faculty is to create an interdisciplinary, collaborative *Engineering in Regenerative Medicine* program that incorporates expertise of engineers, scientists and surgeons where the main research thrusts are 1) engineering of transplantable organ grafts, 2) investigation of matrix-stem cell interactions and 3) transplantation of engineered organ grafts.

(3ae) Hygro-Responsive Membranes for Effective Oil-Water Separation

Arun K. Kota, University of Michigan, Ann Arbor, MI

Several recent events, including the Deepwater Horizon oil-spill in the Gulf of Mexico, have highlighted the difficulty of effective oil-water separation. Efficient, cost-effective processes for oil-water separation, especially in the presence of dispersants (or surfactants), are greatly desired. Surfactant-stabilized mixtures of oil and water are classified by the diameter (d) of the dispersed phase as: i) free oil and water, if $d > 150$ μm, ii) a dispersion, if $20\ \mu m < d < 150\ \mu m$, or iii) an emulsion, if $d < 20$ μm. Conventional gravity separators and skimming techniques are incapable of separating emulsions. Membrane-based technologies are attractive for *demulsification*, i.e., the conversion of an emulsion to a free oil-water mixture, because they are relatively energy-efficient, cost-effective, and applicable across a wide range of industrial effluents. However, for complete oil-water *separation*, demulsification is typically followed by either gravity separation or skimming.

In this work, we have developed superhydrophilic *and* superoleophobic membranes that allow for the separation of a range of different oil-water mixtures including free oil and water, oil-in-water emulsions, water-in-oil emulsions and any combination of these phases in a single unit operation, with greater than 99.9% separation efficiency. Our separation methodology is solely gravity driven and consequently it is expected to be one of the most energy-efficient separation processes for different oil-water mixtures. Further, our dip-coating based membrane fabrication is easily scalable and we demonstrate the separation of several liters of oil-water mixtures using a scaled up apparatus. In addition, we also demonstrate continuous separation of oil-water emulsions for over 100 hours without a decrease in flux. We envision that our separation methodology will be useful for numerous applications including the clean up of oil-spills, wastewater treatment, fuel purification and the separation of numerous commercially relevant emulsions.

(3af) Combining Computation and Experiment to Uncover Environment Friendly Solutions to Energy Problems

Ki Chul Kim, Chemical and Biological Engineering, Northwestern University, Evanston, IL

The world's dependence on fossil fuels has lead to the need for alternate sources of energy as supplies dwindle, as well as a growing need to remove harmful compounds from the air. Solar energy and hydrogen energy are promising candidates for supplanting fossil fuels while novel adsorbents like metal-organic frameworks (MOFs) are promising materials for removing harmful gases. To date, my research has been concerned with both aspects of the fossil fuel problem. Specifically, my M.S. and Ph.D studies were concerned with solar energy and hydrogen energy. In order for solar cells to be a feasible alternative to fossil fuels, they need to have a competitive power conversion efficiency. By adding partially sulfonated ionomer to the active layer of the solar cell, my research led to an increase in power conversion efficiency. Regarding hydrogen energy, my research was focused on understanding the thermodynamics of metal hydride reactions for hydrogen storage applications. Specifically, the goal was to screen thermodynamically promising metal hydride reactions from a full database of metal hydride mixtures using first-principles calculations. The large-scale screening ultimately provided a number of promising single-step or multi-step metal hydride reactions. Currently, my postdoctoral research is investigating MOFs for the removal and separation of harmful gases. I have used computational methods to screen and assess functional groups that could be incorporated into MOF ligands to preferentially adsorb harmful gases under humid conditions. In addition, I am also assessing new methods for characterization of MOFs, as well as assessing MOFs for various gas separations. Inspired by my research background, my future research will be focused on producing energy sources, such as methane or methanol, through carbon dioxide reduction on nanosized clusters or other shapes of metals and metal-based compounds. The nanoscale particles will have different chemical properties from bulk-based surfaces, and thus understanding the effect of the particle size, shape, and type on the carbon dioxide reduction would be an interesting topic for environment-friendly energy applications. Specifically, I have a plan to examine the changes in the mechanisms and activation barriers of the carbon dioxide reduction, increasing the particle size or changing the particle shape and the type of metals.

(3ag) Dynamics and Patterning of Complex Fluids with Energy and Environmental Applications

Cari S. Dutcher, Chemical Engineering and Materials Science, University of California at Davis, Davis, CA

Higher-order fluidic patterning in complex fluids can be engineered for critical contributions in energy and environmental applications, from energy savings for large scale fluid transportation, to improved thermodynamic treatment in climate modeling, to enhanced particulate removal in industrial waste water processing. The patterning used for these applications can be observed in many complex solutions, on many scales, and are induced by such phenomenon as:

1. Viscoelastic effects seen with flow of polymeric solutions in curved streamlines

2. Thermalchemical effects seen in phase partitioning in atmospheric aerosols

3. Electrohydrodynamic effects seen in aggregate formation in colloidal suspensions

My research interests include the study of dynamics, bifurcations, and patterning observed with complex and responsive fluids, using experimental techniques, phenomenological scaling, and rigorous analytic mathematical modeling. Coming from a multi-disciplinary research background, I anticipate establishing a future research program for the study of dynamics and patterning of complex fluids, with the end goal of application-orientated tuning of the fluid behavior and properties.

I completed my Ph.D. in Chemical Engineering at the University of California, Berkeley in May 2009 while working with Professor Susan J. Muller. My doctoral research effort primarily focused on flow instabilities with complex (usually viscoelastic) solutions. The research effort aided in quantitative understanding and optimization of the polymer-induced turbulent drag reduction mechanism, a phenomenon that leads to significant energy savings for large scale fluid transportation. We studied the effects of polymeric additives in Taylor-Couette (TC) flow (flow between concentric, rotating cylinders), rather than in pipe flow, where the flow field transitions directly from laminar to turbulent flow with increasing Reynolds number. In TC flow, as the angular velocities of the inner and outer cylinders are varied, a rich cascade of distinct steady flow patterns, of varied complexity, can be accessed between unidirectional shear flow and fully developed turbulent flow. The use of complex, yet well-characterized, three-dimensional hydrodynamic patterning available in Newtonian TC flow as the background flow field allowed us to probe the effect of the polymeric drag reducing agent on isolated flow features and instability types. The stability planes and equilibrium flow states experimentally established in my PhD thesis, resulting from both centrifugal (inertial) and elastic (polymeric) forces, as well as the phenomenologic scaling models developed for both Newtonian and non-Newtonian fluids in TC geometries, elucidated the role of viscoelasticity in polymer-induced turbulent drag reduction.

Currently, I am a NSF-AGS postdoctoral research fellow at the University of California, Davis, working with Professors Anthony S. Wexler and Simon L. Clegg (Air Quality Research Center) and Professor William D. Ristenpart (Chemical Engineering). My postdoctoral research effort primarily focuses on thermodynamic and electrohydrodynamic properties of multi-component electrolyte containing solutions and aerosols. At the Air Quality Research Center, my work has integrated multilayer adsorption isotherms from gas and solid chemistry, and hydration shell theory and electrostatics from solution chemistry, to form an improved statistical thermodynamic treatment of the Gibbs energy, solvent and solute activity, solute concentration and surface tension for complex solutions and atmospherically relevant aerosols. The resultant thermodynamic model yields unprecedented agreement between the solute concentration and osmotic coefficients for solutions over the entire solute concentration range. Understanding the thermal and chemical forces on the stability of phase partitioning and both the molecular- and micro-scale patterning in atmospheric aerosols is essential for the exact thermodynamic predictions required for accurate climate modeling. Simultaneously, in the department of Chemical Engineering, I have been experimentally investigating colloidal aggregation and patterning, through electrohydrodynamic (EHD) manipulation. Colloidal aggregation through

EHD flow has important energy-saving implications for use of electric fields for large-scale separation processes involving suspended particles.

In addition to being passionate about understanding and tuning complex fluids, I am also passionate about teaching and mentoring graduate and undergraduate students. As a faculty member, I anticipate using research as a teaching opportunity, in the classroom, in the laboratory, and in the broader community.

(3ah) Density-Functional Theories for Solvent-Free Nanoparticle-Organic Hybrid Materials

Hsiu-Yu Yu, School of Chemical and Biomolecular Engineering, Cornell University, Ithaca, NY

Nanoparticle-organic hybrid materials (NOHMs) consist of 10 nm diameter spherical inorganic core particles surface-functionalized with oligomeric organic molecules. Although these systems contain no added solvent, they exhibit fluid behavior with the fluidity provided by the attached oligomers, which in turn mediate the interparticle forces and affect the equilibrium properties and dynamic behavior of the bulk system. The absence of a solvent, the small size of the nanocores and oligomers, and the incompressibility of the tethered oligomeric fluid make the oligomer-mediated interactions non-pairwise-additive. Oligomers from many neighboring cores compete to uniformly fill the interstitial space.

In my PhD research, I developed a classical density-functional approach for model hard spheres with tethered bead-spring oligomers to predict the equilibrium properties and the transport properties of solventless, pure NOHMs. The simple coarse-grained model allows one to have a direct description of the system free energy as a functional of the probability densities of cores and oligomers without assuming pairwise additivity. Under a weak oligomeric-field approximation valid when the oligomer radius of gyration is much greater than the core radius, the equilibrium distribution function of the cores and the concentration field of oligomers are determined semi-analytically. Using the resultant, I obtained the static structure factor of NOHMs that can be utilized to characterize experimental systems as well as the solvent capacity of the suspension as the solute releases the entropic penalty of the space-filling oligomers. This thermodynamic driving force for solute uptake makes NOHMs a candidate for carbon capture showing good CO_2 selectivity over N_2 and CH_4 as the lower affinity of CO
more of the free energy. Meanwhile, since many neighboring particles cooperate in filling the space, it is found that NOHMs can remain disordered even if the core volume fraction is above the freezing transition point of hard-sphere suspensions. Finally, transport properties such as the long-time self-diffusivity and linear viscoelastic behavior are determined by solving for the non-equilibrium probability density function for pairs of particles subjected to a weak applied flow and many-body intercore potential of mean force without hydrodynamic interactions.

(3ai) Advancing Student Engagement in Early Engineering Education

Victoria Goodrich, Department of Chemical and Biomolecular Engineering, University of Notre Dame, Notre Dame, IN

Simply telling students to be engaged in engineering material (or that they will be using the material someday) is rarely enough of a motivating factor for their education. Instead, all professors can practice a variety of techniques to create highly motivated and engaged students. At Notre Dame, we have focused these efforts on large scale projects that require all engineering students to complete open-ended design problems with highly beneficial learning opportunities. In the first-year engineering course, a two-semester series contains three-projects which culminate in an open-ended design project, students create both computational models and physical systems of their own design. While many have not had the requisite classes to complete a highly functioning model, they are able to draw upon early math, chemistry and physics courses in an engineering design environment. In chemical engineering, project based learning continues through the sophomore year in the form of computation projects. Again, students are given freedom to choose projects based on their own interests, thereby designing their own educational goals for the semester. Unlike traditional lecture courses with only exams and homework problems for assessing student progress, these courses allow students to demonstrate understanding of engineering principles across classes and applied to systems that they find personally motivating.

I present here the current progress for project based learning at the University of Notre Dame, including specific projects from the first and second year courses taken by all chemical engineering students. In addition, I present other College of Engineering initiatives towards furthering student engagement and encouraging students to take an active role in their education and career goals. These projects include: creating electronic portfolios, introducing individual learning journals, providing undergraduate research opportunities, and crafting ungraded, computer based study aids.

(3ak) Hard Tetrahedra: Entropy, Geometrical Anisotropy and Structural Complexity

Amir Haji-Akbari, Department of Chemical and Biological Engineering, Princeton University, Princeton, NJ

It is the dream of every material scientist to design materials with desirable tunable properties that spontaneously assemble themselves from smaller bits. Self-assembly of such nano- and colloidal-scale bits is driven by the laws of thermodynamics rather than laborious top-down manufacturing. Recent progress in fabrication techniques has made possible a wide variety of anisotropic nano- and colloidal particles that mimic atoms and molecules in the self assembly of complicated structures. These novel building blocks can be anisotropic in both interaction and shape and can form structures not observed in atomic and molecular systems. A deep understanding of how anisotropy translates into structural complexity is pivotal in devising engineering guidelines for designing self-assembling bits. I am particularly interested in how shape anisotropy alone can give rise to structural complexity and my research plan is to use computational statistical mechanics to understand this relationship.

Hard particles are the most suitable models for studying the exclusive effect of building block geometry on thermodynamic phase behavior. In the realm of atoms and molecules the hard potential mimics excluded volume interactions, and has, for decades, been successfully used for modeling dense states of matters such as liquids and solids. Since all permissible configurations of a hard particle system have the same potential energy, the thermodynamic behavior of such systems is purely determined by entropy. Although entropy is generally associated with 'lack of order', it is a well-known fact that hard particles can assemble into ordered structures that are stabilized by entropy. Such 'entropy-driven disorder-order transitions' have been observed in a wide variety of systems and seem to be ubiquitous in nature [1]. Computational studies of hard particle systems are also interesting due to the intimate relationship that exists between the problem of thermodynamic stability of hard particle systems and the packing problem in geometry. At the limit of infinite pressure, the densest possible arrangement of hard particles is thermodynamically stable. However finding the densest packing(s) of a particular geometric object is often analytically unapproachable, and thus numerical approaches are becoming popular ways of studying packing problems.

Hard tetrahedra, the main focus of my PhD dissertation, are the simple yet perplexing examples of geometrically anisotropic building blocks that self-assemble into complicated structures. The regular tetrahedron is the simplest of the five Platonic solids. In my research, Monte Carlo simulations revealed that these geometrically simple building blocks assemble into a dodecagonal quasicrystal, at sufficiently high densities due to entropy alone [2]. These quasi-periodic arrangements are surprisingly dense. An approximant of the quasicrystal that we constructed manually is an even denser packing and constituted the world record at the time of its discovery. Quasicrystals are an exotic class of solids with long-range order that lack periodicity. In 2011, the Nobel Prize in Chemistry was awarded to Danieh Schechtman for the discovery of the first quasicrystal [3]. We found that the quasicrystal forms robustly from hard tetrahedra, and subsequently found the phase again in a system of hard triangular bipyramids (TBPs), which are dimers of hard tetrahedra [4]. We also confirm that in both systems, the quasicrystal approximant is more stable at intermediate densities than the now known densest packing of these polyhedra [4, 5]. This research suggests the rich and subtle role that entropy plays in stabilizing complex structures and the potential that geometric anisotropy has in inducing complex forms of order.

1. Daan Frenkel. *Entropy-driven phase transitions*. **Physica A**, 263:26–38, 1999.

2. Amir Haji-Akbari, Michael Engel, Aaron S. Keys, Xiaoyo Zheng, Rolfe Petschek, Peter Palffy-Muhoray, and Sharon C. Glotzer. *Disordered, quasicrystalline and crystalline phase of densely-packed tetrahedra*. **Nature**, 462:773-777, 2009.

3. D. Shechtman, I. Blech, D. Gratias, and J. W.Cahn. *Metallic phase with long- range orientational order and no translational symmetry*. **Phys. Rev. Lett.**, 53:1951–1953, 1984.

4. Amir Haji-Akbari, Michael Engel, and Sharon C Glotzer. *Degenerate quasicrystal of hard triangular bipyramids*. **Phys. Rev. Lett.**, 107:215702, 2011.

5. Amir Haji-Akbari, Michael Engel, and Sharon C Glotzer. *Phase diagram of hard tetrahedra*. **J. Chem. Phys.**, 135:194101, 2011.

(3al) Design of Heterogeneous Catalysts for the Conversion of Biomass Into Fuels and Chemicals

Jean Marcel R. Gallo, Chemical & Biological Engineering, University of Wisconsin-Madison, Madison, WI

There has been an increasing interest in renewable sources of carbon for the production of transportation fuels and chemicals due to the diminishing petroleum resources. Cellulose and hemicellulose, components of biomass, can be depolymerized into the monossacharides glucose and xylose, which yields a variety of important platform molecules for the production of fuels and chemicals. For the economical viability and sustainability of biomass conversion processes, development of selective and stable solid catalysts is necessary. My research interest is in the design of novel functional heterogeneous catalysts with appropriate structural and surface properties for the production of biomass-derived compounds. I obtained my undergraduate and Master's degree in Chemistry at the State University of Campinas (Brazil) under the supervision of Prof. Ulf Schuchardt working on the preparation of catalysts for olefin epoxidation. As a graduate student, under the supervision of Prof. Leonardo Marchese at the University of Eastern Piedmont (Italy) and Prof Heloise O. Pastore at the State University of Campinas (Brazil), I studied the surface properties of solid acid catalysts and also developed new electrocatalysts for Methanol Fuel Cell. My post-doctoral research, under the guidance of Prof. James A. Dumesic and Prof. Manos Mavrikakis involves preparing hydrothermally stable solid catalysts and studying methods to improve their selectivities for the conversion of monosaccharides into 5-hydroxymethylfurfural, an important platform chemical. My aim is to create a strong research group focusing on materials science and catalysis, combining the experience I have acquired over the years in various research groups.

(3am) Image-Based Fluid Dynamics for Biomedicine and Beyond

Roman S. Voronov, Chemical and Biomolecular Engineering, University of Pennsylvania, Philadelphia, PA

Research Interests:

Image-Based Modeling is an emerging multi-disciplinary field that finds numerous applications in many diverse industries ranging from Biomedicine to Enhanced Oil Recovery. By using state-of-the-art imaging techniques in conjunction with high performance computer modeling the image-based approach enables researchers to gain insights into systems that otherwise would have been inaccessible. For example, using imaging techniques, such as 3D Computed Tomography or Confocal Microscopy, allows one to obtain exact geometries of complex structures of interest, such as that of living porous media in-vivo or that of inorganic porous media (such as rock formation). Perhaps the main advantage of this modeling approach is that the implementation of the obtained structures in simple robust computer simulations gives practical results that are based on actual experimental data and do not suffer from mathematical complexity or from over-simplifications that are often necessary in other modeling techniques.

Successful Proposals: NIH F32, American Heart Association Postdoctoral Training Grant, XSEDE/Teragrid.

Postdoctoral Project: "*Intra-Thrombus Chemo-Transport and Local Stress Mechanics Under Flow.*"

Under supervision of Scott L. Diamond & Skipp Brass, Chemical and Biomolecular Engineering/School of Medicine, University of Pennsylvania

PhD Dissertation: "*Fluid Shear Stress and Nutrient Transport Effects Via Lattice Boltzmann and Lagrangian Scalar Tracking Simulations of Cell Culture Media Perfusion through Artificial Bone Tissue Engineering Constructs Imaged With MicroCT.*"

Under supervision of Dimitrios V. Papavassiliou & Vassilios I. Sikavitsas, Chemical, Biological and Material Engineering, University of Oklahoma

Research Experience:

My academic career path has been a blend of many fields of science and engineering. My formal training was in simulation, which I have learned on many different physical scales: Density Functional Theory (femto), Molecular Dynamics (nano), and Lattice Boltzmann Fluid Dynamics (meso). Yet, projects that I worked on were always close collaborations with experimentalists: Friction Reducing Surfaces for the NAVY, Bone Tissue Engineering, Enhanced Oil Recovery and Blood Coagulation/Thrombosis. As a result of these collaborations I have acquired invaluable and versatile experience in many research areas: material science, biology, medicine, biochemistry, biomedical imaging, and engineering design, just to name some. Moreover, as a result of intensive training at various supercomputing centers (OU Supercomputing Center for Education & Research - OSCER, Texas Advanced Computing Center - TACC and Lawrence Berkeley National Laboratory – LBNL), I became proficient in high performance computer modeling.

Teaching Experience:

Aside from my research career, I also have extensive teaching experience. I spent a semester teaching basic science to highschool students as a part of a NSF fellowship; I TAed and guest-lectured undergraduate lecture and laboratory courses in Chemical Engineering both the University of Oklahoma and at the University of Texas (for which I won a Campus Wide Best TA Award); I was repeatedly chosen by my department to be the "Pro/II Expert" (duties included lecture undergrads, holding office hours and developing course material); and, lastly, I mentored graduate students in my group.

Future Direction:

As faculty I would like to continue applying Image-Based Modeling in different areas of Biomedicine, as well as in other industries. In particular I would like to unite the knowledge that I obtained from the two topics that I worked on during my Ph.D. and postdoc: Bone Tissue Engineering and Blood Thrombosis, respectively. I believe that the two are closely related, because, Bone Tissue Engineering grafts must interact with the patients' blood post-implantation. Proper blood flow through the artificial tissue must be ensured in order to eliminate waste products from the construct, to mechanically stimulate the bone cells, and to deliver oxygen and nutrients to them. Image-Based Modeling presents itself as a perfect tool for exploring this system, because a) growth of the of the bone tissue can be captured by imaging at different time points and b) effects of the dynamically changing porous structure on the hemo-chemical and mechanical processes within the construct can be readily calculated via high performance computing.

Besides Biomedicine my interests include flows through inorganic porous media (in particular multi-phase flow through porous rock as it is related to enhanced oil recovery), drug delivery in the brain, modeling industrial chemical processes, financial optimization and entrepreneurship. My overall philosophy is to use simple and well-developed simulation methods in order to get practical and realistic results based on actual physical data. Therefore, in the future, I foresee myself collaborating closely with experimentalists and/or having an experimental component in my own laboratory.

Selected Publications:

R. S. Voronov, S. VanGordon, V.I. Sikavitsas and D.V. Papavassiliou, **2010**. "*Computational Modeling of Flow Induced Shear Stresses Within 3D Salt-Leached Porous Scaffolds Imaged via Micro-CT*". J. of Biomechanics, 43 (7), 1279-1286.

R. S. Voronov, S. VanGordon, V.I. Sikavitsas and D.V. Papavassiliou, **2010**. "*Efficient Lagrangian Scalar Tracking Method for Reactive Local Mass Transport Simulation through Porous Media*". Int. J. for Numerical Methods in Fluids. doi: 10.1002/fld.2369

R. S. Voronov, S. VanGordon, V.I. Sikavitsas and D.V. Papavassiliou, **2010**. "*Distribution of Flow-Induced Stresses in Highly Porous Media*" Appl. Phys. Lett. 97, 024101.

R.S. Voronov, D. Papavassiliou and L. Lee, **2006**: "*Boundary Slip and Wetting Properties of Interfaces: Correlation of the Contact Angle with the Slip Length.*" Journal of Chemical Physics, 124, 204701.

Reviews:

R.S. Voronov, D.V. Papavassiliou and L.L. Lee, **2008**: "*Review of Fluid Slip over Superhydrophobic Surfaces and Its Dependence on the Contact Angle.*" Ind. Eng. Chem. Res., 47 (8), 2455-2477. doi: 10.1021/ie0712941

(3an) Design of Functional, Patternable Vapor-Deposited Polymer Thin Films

Christy D. Petruczok and Karen K. Gleason, Chemical Engineering, Massachusetts Institute of Technology, Cambridge, MA

Initiated Chemical Vapor Deposition (iCVD) is a low-energy process for synthesizing functional polymer thin films. Since surface tension and de-wetting effects are absent during vapor-phase processing, iCVD thin films conform to the geometry of the underlying substrate. Additionally, the benign reaction conditions inherent to the iCVD process allow for complete retention of polymer functionality, as well as the ability to deposit on delicate substrates.

iCVD can be used to conformally coat templates defined via electron-beam lithography. The template is removed using a non-solvent for the iCVD layer, yielding well-defined polymeric features that range in size from the micro to the nanoscale. The process is chemically-nonspecific, and can be used with any iCVD polymer, provided an appropriate non-solvent can be determined. The features retain their functionality; this has been confirmed by incorporating poly(4-vinylpyridine) microstructures into sensors for detecting nitroaromatic explosives.

iCVD can also be used to synthesize thin films of conformal photoresist. After deposition, the photosensitivity of the film--typically poly(4-vinylpyridine) or poly(4-aminostyrene)--is enhanced by functionalizing the iCVD polymer with a photoactive diacetylene. Exposure of the functionalized films to UV light induces 1,4 addition photopolymerization of the diacetylene, making the exposed regions insoluble in common solvents. The use of iCVD affords unique attributes to the process, including the ability to pattern bifunctional surfaces as well as polymeric and metal microstructures on non-planar substrates. Importantly, the resolution and sensitivity of the iCVD resist are comparable to those of commercial products.

The properties and chemical composition of iCVD films are easily tuned by adjusting deposition parameters. Divinylbenzene (DVB) is an excellent cross-linking molecule for iCVD copolymers due to its high vapor pressure and vinyl bond reactivity. A quantitative study of DVB vinyl bond reactivity and incorporation of DVB into iCVD copolymers has been performed using FTIR spectroscopy. Experimentally-determined sticking probabilities and monomer reactivity ratios and the effect of DVB incorporation on mechanical properties such as Young's modulus lend insight into the utility of this material.

(3ao) Scalable and Energy Efficient Advanced Separation Devices Through Tunable Materials Chemistry

Ryan P. Lively, Algenol Biofuels, Bonita Springs, FL

My research is aimed towards revolutionizing fluid separation processes critical to the global energy infrastructure via application of process chemistry-inspired materials design. During my PhD at Georgia Tech with Prof. Bill Koros, I introduced the rapid temperature swing adsorption (RTSA) approach for post-combustion CO_2capture, whereby I adapted knowledge developed in membrane science to design unique nanoscale composite adsorbent/heat exchangers. Through collaboration with faculty at Georgia Tech, I successfully guided this RTSA approach towards a new $3.0M research program at Georgia Tech funded by the Department of Energy, General Electric, Algenol Biofuels, and Southern Company.

Currently, I am a post-doctoral fellow working for Algenol Biofuels, where I am expanding my expertise in gas and liquid separations by designing new materials and processes for gas and liquid management in enclosed algal bioreactors. This position has allowed me to interact first-hand with realistic separation challenges and develop workable solutions rooted in materials science. With an emphasis on materials research and a "top-down," process-guided materials development strategy, I have created membranes, sorbents and barriers that are tailored specifically for industrially relevant applications. By using an array of readily tunable microporous materials, functional polymers, tunable copolymers, and a stable of commercially available polymers, I have created composite materials (e.g., mixed matrix membranes, hollow fiber sorbents, micro-capillary heat exchangers, etc.) that can be integrated into demanding industrial systems.

My future research will address critical issues in separation science via an approach that combines innovative materials design with deep expertise in developing practical, scalable processes. Whereas many research programs offer expertise in materials or process chemistry, my program will uniquely unite these two topics with a strong focus on sustainability and manufacturable materials to allow for revolutionary advances in separations to address modern problems in energy, water supply, and pollution control. Inspired by the process challenges and opportunities, I will endeavor to develop a fundamental understanding of hybrid material formation mechanisms and subsequent separation-related properties. By transitioning from the "thermal age" of separations towards an era of efficient, low energy membrane- and adsorption-based separations, significant advances in both energy conservation and CO_2 footprint reductions will be achieved.

Select Publications (18 total, 11 first author, 145 total citations):

1. Lively RP, Mysona JA, Chance RR, Koros WJ. Formation of defect-free latex films on porous fiber supports. *ACS Appl. Mater. Interfaces.* **2011**;3(9):3568-82.

2. Lively RP, Dose ME, Thompson JA, McCool BA, Chance RR, Koros WJ. Ethanol and water adsorption in methanol-derived ZIF-71. *Chem. Commun.* **2011**;47:8667-8669.

3. Lively RP, Chance RR, Kelley BT, Deckman HW, Drese JH, Jones CJ, Koros WJ. Hollow Fiber Adsorbents for CO_2 Removal from Flue Gas. *Ind. Eng. Chem. Res.* **2009**;48(15):7314-7324.

4. Lively RP, Dose ME, Xu L, Johnson JR, Zhang K, Vaugh JT, Lydon ME, Thompson JA, Lee JS, Liu L, Hu Z, Realff M, Koros WJ. A high-flux polyimide as a contender for CO_2 recovery from flue gas. *J. Memb. Sci.* **2012**; *In submission*.

5. Drese JH, Choi S, Lively RP, Koros WJ, Fauth DJ, Gray ML, Jones CW. Synthesis-Structure-Property Relationships for Hyperbranched Aminosilica CO_2 Adsorbents. *Adv. Funct. Mater.* **2009**;19:3821-3832.

(3ap) Ionic Liquid Pretreatment Technology: Effects of Pretreatment Conditions On Biomass Structure, Composition, and Enzymatic Digestibility

Christopher J. Barr, Chemical and Environmental Engineering, University of Toledo, Toledo, OH

Rising fuel costs due to the high costs of oil has driven the demand for alternative energy. One of the many alternative energy sources that have been studied over the past years has been biofuels derived from lignocellulosic biomass instead of from corn. This change in biofuel feedstock is driven by opportunity costs associated with corn as fuel or food. Lignocellulosic biomass is an abundant source of carbon consisting of three major components: cellulose, hemicellulose, and lignin. Because it is non-food based feedstock, lignocellulosic biomass serves as a renewable carbon source while mitigating impact on food supply. Monomeric sugars can be produced through a biochemical process which includes pretreatment of the biomass to produce a more easily hydrolyzed substrate and enzymatic deconstruction of polysaccharides to monomeric sugars. These sugars can then be used in the production of fuels and value-added chemicals through fermentation or chemical conversion.

During my tenure at The University of Toledo, I have been examining the mechanisms of the ionic liquid (IL) pretreatment technology as well as enzymatic saccharification of pretreated biomass as they relate to monomeric sugar production for future chemical and biofuel processing. As part of this research, we have used X-ray powder diffraction and fiber diffraction crystallography to determine the post-pretreatment cellulose structure in order to determine its effect on enzymatic digestibility. Additionally, we examined the roles of different enzymatic activities during the deconstruction of IL pretreated biomass by using tailored enzyme mixtures containing known quantities of purified and characterized mono-component enzymes. This knowledge of required enzymatic activity can lead to a reduction in the amount and, ultimately, costs of the enzymatic portion of the biofuels process.

(3aq) Catalytic Conversion and Kinetics Study of Biomass to Biofuels and Chemicals

Nafiseh Rajabbeigi and Michael Tsapatsis, Department of Chemical Engineering & Materials Science, University of Minnesota, Minneapolis, MN

Environmental concerns regarding the use of crude oil and increasing oil prices continue to promote advances in new sources of energy, chemical, and fuel production. In particular, the catalytic conversion of abundant cellulosic sugars such as glucose, fructose, and xylose into chemical intermediates attracts attention as an alternative to the catalytic cracking of petroleum hydrocarbons. The design of sustainable processes involving catalytic conversion of biomass currently is a challenge due to the costly process of production of biofuels.

My research involves materials synthesis and catalytic science, with an emphasis on catalytic conversion of biomass into biofuels and chemicals. I am studying the isomerization of glucose into fructose, kinetics study of biofuel formation from biomass catalyzed by different zeolites, and comparing the activities of Sn-containing zeolites for certain catalytic reactions. In addition, the separation of products from reaction mixtures by adsorption process is another complementary aspect of my research.

My future research direction aims to develop a new class of nanostructured materials as heterogeneous catalysts using hierarchical approach to increase the bioconversion efficiency.

(3ar) Current Developments in Wastewater Treatment Facility-Based Biorefinery Concept Producing Biofuel Feedstock Biocrude

Andro Mondala, Dave C. Swalm School of Chemical Engineering, Mississippi State University, Mississippi State, MS

This poster will present a summary of the most recent developments in our group's ongoing study on developing a wastewater treatment plant–based biorefinery concept producing biofuel feedstock lipids and other potentially high value bioproducts. These include (1) the use volatile fatty acids obtained via anaerobic digestion of lignocellulose biomass as platform chemical/fermentation substrates for biocrude production by activated sludge cultures; (2) the evaluation of high-strength industrial wastewaters originating from pulp and paper mills, poultry processing plants, and landfill leachates as fermentation media and nutrient source for the fermentative biocrude production process; and (3) the evaluation of the pilot-scale facility containing twin 500-L bioreactor units with downstream processing equipment for a larger-scale production of biocrude. These studies will aim to further improve the cost-effectiveness of the proposed biocrude production process using existing wastewater treatment facility infrastructure by using low-cost and abundant carbon and nutrient sources and optimizing the downstream recovery processes to minimize utilities and operational costs.

(3as) Electrostatics in Non-Polar Systems: Transitions From Unstable Colloids to Molecular Dissolution Via Polymeric Functionalization

Sara M. Hashmi, Chemical & Environmental Engineering, Yale University, New Haven, CT

Asphaltenes, the most aromatic and largest molecular weight component of petroleum fluids, tend to phase separate under a variety of conditions, causing many problems for hydrocarbon energy production in the petroleum industry. While large amounts of aromatic solvents provide a thermodynamic mechanism for addressing asphaltene precipitation, their use is neither economical nor environmentally friendly. Methods requiring smaller doses of chemicals are highly preferred. However, chemical design principles for asphaltene dispersants require detailed understanding not only of asphaltene chemistry on the molecular scale, but also of the interactions governing asphaltene dynamics on colloidal length scales. One additional challenge for asphaltene stabilization lies in the non-polar nature of petroleum itself. Electrostatic interactions, ubiquitous for stabilizing aqueous suspensions, provide a double-edged sword in non-polar media: they are both strong and long-range. While mixtures of homogeneously charged molecules or particles will remain stable, even small amounts of heterogeneous surface charging can lead to destabilization and separation.

During my postdoctoral research, I seek to understand the fundamentals of asphaltene stabilization by dispersants by using a variety of experimental methods, including static and dynamic light scattering, UV-visible spectroscopy, and measurements of bulk conductivity and sedimentation dynamics. Both non-ionic and ionic dispersants can stabilize asphaltenes at different stages in the precipitation process, albeit through very different mechanisms: the first involving polymeric adsorption onto colloidal surfaces and the second relying on acid-base interactions at the molecular level. My work has uncovered several unique aspects of asphaltenes from a materials science perspective, including the presence of both positive and negative surface charges on unstable asphaltene colloids and the importance of heteroatomic content in tuning the solubility characteristics of these π-conjugated materials. Furthermore, the different mechanisms of action of the non-ionic and ionic dispersants at the molecular and colloidal scales have consequences for their utility in addressing the main industrial concern: asphaltene precipitation in wellbores and pipelines. I will also describe measurements assessing dispersant effectiveness for both inhibiting and reversing asphaltene deposition from bulk flows.

(3at) Multi-Scale Approaches in Systems with Nanoscale Phenomena and Novel Materials

Pil Seung Chung, Chemical Engineering, Carnegie Mellon University, Pittsburgh, PA

During the past few decades, along with remarkable scientific and engineering achievements, the world has faced immense challenges in efficiency and system size, which motivates the investigation of novel materials and nanoscale systems. Analytic approaches have been widely studied to satisfy the demand by tuning physical properties, chemical structures, and functionalities of the new materials as a part of the eye-opening progress. The broad progress in nanotechnology provides us the understanding of nanoscale physics to control the system properties. However, advances in nano materials and analysis methods direct us to the critical issue of the interaction to meso/macroscopic scales since most systems comprise phenomena at different time and length scales, and often described via a hierarchy of scale-specific models. Consequently, the multi-scale approach based on a clear understanding of nanoscale phenomena becomes an essential multidisciplinary analysis paradigm in science and engineering fields in order to realize the new nano materials and systems. Via this approach, the mechanism of information communication can be obtained from one scale and passed to another scale allowing one to predict the system scale behavior from first principles in a bottom-up approach. The device scale properties, nonetheless, can be obtained with nanoscale resolution using the top-down approach.

My research interest covers the broad area of both theoretical and experimental multi-scale approaches in novel materials and nanoscale phenomena including tribology, energy, and nano electronics applications by using novel atomistic/molecular and mesoscale/continuum theory and developing methodologies for hierarchical integration of the systems at the different scales, which will be expanded to multi-phenomena/physics related issues. The device scale properties will be coupled with optimization for the process system integration. I also intend to perform experiment and will vigorously collaborate with experimentalist to complement the full-scale simulation. In this poster session, I will introduce and discuss the details of my research plan.

(3au) Generation of High-Value Products From Biomass – the Bioseparation Route

Abhijit Tarafder, Department of Chemistry, The University of Tennessee, Knoxville, TN

Biorefineries are one of the most promising technological alternatives available in shifting from fossil-based to bio-based industries. One of the primary criteria to make biorefineries economically viable is to produce high-value co-products along with low-value bio-fuels. Various routes for manufacturing high-value products from biomass are being investigated, where a more common approach is to convert the non-fuel producing components of the biomass to high-value chemicals or materials, through synthetic routes.

While this approach is certainly useful, I propose to develop separation routes for retrieving high-value biomolecules from the biomass, without involving any reaction process. Biomolecules have attained significant attention as non-synthetic natural products, especially in areas like fine chemicals, pharmaceuticals and nutraceuticals, and there is a strongly growing interest in applying biomolecular products as various other consumer items. Retrieval of high-value biomolecules from the biomass can add significant value to the overall process. For example separation of compounds like astaxanthin ($2500/kg) or eicosapentaenoic acid (EPA) ($4600/kg) from algal biomass will be significantly beneficial compared to the scheme of treating them similarly with the other compounds present in the algal biomass.

Separation of such products, however, can be challenging and costly. My primary tool for implementing the separation route is the chromatographic processes. Chromatography refers to a set of separation techniques known for their ability to separate mixtures of very similar molecules, especially the biomolecules. They are universally used for chemical and biological analyses and play a pivotal role in the downstreams of biopharmaceutical and fine chemical industries. Chromatography, for their ability to adapt to the changes in feed conditions, their strength in multicomponent separation and the possibility of operating them with benign operating conditions, are uniquely competent to handle the separation challenges of biorefineries. This process found limited applications in the fossil-based chemical industries, but in a bio-based industrial setup and with our growing interests in manufacturing natural products will make chromatography a crucial process option in the future industries.

In this session I will discuss the research plans I have to develop separation schemes for high-value biomolecules from different biomasses and how I plan to design economically and environmentally sustainable chromatographic processes for such separation.

(3av) In Silico Design of Nanoporous Materials for Energy Storage and Environmental Remediation Applications

Jeremy C. Palmer, Department of Chemical and Biological Engineering, Princeton University, New Jersey , NJ

Confinement of guest phases within the pores of nanostructured materials, such as porous carbons, zeolites and metal-organic frameworks, can signficantly alter phase behavior, chemical equilibria, reaction rates and mechanisms, and transport properties. In order to develop nanoporous materials with desired performance characteristics, these effects may be finely tuned by modulating the structural features (e.g., porosity, pore size and surface area) of the materials through judicious choice of the precursors and synthesis conditions. As a result, nanoporous materials are playing an increasingly prominent role in the development of many emerging energy- and environment-related technologies, including electric double-layer supercapacitors, membrane systems for water purification and desalination, and adsorbents for CO_2 capture and sequestration. However, the full potential of these materials has not been realized due to the absence of rational design principles to guide their synthesis and application. Development of such guidelines for porous materials requires a complete understanding of the relationship between their nanostructure and functional properties. At present, this level of understanding is difficult to achieve through empirical investigation alone, due to the limited ability of existing experimental techniques to fully resolve the nanoscopic features of these materials and the mechanistic origins of their performance characteristics.

One of my main research interests is to establish a theoretical framework based in statistical mechanics and molecular simulations to assist in the development of rational design principles for nanoporous materials. These approaches are well-suited for studying phenomena such as adsorption, transport and chemical reactions at the nanoscale, and can provide information that is complementary to experimental measurements and is vital to understanding confinement effects in nanostructured materials. Specifically, molecular simulation techniques can be used to: (1) resolve the structural features of materials that are challenging to characterize with experiment, (2) elucidate the molecular mechanisms that are responsible for confinement effects, (3) combinatorially screen materials to identify structural features that give rise to enhanced performance attributes, and (4) make quantitative predictions regarding material performance. As an example, I will discuss the application of molecular simulation techniques to model disordered nanoporous carbons (DNCs), which have gained recent attention due to their exceptional performance as supercapacitor and battery electrode materials. Although these materials are among the most challenging to experimentally characterize and model due to their amorphous structure, I will demonstrate that simulation methods can be used to help achieve a fundamental understanding of the relationship between the structure and function of DNCs and lay the foundation for establishing rational design principles.

(3aw) Multiscale Simulations of Soft Materials and Complex Fluids

Li Xi, Department of Chemical Engineering, Massachusetts Institute of Technology, Cambridge, MA

My general research interest lies in the study of soft matter using multiscale computational approaches. The scope of soft matter covers a wide range of materials of practical interest, such as polymers, surfactants, colloids, gels, biomolecules, liquid crystals and complex fluids. A special focus of my future research is to couple simulation techniques at different length scales, and address problems inspired by practical applications in the areas of energy, environment and biomaterials. My past experience in simulations at both small and large scales prepares me for this plan. During my Ph.D. study at University of Wisconsin-Madison, I worked on projects involving polymeric flows at different limits of Reynolds number, typical of applications from microfluidics to large scale fluid transportation. Among my achievements is a new framework of understanding a long-standing puzzle in the area of viscoelastic turbulence. Recognizing the importance of multiscale dynamics in soft matter, I decided to enter a new field and obtain training in microscopic scale simulation after graduation. My current postdoctoral research at MIT focuses on the understanding and prediction of transport performance of glassy amorphous polymer materials. In terms of simulation techniques, I obtained training at both limits of scales relevant to the study of soft matter: from the full atomistic scale to the continuum level.

(3ax) Nanoscale Understanding On the Structure and Dynamics of Biomembranes and Biomacromolecules

Mohan B. Boggara, Center for Biotechnology and Interdisciplinary Studies, Rensselaer Polytechnic Institute, Troy, NY

Using a combination of biophysical, computational and biological approach, I will demonstrate in detail that nanoscale insights play a major role in number of biologically relevant problems. In particular I will show examples from my research work on understanding (1) interactions between amphiphilic drugs and membranes (2) interactions between proteins and peptide-inhibitors (3) self-assembly of misfolded proteins and (4) diffusion of membrane-receptor clusters in live cell membranes. Based on my doctoral and post-doctoral research work, I aim to develop a research program focusing on various open questions in above areas and further expand upon them. In this poster, I plan to discuss in detail about specific questions I am interesed in, both short term as well as long term, and identify potential funding sources. My research program is expected to contribute to various areas including, but not limited to, therapeutic drug delivery, design of bio-inspired nanomaterials, structure-based drug design, protein misfolding and cancer therapeutics. Throughout my doctoral and postdoctoral studies, I have also been actively involved in mentoring undergraduate and graduate students. Further, I am interested in teaching various core chemical engineering courses as well as advanced graduate level courses.

(3ay) Solution Phase Self-Assembly of Soft Materials

Manickam Adhimoolam Arunagirinathan, Materials Science and Engineering, Stanford University, Stanford, CA; SLAC National Accelerator Laboratory, Stanford Institute for Materials and Energy Sciences, Menlo Park, CA

Self-assembly of surfactants, phospholipids, block copolymers and biopolymers in solution to form diverse equilibrium and non-equilibrium microstructures have great significance from basic evolution of life to development of engineered advanced nanomaterials. Understanding the mechanism of microstructure formation at different length scales necessitates appropriate model systems and analytical tools. In this context, I describe (i) the self assembly behavior of a simple dodecylsulphate surfactant influenced by cesium counterions, (ii) the lamellar phase immiscibilty of the swelling amphiphile polyethoxysorbitan trioleate (Tween85) to form concentric cylinders, and (iii) the structural reorganization of clathrin protein triskelia to form 3D cage-like assemblies, as studied using Optical Microscopy, Dynamic Light Scattering (DLS), Cryo Transmission Electron Microscopy (cryo-TEM), cryo Electron Tomography (cryo-ET), and Freeze Fracture Transmission Electron Microscopy (FFTEM).

Charged surfactants self-assemble to form elongated structures such as worm-like micelles and crystalline precipitates depending on the extent of electrostatic screening of head group repulsion by electrolytes, oppositely charged surfactants, or hydrotropes. Here, the effect of screening the electrostatic repulsion between anionic headgroups of dodecylsulphate by hydrated cesium counterions to form worm-like protrusions and tubular structures of complex hexagonal phase was explored using Cryo-TEM in correlation with earlier scattering studies. Compared to sodium dodecylsulphate, the concentration of cesium dodecylsulphate required to form extended microstructures has significantly decreased. These data demonstrate that the nature of counterion influences self-assembly and phase behavior of charged surfactant systems.

To understand the lamellar phase immiscibility exhibited by swelling amphiphiles, the formation of tubular microstructures made of concentric lamellar cylinders via dissolution of polyethoxy sorbitan trioleate (Tween 85) in water was studied using optical microscopy and FFTEM. The dissolution characteristic of lamellar phase indicated that the gel phase coexists with fluid lamellar bilayers. Further investigation at elevated temperatures revealed the existence of a monotectic phase transition leading to fluid phase immiscibility. FFTEM identification of the focal eddies evolved during diffusion of water into the concentrated surfactant, the lamellar mesophase formation, and the isotropic fluid phase nucleation helped in understanding the complex monotectic phase behavior of the Tween85/water system.

Clathrin is a ~630 kDa protein that resembles a three-legged pinwheel, *i.e.,* a triskelion, with each leg having an extended length of ~50 nm and width of ~3 nm. Due to molecular recognition interactions between legs and the non-planar geometry of the basic building block, clathrin self-assembles to form 3D assemblies resembling fullerene structures. Here we use cryo-TEM, cryo-ET and DLS to explore the influence of pH on the kinetics and morphology of clathrin self-assembly and to discover potential new nanostructures that clathrin may form. Our results demonstrate that clathrin has strikingly different assembly kinetics at pH 5 (below the isoelectric point) where discrete spherical cages (~60-100 nm) form over the course of several days via the rearrangement of large disordered aggregates versus at pH 6 (above the isoelectric point) where stable spherical cages (~60-100 nm) form within 30 minutes. This corroborates DLS data that indicate clathrin self-assembly is a function of both pH and time, with large length-scale structural rearrangements occurring over long time-scales (*i.e.,* days).

(3az) Designing of Novel Porous Materials for Functional Applications

Dipendu Saha, Materials Science and Technology Division, Oak Ridge National Laboratory , Oak Ridge, TN

Advanced and functional nanoporous materials possess an important role in the energy and gas storage, separation and purification, drug delivery, catalysis and wide variety of benign environmental applications. My primary research will focus on the synthesis, characterization and usage of application-specific novel adsorbents accompanied by the fundamental studies to understand the adsorbent and their role in adsorption processes. I like to segregate the research on nanoporous based materials on to three broad categories, *(i)* Surfactant induced mesoporous carbon from sustainable and natural precursor *(ii)* Porous Media based targeted drug delivery system and *(iii)*Role of novel porous materials in the remedy of environmental problems including gas separation and water purification.

Selected peer-reviewed publications (out of 30)

1) N. Gallego, L. He, D. Saha, Y. Melnichenko, C. I. Contescu, ***J. Am. Chem. Soc.***vol 133, no. 35, pp 13794-13797, 2011.

2) D. Saha, S. Deng, ***J. Physical Chemistry Letters***, vol.1, no. 1, pp 73-78, 2009.

3) D. Saha, Z. Bao, F. Jia, S. Deng, ***Environmental Science and Technology***vol. 44 no. 5 pp 1820-1826, 2010.

4) D. Saha, C.I. Contescu, N.C. Gallego, ***Langmuir***Vol. 28, no. 13, pp. 5669-5677.

5) D. Saha, S. Deng, ***Langmuir***, vol. 25, no. 21, pp. 12550-12560, 2009.

6) D. Saha, S. Deng, ***Langmuir***, vol. 26 no. 11 pp 8414-8418, 2010.

7) D. Saha, S. Deng, ***Langmuir*** vol. 27, no. 11, pp 6780-6786, 2011.

8) D. Saha, C.I. Contescu, N.C. Gallego, ***Microporous and Mesoporous Materials***vol. 155, pp 71-74, 2012.

9) D. Saha, S. Deng, ***Carbon***, vol. 48, no. 12, 3471-3476, 2010.

10) C. I. Contescu, D. Saha, N. Gallego, A. Kolesnikov, E. Mamontov, ***Carbon***, vol. 50, no. 3, pp 1071-1082, 2011.

11) D. Saha, S. Deng, ***J. Colloid and Interface Science*** vol. 345, no. 2, pp 402-409, 2010.

12) L. Camacho, A. Torres, D. Saha, S. Deng, ***J. Colloid and Interface Science***, vol. 349, no. 1, pp. 307-313, 2010.

13) D. Saha, Z. Wei, S. Deng, ***Separation and Purification Technology***vol. 64, no. 3. pp. 280-287, 2008.

14) D. Saha, Z. Wei, S. Deng, ***Intl. J. Hydrogen Energy***, vol. 33, no. 24, pp. 7479-7488, 2008.

15) D. Saha, S. Deng, ***Journal of Chemical and Engineering Data*** vol. 54, no. 8, pp. 2245-2250, 2009.

(3b) Confined Syntheses for Hierarchical Catalysts and Membrane Fabrication for Separation: Key Components for Biorefinery Processes

Won Cheol Yoo, University of Minnesota, Minneapolis, MN

The development of sustainable energy resources is imperative due to continuously increasing energy demands in a fragile global market, in which plant biomass is the only sustainable resource that can substitute the use of petroleum sources. In this respect, lignocellulose decomposition to monomeric sugars, control of the isomerization of sugars and biomass derivatives, and energy-efficient selective separations have challenged the commercial realization of processes that can generate liquid fuels and chemicals. For the production of sustainable energy from biomass, my expertise in the areas of zeolite/solid catalyst design, membrane fabrication and separation applications, catalytic conversion of biomass and photocatalytic reactions, and nanostructured materials assembly via surface and colloidal chemistry can have an immediate impact in this emerging area.

My graduate research devoted to designing a wide range of porous materials (micro/meso/macropores) and testing them for a range of applications, with a focus on molecular sieve membranes for separation of vapor/gas isomers (ie., xylene and butane isomers). To accomplish this, the concept of reactor engineering of confined systems has been applied to grow zeolite nanocrystals inside confinement while studying their growth mechanism. Controlled morphologies of zeolite nano crystals can be produced, which are used as a seed layer for zeolite membrane growth. In addition, design of zeolite catalysts with controlled extra-porosity was achieved for overcome the diffusion limitations, which have restricted the catalytic activity of zeolites. Furthermore, the ability to control the porosity was applied to different materials (e.g., silica and carbon materials) and self-assembly of nanostructured materials was achieved via surface chemistry.

My recent focus in the postdoc course is related with generation of renewable energy from biomass using synthetic solid materials. Hierarchically designed and multi-functional catalysts have been synthesized for biomass conversion (i.g., cellulose to monomeric sugars) as well as for photocatalytic reactions. In addition, separation of biomass derivatives by selective adsorption using synthetic carbon materials has been investigated.

My proposed research will focus on the design of multi-functional catalysts for biorefinery and catalytic reactions. Hierarchically designed solid catalyst (e.g., zeolites, metal oxide catalysts, functionalized carbon materials) will be used as a template for another catalyst growth, where I will investigate growth mechanism of composite materials as well as sequential catalytic reactions. In addition, zeolitic and composite (zeolite + polymer) membranes will be fabricated for gas separation (e.g., CO_2, N_2, H_2 and isomers) and pervaporation for selective alcohol separation from water mixture. As well, separation of biomass derivatives and water purification will be developed using porous material (i.e., carbon, zeolite and mesoporous materials) with control of porosity, surface modification for specific chemical interaction and continuous reactor design.

(3ba) Macromolecular Drug Delivery for Cancer Therapy

Sutapa Barua, Chemical Engineering, University of California, Santa Barbara , Santa Barbara, CA

The use of macromolecular therapeutic agents e.g., monoclonal antibodies and nucleic acids has come to the forefront in recent years for the treatment of cancer. However, several challenges, including cell target specificity, intracellular delivery, low therapeutic payload, biologically active drug concentration and toxicity, still prevail for eliciting effective responses from cancer cells to therapy. My post-doctoral research with Professor Samir Mitragotri at University of California, Santa Barbara focuses on synthesis of advanced engineered nanoparticles with controlled size, shape and surface chemistry for targeted delivery of therapeutic drug macromolecules. My Ph.D. thesis research under the guidance of Professor Kaushal Rege at Arizona State University, Tempe focused on combination treatment strategies for non-viral gene delivery to induce apoptotic cancer cell death. In this poster session, I will highlight examples from my graduate and postdoctoral research which demonstrate the ability of synthetic nano-biomaterials in delivery of drugs, antibodies, and nucleic acids. I will present my vision for my independent research program at the interface of nanotechnology and cancer therapy that encompasses: 1) development of new carrier systems with specific cellular and sub-cellular target effects for cancer therapy, and 2) detailed mechanistic investigations of sub-cellular interactions between synthetic materials and sub-cellular organelles.

EDUCATION:

B.Sc., Chemical Engineering, 2003, Bangladesh University of Engineering & Technology (BUET), Dhaka, Bangladesh

M.Sc. in Chemical Engineering, 2007, Montana State University (MSU), MT, USA

Ph.D. in Chemical Engineering, 2011, Arizona State University (ASU), AZ, USA

EXPERIENCE:

Post-doctoral Researcher, Chemical Engineering, University of California, Santa Barbara, Santa Barbara, CA; 07/2011-present

Graduate Research Assistant/Associate

Chemical Engineering, Arizona State University, AZ; 08/2007- 05/2011

Chemical, Biological Engineering, Montana State University, MT; 08/2005- 08/2007

Chemical Engineering, Washington State University, WA; 08/2004- 08/2005

Teaching (Lecturer), Chemical Engineering, Bangladesh University of Engineering & Technology (BUET), Dhaka, Bangladesh; 02/2003- 07/2004

SELECTED PUBLICATIONS:

Barua, S.; Ramos, J.; Potta, T.; Taylor, D.; Huang, H.C.; Montanez, G.; Rege, K., *Review on Discovery of Cationic Polymers for Non-viral Gene Delivery*. Combinatorial Chemistry & High Throughput Screening, 2011, 14(10), 908-24

Barua, S.; Rege, K., *The influence of Mediators of Intracellular Trafficking on Transgene Expression Efficacy of Polymer-plasmid DNA Complexes*. Biomaterials, 2010, 31 (22), 5894-5902

Barua, S.; Joshi, A.; Banerjee, A.; Matthews, D.; Sharfstein, S. T.; Cramer, S. M.; Kane, R. S.; Rege, K., *Parallel*

Synthesis and Screening of Polymers for Nonviral Gene Delivery. Molecular Pharmaceutics, 2009, 6 (1), 86-97

Barua, S.; Rege, K., *Cancer-Cell-Phenotype-Dependent Differential Intracellular Trafficking of Unconjugated Quantum Dots*. Small, 2009, 5 (3), 370-376

MANUSCRIPTS IN PREPARATION

Barua, S.; Voelkel-Johnson, C.; Von Hoff, D.D.; Han, H.; Rege, K., *Inhibition of Aurora Kinases Enhances Transgene Expression and Cancer Cell Apoptosis following delivery of Polymer-Plasmid DNA Complexes. In preparation, 2012*

Barua, S.; Yoo, J.W.; Mitragotri, S., *Endocytosis of Nano- and Microparticles in Breast Cancer Cells. In preparation, 2012*

AWARDS/ HONORS:

3rdplace, poster competition, National Nanotechnology Infrastructure Network Symposium, Arizona State University, 2011

Outstanding Research Assistant Award, Chemical Engineering, Arizona State University, 2009

Travel Grant Awards, Arizona State University, 2009 and 2010

Participant, 57th Lindau meeting with 18 honorable Nobel Laureates in Lindau, Germany, 2007

2ndplace, poster competition of the Intermountain Systems Biology Symposium, 2007

INRA (Inland Northwest Research Alliance) fellowship, Montana State University, 2005-2007

2ndplace poster competition, William R. Wiley Exposition at Washington State University, 2005

Dean List Awards, Bangladesh University of Engineering & Technology (BUET), 1999-2002

Scholarship, Bangladesh Scholarship Council and the Nippon Foundation, Japan, 1998 and 2000

(3bb) Engineering in the Microvasculature: The Mechanical Microenvironment's Control of Systemic Metabolism

Joseph M. Rutkowski, Touchstone Diabetes Center - Department of Internal Medicine, UT Southwestern Medical Center, Dallas, TX

Lying between the endothelial beds of the body's circulation is the interstitial microenviroment wherein the complex biohphysical and biochemical interplay of cells, extracellular matrix, and the molecules of interstitial fluid control both immediate tissue homeostasis and systemic functions. The modulators of interstitial flow rates and shear, tissue hydration and composition, and cellular trafficking are the endothelial microvasculature of the blood and venous capillaries and the lymphatic system. I hypothesize that the blood and lymphatic microvasculature, as gatekeepers to the interstitium and governors of molecular and fluid fluxes, regulate tissue homeostasis and modulate metabolism through their endothelial physiology.

In my doctoral studies under Prof. Melody Swartz at the EPFL, we demonstrated that the primary modulator of interstitial drainage, the initial lymphatic vasculature, both controls the interstitium and cells found within and is itself modulated by the tissue mechnical environment. Generation of new functional lymphatic capillaries, lymphangiogenesis, was found to be directional and fully dependent on interstitial flow. Blocked lymphatic drainage in lymphedema resulted in hyperproliferation, yet poor organization and decreased function of lymphatic capillaries. Where interstitial flow rates slowed towards zero, inflammatory chemokines accumulated, interstitial immune cells failed to traffick, and adipose tissue depots expanded in adipogenesis. Indeed, we quantified lymphatic clearance rates in genetic models of lymphatic dysfunction and found a correlation with fluid transport and remodeling of the extracellular matrix and adipose expansion on tissue hydraulic conductivities and subsequent treatment and potential resolution of the disease. Blood capillary extravasation, as controlled by Starling's Law, was reduced in these instances due to the increased hydraulic and osmotic pressures of the interstitium. Concurrently, adipose expansion or hypercholesterolemia strongly reduced lymphatic drainage capability by inducing vessel degeneration. As lymphatic vessels are the primary route of lipid transport throughout the body, these studies combined to demonstrate the potential that the interplay of microvasculature cells with their surrounding tissues influences systemic metabolism.

No longer considered to be a passive reservoir, adipose tissue is an active metabolic player and dynamic endocrine organ. My current postdoctoral research at UT Southwestern Medical Center focuses on adipose tissue physiology and how the microvasculature may control its metabolism and inflammation via adipokine and metabolite transport. The adipokine adiponectin, intruiging in that it is secreted in multiple oligomer sizes, has exhibited positive effects in regulation of systemic metabolism and insulin sensitivity. In the laboratory of Prof. Philipp Scherer I have demonstrated that oligomer size, particularly that of the high molecular weight form, presents transport limitations across endothelial barriers with low intercellular permeabilities. I have quantified size-dependent exclusion of adiponectin oligomers in model endothelial cells *in vitro*, and used genetically- and metabolically-altered mouse models to discern that oligomer function likely depends on microvasculature permeabilities *in vivo*. Adiponectin transport, and potentially function, are pharmacologically and pathologically limited by endothelial function. I have demonstrated that metabolic stresses modulate the permeability of the vasculature for increased or decreased lipolytic fluxes with feeding and fasting, and endothelial dysfunction regulates systemic adipokine effects. The critical barrier function of the endothelium thus modualtes multiple aspects of adipose function.

Ongoing studies are aimed at discerning the interplay between adipose tissue and its vasculature. One concomitant pathology with obesity is chronic kidney disease; changes in renal function alter transvascular fluxes and protein concentrations in the periphery, resulting in excess fluid accumulation. In a recently developed inducible model of renal podocyte apoptosis, adiponectin altered renal recovery. Altering systemic fluid balance and interstitial fluxes pharmacolgically or by controlling renal physiology drives either adipogenesis or adipose wasting. Changes in lipid uptake or secretion coincide with altered adipokine profiles: all of which alter or are controlled by endothelial transport.

I propose that the biophysical environment of adipose tissue, specifically matrix composition, interstitial flow, and lymphatic drainage, influence adipocyte behaviors and systemic metabolism and that changes in cell infiltration and chemokine levels in the interstitial fluid have a direct effect on adipogenesis, adipokine secretion, and metabolic lipid fluxes. My laboratory will take an integrative biomedical engineering approach to the important and relevant problem of obesity and metabolic dysfunction by combining and contributing to fundamental mechanistic insights in adipose tissue physiology, microvascular biology, and interstitial fluid mechanics. My laboratory will develop and utilize *in vitro* tissue engineered models to study the adipose biophysical environment and adipocyte interactions with blood and lymphatic endothelial cells. We will concurrently utilize relevant, targeted transgenic *in vivo* models to demonstrate the physiologic relevance of our *in vitro* advances.

Accepted Peer-Reviewed Publications

UT Southwestern Medical Center

Wernstedt Asterholm I, McDonald J, Blanchard PG, Sinha M, Xiao Q, Mistry J, **Rutkowski JM**, Dehaies Y, Brekken RA, Scherer PE. Lack of immunological fitness during fasting in metabolically challenged animals. *J Lipid Res*. **2012**Apr 13.

Fischer-Posovszky P, Wang QA, Asterholm IW, **Rutkowski JM**, Scherer PE. Targeted deletion of adipocytes by apoptosis leads to adipose tissue recruitment of alternatively activated m2 macrophages. *Endocrinology*. **2011**Aug;152(8):3074-81.

Holland WL, Miller RA, Wang ZV, Sun K, Barth BM, Bui HH, Davis KE, Bikman BT, Halberg N, **Rutkowski JM**, Wade MR, Tenorio VM, Kuo MS, Brozinick JT, Zhang BB, Birnbaum MJ, Summers SA, Scherer PE. Receptor-mediated activation of ceramidase activity initiates the pleiotropic actions of adiponectin. *Nat Med* **2011**Jan;17(1):55-63.

Rutkowski JM, Davis KE, Scherer PE. Mechanisms of obesity and related pathologies: the macro- and microcirculation of adipose tissue. *FEBS J*. **2009**Oct;276(20):5738-46.

École Polytechnique Fédérale de Lausanne (EPFL)

Miteva DO, **Rutkowski JM**, Dixon JB, Kilarski W, Shields JD, Swartz MA. Transmural flow modulates cell and fluid transport functions of lymphatic endothelium. *Circ Res*. **2010**Mar 19;106(5):920-31.

Rutkowski JM, Markhus CE, Gyenge CC, Alitalo K, Wiig H, Swartz MA. Dermal matrix remodeling and fat accumulation control tissue swelling and hydraulic conductivity during murine primary lymphedema. *Am J Pathol.* **2010**Mar;176(3):1122-9.

Lim HY, **Rutkowski JM**, Helft J, Reddy ST, Swartz MA, Randolph GJ, Angeli VA. Hypercholesterolemic mice exhibit lymphatic vessel dysfunction and degeneration. *Am J Pathol.* **2009** Sep;175(3):1328-37.

Goldman JG*, **Rutkowski JM***, Shields JD*, Pasquier M, Cui Y, Schmökel HG, Pytowski B, Swartz MA. Co-operative and redundant roles of VEGFR-2 and VEGFR-3 signaling in adult lymphangiogenesis. *FASEB J.* **2007**Apr;21(4):1003-12. (* equal contribution)

Goldman J, Conley KA, Raehl A, Bondy DM, Pytowski B, Swartz MA, **Rutkowski JM**, Jaroch DB, Ongstad EL. Regulation of VEGF-C by interstitial flow. *Am J Physiol Heart Circ Physiol.* **2007**May;292(5):H2176-83.

Rutkowski JM and Swartz MA. A driving force for change: Interstitial flow as a morphoregulator. *Trends Cell Biol.* **2007**Jan;17(1):44-50.

Rutkowski JM, Moya M, Johannes J, Goldman J, Swartz MA. Secondary lymphedema in the mouse tail: lymphatic hyperplasia, VEGF-C upregulation, and the protective role of MMP-9. *Microvasc Res.* **2006**Nov;72(3):161-71.

Rutkowski JM, Boardman KC, Swartz MA. Characterization of lymphangiogenesis in a model of adult skin regeneration. *Am J Physiol Heart Circ Physiol.* **2006**Sep;291(3):H1402-10.

Penn State University

Rutkowski JM, Santiago LY, Ben-Jebria, Ultman JS. Comparison of ozone-specific (OZAC) and oxygen radical (ORAC) antioxidant capacity assays for use with nasal lavage fluid. *Toxicol In Vitro.* **2011**Oct;25(7):1406-13.

Rutkowski JM, Santiago LY, Ben-Jebria, Ultman JS. Development of an assay for ozone-specific antioxidant capacity. *Inhal Toxicol.* **2003** Nov;15(13):1369-85.

(3bc) Green Chemistry: Metal Organic Frameworks (MOFs) for CO2 Separation and Energy Storage

Sangil Han, Chemical Engineering, Massachusetts Institute of Technology, Cambridge, MA; School of Chemical & Biomolecular Engineering, Georgia Tech, Atlanta, GA

The capture of CO_2 from fossil fuel combustion flue gas is the focus of significant research efforts. The existing technologies are extremely energy intensive, including amine absorption and pressure-swing adsorption. Metal-organic frameworks (MOF) are strong candidates as size-selective, high-capacity materials for separation of CO_2 and other gases. MOFs exhibit nanoporous crystalline structure in which organic linker molecules self-assemble with metal cations to form extended crystals with well-defined pore size, high surface area, low framework density, and thermal stability. These properties make MOFs an attractive alternative to traditional zeolite materials for gas separation in membrane separation technologies.

In principle, a potentially infinite number of MOFs with various combinations of organic linker and metal cation present opportunity to tailor pore structure, size, and functionality for CO_2 capture from flue gas. many MOFs are unstable in water vapor and acidic gases and the principles for designing water-stable MOFs are unknown. Most previous work reports only single gas adsorption capacity with a small number of samples, which limits the fast selection of promising candidates among the huge number of the available MOFs. Our experimental approach assesses relevant adsorption properties in a high-throughput mode and also directly examines the stability of materials with respect to humidity and acid gas exposure.

In addition to the above postdoc research, this poster will include my phd research with regard to molecule transport through anisotropic phase of liquid crystals, especially liquid crystal supported membranes for enantiomer separation.

(3bd) Simulation of Materials for Alternative Energy Applications

Karl D. Hammond, Nuclear Engineering, University of Tennessee, Knoxville, TN; Chemical Engineering, University of Massachusetts, Amherst, MA

My research addresses the challenging yet vitally important materials science associated with alternative energy, specifically porous/catalytic materials and nuclear fusion/plasma-facing materials. Such materials research is inherently multidisciplinary, involving solid-state physics, chemistry, surface science, catalysis, and materials science. The impact is similarly broad, with potential effects on petrochemical separations, biomass conversion, catalytic processes, and nuclear energy production (both fission and fusion). This poster highlights my research on catalytic and nuclear materials, which uses both experimental and computational approaches. The first half, on catalytic materials, focuses on zeolites that are post-synthetically treated with ammonia to create base-catalytic sites. Specifically, we establish some spectroscopic "signatures" for nitrogen incorporation in zeolites that can be used as diagnostic criteria for the presence of nitrogen in the framework. The second half of the poster presents research into materials for nuclear fusion power applications, specifically for plasma-facing components in the divertor of tokamak reactors. This work uses atomistic and Monte Carlo simulation techniques to study the evolution of tungsten surfaces under plasma exposure.

(3be) First-Principles Multiscale Modeling of Materials for Energy and Environmental Applications

Giannis Mpourmpakis, Catalysis Center for Energy Innovation, Chemical and Biomolecular Engineering, University of Delaware, Newark, DE

The rapid increase in global energy demand, in conjunction with the declining fossil fuel reserves make it necessary to adopt new strategies for energy production and utilization. In addition, emissions from combustion processes, pose serious environmental and health challenges. Future energy production and storage will employ a diverse suite of technologies, including renewables, such as biomass and processes with improved energy efficiency. Heterogeneous nanocatalysts will play a keyrole in these technologies. Increased precision in molecular architecture over multiple length scales, and/or tailored multi-functionality will often be needed in designing such catalysts. Computational methods have and will continue to play an important role in overcoming these challenges, as the demand for a fundamental understanding of realistic catalyst structures necessitates novel simulation methodologies and multiscale theoretical approaches.

My research addresses these challenges by following a bottom-up approach. Using quantum chemical methods in my core research, I elucidate the underlying pathways for chemistries of practical interest, calculate reaction energies and barriers with high accuracy, and provide insights into complicated physicochemical processes (i.e. charge transfer from support-vacancies to nanoparticles, "magic number" stability of nanoparticles). In addition, I integrate this molecular-scale information into larger scale simulation frameworks to model the behavior of complex materials under realistic reaction conditions. In this poster, I will give an overview of examples which i) reveal the mechanisms of colloidal metal nanoparticle growth (combining quantum with molecular dynamics), ii) unravel complex catalytic phenomena on the CO oxidation reaction on metal oxide supported Au nanocatalysts (combining quantum with Kinetic Monte Carlo simulations) and iii) predict novel materials for Hydrogen storage applications (combining quantum with Grand Canonical Monte Carlo Simulations).

(3bf) Nanomedicines That Overcome Extra- and Intracellular Barriers

Anthony J. Kim, The Center for Nanomedicine, Johns Hopkins University School of Medicine, Baltimore, MD

No gene delivery system, viral or non-viral, has shown evidence of efficiently penetrating the human mucus barrier; this reality may account in part for the limited success in clinical trials for CF gene therapy to date. Gene vectors that fail to penetrate airway mucus layers are ultimately cleared from the lungs via mucus clearance mechanisms. In addition, poor gene transfer to the airway epithelium has been attributed to a number of cellular barriers, including poor cellular uptake across the apical membrane, unproductive intracellular trafficking, and inefficient nuclear import.

To potentially overcome these formidable barriers, we developed a polymeric nanoparticle platform capable of penetrating sputum expectorated from the lungs of patients with cystic fibrosis (CF), while maintaining an ability to provide efficient airway gene transfer in mice and in human airway epithelial cells grown at air-liquid interface. Mucus penetrating nanoparticles markedly enhanced particle diffusion in CF sputum, leading to enhanced particle distribution, retention and gene transfer in the mouse airways. These results represent an important step toward the rational development of an efficient nanoparticle platform for the lung diseases.

Ph.D. advisor:

John C. Crocker (University of Pennsylvania)

Postdoctoral advisor:

Daniel A. Hammer (University of Pennsylvania)

Justin Hanes (Johns Hopkins University)

Funding: NIH Fellowship (F32 HL103137)

(3bg) Membranes for Hydrogen Separation and Adsorbents for Sulfur Removal: Application to IGCC Plants

Bahman Elyassi, Chemical Engineering & Materials Science, University of Minnesota, Minneapolis, MN

Meeting the world's rising energy demands in ways that are sustainable and environmentally compatible highlights the importance of developing advanced energy efficient separation and emission control technologies. While membrane separation is widely considered as one of the most promising alternatives to conventional energy-intensive industrial separations, high cost and difficulties in scale-up constrain their commercial application for high-temperature separations. To address these issues, we recently demonstrated a new concept to synthesize zeolite membranes for hydrogen separation using layered (2D) zeolites. The fabrication method consists of exfoliation of lamellar zeolites to the constituent nanosheets and forming well-packed monolayers on porous substrates. The pores that are perpendicular to the monolayers selectively transport hydrogen through the membrane. The proposed approach lays the foundation for the fabrication of viable membranes using a range of layered zeolites. My other work focused on sulfur removal using solid sorbents. In this project, I investigated the synthesis of mixed metal oxides with high H_2S adsorption capacity.

These two projects aim to synergistically improve the economy of integrated gasification combined cycle with carbon capture and sequestration (IGCC/CCS) technology. IGCC/CCS is of particular interest to coproduce electricity, hydrogen, fuel, and chemicals with near-zero emissions. However, the challenge associated with CCS technology is to improve the economy of the process to close the cost gap between IGCC/CCS and conventional coal-fired power plants. The Department of Energy (DOE) has reported that by integrating warm-gas cleanup technologies into IGCC plants, an improvement of 8% in thermal efficiency over conventional cold-gas cleanup methods is achievable. In this context, two aspects of IGCC/CSS plants will be discussed that both aim to eliminate the energy penalty associated with carbon capture and sulfur removal: hydrogen purification using membranes at high temperatures and desulfurization of fuel gas using solid sorbents.

(3bh) Understanding Nature's Catalysts: The Theoretical Description of Metallocluster Containing Enzymes

Sandeep Sharma, Department of Chemistry, Princeton University, Princeton, NJ

Metalloclusters are transition metal containing prosthetic groups found in all forms of life and facilitate some of the most remarkable chemistry on earth. For example, iron-sulfur containing enzymes are able to reduce atmospheric N_2 to ammonia at ambient conditions, as compared to Haber-Bosch process that is carried out at 400 C and 200 bar and which uses more than 1% of global energy production. As another example, nickel and molybdenum containing carbon monoxide dehydrogenase (CODH) can electrochemically convert CO_2 to CO with essentially no over-potential, whereas the best known synthetic catalysts, made of gold, are not only expensive but also require significant over-potential. Given the remarkable properties of metalloclusters, the detailed theoretical understanding of their action is clearly central not only to problems of biology, but also to inform new paradigms for designing materials for catalysis, energy, and beyond.

In my phd I studied the complex reaction mechanisms of soot formation and proposed pathways for formation of soot precursors using quantum chemical methods. My work demonstrated that it is possible to understand and even quantitatively predict the kinetics of highly complex, real-life systems, so long as reliable quantum chemical methods are available [C&F, 2010]. However, although commonly used quantum chemistry methods work well for the kinds of organic molecules that I had studied, they typically fail for metalloclusters, sometimes catastrophically, due to the strong electronic correlations in transition metal electronic structure.

Recently, with the advent of the Density Matrix Renormalization Group (DMRG) method, quantum chemistry now has a method that is in principle capable of treating the kinds of strongly correlated electronic structure found in metalloclusters. In my postdoctoral work I have made significant contributions to the development and application of the DMRG to the quantum chemical systems [ARPC, 2011]. To the best of my knowledge I have written the most efficient implementation of DMRG, which is furthermore capable of targeting specific spin states in molecules [JCP, 2012]. I have also demonstrated the power of the DMRG to model small molecules containing Fe_xS_y groups, which will be described in greater detail in the poster.

During my faculty appointment I intend to combine my expertise in advanced electronic structure methods and reaction dynamics to explore the thermochemistry and kinetics of realistic metallocluster containing enzymes. In particular, I will work to bridge the gap between the kinds of model problems that new electronic structure methods are developed for, and the challenges of describing kinetics in real materials and biological problems. This will require not only the applications of the current tools, but also development of new methodologies. A parallel direction will be to use our theoretical understanding of biological motifs to guide the design of synthetic metalloclusters for catalysis.

(3bi) Atomistic-Level Investigation of Efficient Energy Conversion and Storage

Dong-Hee Lim, Energy Resources Engineering, Stanford University, Stanford, CA

A first-principles approach based on density functional theory (DFT) calculations coupled with thermodynamics greatly facilitates finding solutions to energy-related challenges, providing knowledge of atomic structures, reaction mechanisms, and electronic properties. My research interests focus on atomic and molecular-level investigations of efficient energy conversion and storage using approaches based upon quantum mechanics and thermodynamics. I am specifically interested in understanding and enhancing the electrochemical CO_2 conversion to hydrocarbon-based fuels, the oxygen reduction reaction (ORR) in polymer electrolyte membrane (PEM) fuel cells, and the phase stability of metal complex hydrides for hydrogen storage.

The main challenge for advancing CO_2 reduction is to improve the energy efficiency. Although copper has been widely accepted as a promising metal for the reduction of CO_2 to hydrocarbon fuels, its conversion is primarily hindered by the key step of CO protonation since CO binds only weakly to copper. This large kinetic barrier may be overcome by utilizing nanostructures of copper on graphene supports due to changes associated with nanoparticle size that may enhance the surface activity of copper for improved stabilization of the intermediates of CO_2 reduction. Furthermore, graphene support may not only prevent the sintering of copper nanoparticles, but may also provide a balance in the intermediate binding strength that may allow for enhanced catalyst turnover.

The ORR is one of the central focuses amidst ongoing studies of electrode reactions in polymer electrolyte membrane (PEM) fuel cells due to the slow kinetics taking place at the cathode electrode. To improve ORR kinetics, Pt and Pt alloy nanocatalysts supported on graphene, graphene nanoplatelets, and nanoscale graphite have been recently utilized. These supports reduce the high cost of the precious metal, improve the kinetics, and increase the durability of Pt support materials. Despite efforts on experimental investigations of the ORR mechanisms, few theoretical studies of the ORR mechanisms on graphene-supported Pt nanoparticles have been conducted. We provide details of the mechanisms of ORR on defective graphene supported-Pt nanoparticles.

Complex hydrides including alanates ($[AlH_4]^-$) have recently gained attention as alternative hydrogen storage materials. Many of these materials have been known to release H_2 upon contact with water; however, the hydrolysis reactions are highly irreversible, a process known as "one-pass" hydrogen storage. Nanostructuring and nanocatalysis have been accepted as promising methods to overcome the irreversible hydrogenation process. Predicting which phases may be more stable as a function of nanoparticle size may contribute to nanostructuring complex hydrides for hydrogen storage. We provide phase stability diagrams of the nanoparticle of $Mg(AlH_4)_2$ as a function of particle size and temperature using DFT calculations, cluster expansion, and Monte Carlo simulations.

(3bj) Optical Nanoscopy Will Enable the Creation of New Materials

Chaitanya K. Ullal, Max Planck Institute for Biophysical Chemistry, Goettingen, Germany

Research Motivation

Much of the success in the design of material properties is due to the control of the structure of matter. Simultaneously, contemporary challenges in Energy, Health and Computing demand novel materials with previously unattainable, multifunctional or maximized properties. My research experience and interests broadly reflect the importance of structure in the creation of new materials. My specific interests lie in the study of three dimensional (3D) nanoscale morphology. In particular, the nascent ability of Optical Nanoscopy to image deep within the bulk of nanostructured materials will be exploited. The research dream is to "dial-in" material properties by controlling 3D morphology on the nanoscale with exquisite precision. Soft Matter, Optics and Unconventional Lithography are the primary themes of my work. Short term projects in self assembled polymers, multifunctional materials and transformation optics that target immediate needs in Energy, Health and Computing have been identified.

Optical Nanoscopy

This poster presents far-field optical nanoscopy as a viable candidate for the 3D imaging of polymeric nanostructures. Non-invasive imaging with nanometric resolution that collects local, dynamic, molecular specific and 3D structural information can shorten the process of establishing structure–property relationships. Conventionally, real space imaging (as opposed to scattering based methods) at the nanoscale has been performed with electron or scanning probe microscopes. However, these techniques do not provide easy access to truly 3D information, either yielding 2D projections of thin sections or being restricted to near surface morphologies. Although fluorescence microscopy meets the demands of in situ 3D imaging, the width of the focal spot, and thus the resultant resolution, was till recently limited by diffraction of the light used. This barrier has now been circumvented by exploiting the on–off switching of the fluorescence ability of markers.

Stimulated emission depletion (STED) microscopy, which produces tiny focal spots by deactivating molecules on the outer rim of the diffraction-limited spot is a powerful implementation of this approach. The singular strengths of fluorescence microscopy that are retained by STED are particularly relevant to polymeric nanostructures. This includes the ability to non-invasively image within the volume of a material, capture dynamic phenomena, and the ease and specificity with which target phases can be tagged. Specific examples that would be challenging to image with contemporary techniques, such as the non-destructive nanoscale imaging of colloidal crystals and block copolymers that consist of 3D data stacks; the recording of assembling colloidal 2D nanostructures at speeds as high as 200 Hz; the in situ imaging of solvent swollen block copolymers and the development of a flexible microdomain specific tagging technique are presented. The ability of STED to "see" within polymers with nanometric resolution highlights the enormous potential of subdiffraction far-field fluorescence microscopy for the polymer sciences.

(3bk) Biomolecular Recognition of Receptor Proteins and Their Roles in Tumor Cell Adhesion in the Vasculature

Luthur Siu-Lun Cheung, Chemical and Biomolecular Engineering, Johns Hopkins University, Baltimore, MD

Cell-cell adhesive interactions play a pivotal role in major pathophysiological vascular processes, such as inflammation, thrombosis, and cancer metastasis, and are regulated by hemodynamic forces generated by blood flow. Elucidating molecular and biophysical properties of membrane receptor-ligand interactions may provide guidelines for developing promising therapeutic strategies to combat these fatal diseases. Membrane receptor-ligand interaction is not a simple lock-and-key problem in which only the right key can operate the lock. The protein-protein interface is constructed by a series of amino acid residues; therefore, the binding free energy may not be uniformly distributed among these interfacial residues. The goal of my research career is to elucidate the biomolecular recognition of receptor proteins and their roles in tumor cell adhesion in the vasculature.

My research expertise is in the area of microfluidics and protein-ligand interactions. I received Ph.D. at the University of Arizona, where my dissertation focuses on the effect of shear flow on the attachment, deformation and detachment of tumor cells in microchannels. In addition to research training, I also had a precious opportunity to participate in the development of a new laboratory. At the Johns Hopkins University, I continue my research with Prof. Konstantinos Konstantopoulos to investigate the complex interplay and structure-activity relationship of receptor-ligand interactions pertinent to inflammation and cancer metastasis. In 2011, I received a postdoctoral fellowship award from the American Heart Association.

There are three core topics in my research as follows: (1) Lab-on-a-chip devices: Design and fabricate microfluidic devices for studying the cell adhesion in pathophysiological vascular processes, such as inflammation and cancer metastasis. (2) Bio-molecular recognition: Quantify the kinetic and micromechanical properties (such as dissociation rate, reactive compliance and tensile strength) of membrane protein-ligand interactions under dynamic loading. (3) Tumor cell mechanics: Develop models and experimental techniques to study the effect of shear flow on the cell deformation and migration during intra- and extra-vasation processes. Cumulatively, these studies will provide scientific insights on how the tumor cell receptors interact with their target molecules in the vasculature.

Selected peer-reviewed publications:

1. Z. Tong*, **L.S.L. Cheung***, K.J. Stebe and K. Konstantopoulos, "Selectin-mediated Adhesion in Shear Flow Using Micropatterned Substrates: Multiple-Bond Interactions Govern the Critical Length for Cell Binding", *Integr Biol*, 2012 (accepted). *Equally contribution authors

2. **L.S.L. Cheung**, M. Kanwar, M. Ostermeier and K. Konstantopoulos, "A Hot-Spot Motif Characterizes the Interface Between a Designed Ankyrin-Repeat Protein and Its Target ligand". *Biophys J*, 102(3), 407-416, 2012.

3. X.J. Zheng, **L.S.L. Cheung**, J.A. Schroeder, L. Jiang and Y. Zohar, "Cell Receptor and Surface Ligand Density Effects on Dynamic States of Adhering Circulating Tumor Cells", *Lab Chip*, 11(20), 3431-3439, 2011.

4. X. Zheng, **L.S.L. Cheung**, J.A. Schroeder, L. Jiang and Y. Zohar, "A High-Performance Microsystem for Isolating Circulating Tumor Cells", *Lab Chip*, 11(19), 3269-3276, 2011.

5. **L.S.L. Cheung**, X. Zheng, L. Wang, J.C. Baygents, R. Guzman, J.A. Schroeder, R.L. Heimark and Y. Zohar, "Adhesion Dynamics of Circulating Tumor Cells under Shear Flow in a Bio-Functionalized Microchannel", *J Micromech Microeng*, 21(5), 054033, 2011.

6. P. Sundd, M.K. Pospieszalska, **L.S.L. Cheung**, K. Konstantopoulos and K. Ley, "Biomechanics of Leukocyte Rolling", *Biorheology*, 48, 1-35, 2011.

7. **L.S.L. Cheung**, K. Konstantopoulos, "An Analytical Model for Determining Two-Dimensional Receptor-Ligand Kinetics", *Biophys J*, 100(10), 2338-2346, 2011.

8. **L.S.L. Cheung**, P.S. Raman, E.M. Balzer, D. Wirtz and K. Konstantopoulos, "Biophysics of Selectin–Ligand Interactions in Inflammation and Cancer", *Phys Biol*, 8(1), 015013, 2011.

9. T. Gudipaty, M. T. Stamm, **L.S.L. Cheung**, L. Jiang and Y. Zohar, "Cluster Formation and Growth in Microchannel Flow of Dilute Particle Suspension", *Microfluid Nanofluid*, 10(3), 661-669, 2010.

10. **L.S.L. Cheung**, X. Zheng, L. Wang, R. Guzman, J.A. Schroeder, R.L. Heimark, J.C. Baygents and Y. Zohar, "Kinematics of Specifically Captured Circulating Tumor Cells in Bio-Functionalized Microchannels", *J Microelectromech S*, 19(4), 752-763, 2010.

11. **L.S.L. Cheung**, X. Zheng, A. Stopa, J.C. Baygents, R. Guzman, J.A. Schroeder, R.L. Heimark and Y. Zohar, "Detachment of Captured Cancer Cells under Flow Acceleration in a Bio-functionalized Microchannel", *Lab Chip*, 9, 1721-1731, 2009.

12. M. Lee, **L.S.L. Cheung**, Y.K. Lee and Y. Zohar, "Height Effect on Nucleation-Site Activity and Size-Dependent Bubble Dynamics in Microchannel Convective Boiling", *J Micromech Microeng*, 15, 2121-2129, 2005.

Book chapters:

13. **L.S.L. Cheung**, P.S. Raman, D. Wirtz and K. Konstantopoulos, "Biophysics of Selectin-Mediated Cell Adhesion", *Comprehensive Biophysics: Cell Biophysics,* Elsevier, 7, 2012 (in press May 31st).

14. L.M. Lee, **L.S.L. Cheung** and Y. Zohar, "Microfluidics: Device Science and Technology", *CISM Courses and Lectures, International Centre for Mechanical Sciences*, Springer, 428, 157-211, 2006.

(3bl) Design of Advanced Catalytic Systems Through Computational Methods

Diego A. Gomez Gualdron, Chemical Engineering, Texas A&M University, College Station, TX

Catalysts are materials pervasive to many areas such as petroleum refining, fuel cell technology, hydrogen production, fine synthesis, food processing, and pollutant elimination, among others. Catalysts typically work at the nanoscale, which creates huge opportunities to design application-driven new catalytic systems through rational manipulation and design of nanoscale structures. On the other hand, recent advancements in computation have solidified *molecular simulations* and *computational chemistry* as powerful tools to investigate phenomena at the nanoscale, and to design materials using a bottom-up approach.

As a materials scientist possessing an expertise in molecular simulation, my research interests encompasses the utilization of simulation methods such as density functional theory (DFT), molecular dynamics (MD), and multiscale methods to the design of advanced catalytic systems for: 1) Synthesis of advanced materials (e.g. synthesis of single-walled carbon nanotubes (SWCNTs)), 2) Energy-efficient (e.g. photocatalysis) synthesis of chemical compounds, 3) Synthesis of alternative fuels (e.g. hydrogen), and 4) bio-inspired heterogeneous catalysis. In planning my research I always consider the experimental component that in a multidisciplinary context I expect from multiple collaborations at the departmental, college, and other institutions levels.

Here I show research performed during my doctoral studies in the Materials Science & Engineering Program at Texas A&M University, under the supervision of Dr. Perla Balbuena in the Artie McFerrin Chemical Engineering Department. I focus on the design of a catalytic system for the chiral-selective synthesis of SWCNTs. The chiral-dependent electronic and optical properties of SWCNTs promise to revolutionize the fabrication of power-efficient electronic devices, and ultrasensitive bio- and chemo-sensors, as well as introduce new diagnosis and therapeutic strategies. However, with the availability of commercial quantities of nanotubes of predetermined properties as a prerequisite for the exploitation of these technologies, the need for chiral-selective nanotube synthesis is evident.

Using the interaction of the dispersed catalyst nanoparticles with the support as a means of controlling the nanoparticle structure/dynamics we examine the conditions to create a template effect that helps guide the nanotube growth toward a desired chirality. Using molecular simulation we investigate the mechanism that establishes a correlation between the nanoparticle and nanotube structures, and the impact of nanoparticle size and strength of interaction with the support on the template effect effectiveness. We have found that this effectiveness is affected, because the latter factors affect the rate of defect annealing, nanoparticle/nanoparticle contact, the heterogeneity of the nanoparticle surface, and the mobility of the nanoparticle. We have also obtained valuable information about the nanotube growth mechanism, and examined the implications for chirality control and the accuracy of the vapor-liquid-solid (VLS) mechanism to describe nanotube growth.

(3bm) In Vivo in Vitro and in Situ Biosensor Applications

Kevin J. Cash and Heather A. Clark, Pharmaceutical Sciences, Northeastern University, Boston, MA

Recent years have shown the incredible versatility and utility of polymeric nanosensors. Sensors have been developed for a wide variety of ionic analytes (e.g. sodium, potassium, pH) as well as nonionic analytics (e.g. glucose) and have been applied for challenging research fields such as detection of sodium sparks in live cardiomyocytes. This field of research has yielded new knowledge on cellular metabolism and function. Additionally, these sensors can be applied in vivo, through implantation in or under the skin in animals; showing potential as a research tool for continuous monitoring of analyte concentrations without the need for direct samples to be taken from the animal.

To date the choice of recognition element has limited the range of targets which can be measured with these polymeric nanosensors. This stems from the need for small molecule recognition elements (e.g. ionophores) which can be incorporated into the polymer matrix. Through the combination of these nanosensors with enzymatic recognition elements we can now detect a far wider variety of biologically relevant targets through detection of metabolic byproducts of enzymatic reactions. This poster will discuss continuous monitoring of histamine concentrations in small animals as a proof of concept for eventual monitoring of a suite of physiological parameters. This will also form the foundation of my research program which will focus on the development of novel biomedical diagnostic approaches for *in vivo* and *in vitro* monitoring, and will also leverage the advances made in biomedical research applications toward improved process monitoring and analytics.

Postdoctoral Research Advisor

Heather A. Clark, Department of Pharmaceutical Sciences, Northeastern University, Boston MA

Graduate Research Advisor

Kevin W. Plaxco, Department of Chemistry and Biochemistry, University of California, Santa Barbara

Funding

KJC: NIH/NIBIB F32 EB015270-01

HAC: DARPA, NIH/NIGMS R01 GM084366

(3bn) Application of Systems Biology Tools to Investigate Anti-Angiogenic Cancer Therapies

Stacey D. Finley, Biomedical Engineering, Johns Hopkins University, School of Medicine, Baltimore, MD

Angiogenesis is the formation of new blood vessels from pre-existing vasculature. This tightly regulated biological process is involved in physiological function such as wound healing and exercise, as well as in pathological conditions, including preeclampsia, ischemic heart disease, and cancer. Inducing angiogenesis is a hallmark of cancer, as tumors cannot grow beyond 1 mm in diameter without eliciting the formation of blood capillaries to supply oxygen and other nutrients. Vascular endothelial growth factor (VEGF) is a key regulator of angiogenesis and its role in cancer biology has been widely studied. Given the action of VEGF in promoting angiogenesis, it has been targeted in various cancer treatments.

Systems biology approaches, including computational models, are useful in gaining insight into the complexity of tumor angiogenesis. These models provide a framework to test biological hypotheses and optimize effective therapies that aim to inhibit tumor vascularization and growth. In this poster, I describe the development of whole-body compartment models of VEGF kinetics and transport in mice and humans and the application of these models to predict the effect of various anti-angiogenic therapies that target VEGF. The mouse model has been optimized using experimental data and complements pre-clinical drug studies. The human model is applied to interpret clinical data. In both instances, the models predict how properties of the tumor microenvironment influence the response to anti-angiogenic agents, which is difficult to quantify clinically. The models aid in the development and optimization of personalized cancer treatment strategies that target the VEGF pathway.

During my doctoral research at Northwestern University, I worked with Profs. Linda Broadbelt and Vassily Hatzimanikatis to apply computational tools to predict and characterize novel metabolic pathways. My current postdoctoral work with Prof. Aleksander Popel at Johns Hopkins University applies a molecular-detailed model of VEGF kinetics and transport to understand the effect of therapeutic agents that inhibit VEGF signaling. In the next phase of my career, I would like to establish a research lab that applies engineering and systems biology tools to study relevant biological questions related to the development of optimal cancer therapies. This work will build upon my training in Chemical Engineering and biomedical research. As a starting point, I aim to develop integrative computational models of cancer metabolism and VEGF-mediated angiogenesis, two processes that characterize cancer. These models will be applied to investigate how tumor cell metabolism and VEGF activation lead to tumor angiogenesis and how these processes can be targeted in order to inhibit the development of cancer. The poster will present the aims of this proposed research.

(3bo) Design and Engineering of Novel Nanomaterials for Energy and Environmental Sustainability

Camille Petit, Department of Earth and Environmental Engineering and Department of Chemical Engineering, Columbia University, New York, NY, Teresa Bandosz, Department of Chemistry, The City University of New York, New York, NY and Ah-Hyung Alissa Park, Earth and Environmental Engineering & Chemical Engineering, Columbia University, New York, NY

Energy, water, and the environment have been identified as part of the "*Top Ten Problems Facing Humanity Over the Next 50 Years,*" as reported by the National Academy of Sciences. Addressing those challenges is thus clearly part of the mission of the scientific and engineering community today and in the years to come. Innovative and transformative approaches and solutions must be developed and nanotechnology has a critical role to play towards the path to sustainability.

This poster will provide an overview of my research in the synthesis and characterization of nanoscale materials for various energy and environmental technologies. Specifically, my graduate work consisted of the design and physical and chemical tuning of reactive carbon-based adsorbents for the gas separation of small molecules, such as ammonia. The solid adsorbents developed include modified activated carbons and graphite oxide and metal-organic framework/graphene composites. The nanostructure and chemistry of these materials were designed and controlled to provide strong adsorption forces that allowed these adsorbents to greatly surpass the efficiency of conventional physical adsorbents. During the last year, my postodoctoral work focused on the synthesis, characterization, and evaluation of novel liquid-like nano-scale inorganic-organic hybrid materials for CO_2 capture. These materials consist of an inorganic core tethered with a polymeric canopy that acts as the solvating medium. CO_2 capture by these materials is expected to arise from an enthalpic effect via selection of task-specific groups, as well as an entropic effect via the specific structural arrangement of the polymer chains. This unique feature in concert with the negligible vapor pressure and high thermal stability of these materials make them compelling candidates to replace the conventional amine-based solvents. The research efforts are directed towards the structural characterization of these novel materials as well as the engineering of the canopy (functionalization and structural design) to understand fundamental aspects of CO_2 capture by those materials and optimize their capacity. The findings from all these studies have also provided important insight into a wide range of novel nanoscale materials that could be critical for a number of research areas (e.g., gas separation, energy storage, heterogeneous catalysis, sensing).

Building on my expertise in materials science, environmental chemistry, and separation processes, I would like to focus my academic career on the design, synthesis and investigation of nanomaterials as a platform to address critical issues for environmental, water, and energy sustainability.

(3bp) Large Amplitude Oscillatory Flow, a Microstructural Perspective

James W. Swan, Department of Chemical and Biomolecular Engineering, University of Delaware, Newark, DE

A microstructural understanding of the origin of the large-amplitude rheological response of materials is an on-going challenge. I have conducted a detailed study of the response to large amplitude oscillatory flows (LAOF) from a micromechanical perspective -- the microrheology of colloidal dispersions. The solution to this problem combines the micro-viscoelasticity with the material microstructure and provides considerable insight into the physical processes involved in large amplitude flows. In particular, the relative importance of hydrodynamic and thermal forces determines the amount of energy stored and lost during the process. Interestingly, while hydrodynamic interactions are largely responsible for the behavior of strongly driven materials, I show that hydrodynamics cannot contribute to net energy storage. A comparison of the response for driven suspensions with and without hydrodynamic interactions reveals that a detailed understanding of the material microstructure is necessary for meaningful interpretation of large amplitude oscillatory flows. This simple model provides such details and recovers the rich behavior observed experimentally in a broad class of materials -- ranging from colloidal dispersions to polymer melts and even slug mucus.

(3bq) Microrheological Characterization Techniques for Biological Applications and Soft Material Design

Kelly M. Schultz, Department of Chemical and Biological Engineering, University of Colorado at Boulder, Howard Hughes Medical Institute, Boulder, CO, Eric M. Furst, Department of Chemical and Biomolecular Engineering, University of Delaware, Newark, DE and Kristi S. Anseth, Department of Chemical and Biological Engineering, University of Colorado, Howard Hughes Medical Institute, Boulder, CO

Synthetic scaffolds are designed to be remodeled and degraded by migrating cells for applications such as tissue engineering, wound healing and stem cell culture. Although there are distinct advantages of these materials, little is known about the interactions cells have with these synthetic microenvironments, especially in the region immediately surrounding the cell, the pericellular region. Development of characterization methods that will measure the rheological properties of the hydrogel material, both with and without the presence of encapsulated cells will enable further advances within the field. Specifically, microrheological characterization techniques of synthetic hydrogel scaffolds have been developed to quantify the rheological properties of equilibrated scaffold materials and through transient conditions, such as those occurring during gelation and degradation.

My doctoral research focused on developing high-throughput characterization techniques using multiple particle tracking microrheology (MPT) that would rapidly measure material rheological properties while conserving both material and time. MPT is a passive microrheological technique that measures the Brownian motion of embedded probe particles. The trajectories of these particles are directly related to rheological properties, such as viscosity and creep compliance, using the Generalized Stokes-Einstein Relation (GSER). This technique requires small sample volumes (4-40 μL) while enabling rheological measurements of both the hydrogelation reaction and the final material properties. During the hydrogelation reaction, the material is monitored with time and the sol-gel transition is measured, yielding the critical gelation time and critical relaxation exponents. High-throughput microrheological measurements in a microfluidic device, μ^2rheology, are then used to screen the equilibrated materials, significantly reducing the time of sample preparation and increasing the amount of samples evaluated. From the data, gelation state diagrams identify compositions where hydrogels form. Overall, our methods can be used to thoroughly screen the material rheological properties, yielding a high information density, but requiring only small volumetric amounts.

Building on my expertise in microrheological characterization, my postdoctoral research focuses on the characterization of hydrogelators during enzymatic and cellular mediated degradation. Our approach has been to measure the enzymatic degradation of a synthetic scaffold as the amount of matrix metalloproteinases (MMP) degradable peptide cross-links (KCGPQG*IWGQCK) is varied. The ability to tailor the amount of degradation in the material enables fundamental studies of the material properties with degradable cross-links close to the critical fraction of cross-links required to form a gel, $p_{c,A}$, calculated using Flory-Stockmayer theory. The material degrades at slightly higher concentrations of degradable cross-links than predicted with Flory-Stockmayer theory, due to nonidealities present in the matrix and not accounted for in the theory. With these investigations we are able to identify the gel-sol transition and spatially characterize material properties during degradation. These materials and results can then be used to study the pericellular region of encapsulated cells during remodeling and degradation of the material. These experiments will enable the identification of the rheological properties and spatial variations that are required to enhance cell motility.

(3br) Characterization of Fractal-Like Aerosols During Sintering

Max Eggersdorfer, Particle Technology Laboratory, Swiss Federal Institute of Technology (ETH Zurich), Zurich, Switzerland

Characterization of Fractal-like Aerosols during Sintering

M.L. Eggersdorfer

Particle Technology Laboratory, Institute of Process Engineering, Department of Mechanical and Process Engineering, ETH Zurich, Sonneggstrasse 3, CH-8092 Zürich, Switzerland.

Nanoparticle production in the gas phase is a well-established route for carbon black, pigmentary titania and fumed silica and a promising method for many new, sophisticated materials like catalysts, sensors, phosphors, biomaterials and even nutritional products. Such particles grow by gas and surface reaction, coagulation and sintering and tend to form fractal-like aggregates and agglomerates affecting their transport, light scattering, effective surface area and density [1]. The (real-time) characterization of these structures and their constituent primary particles is necessary for continuous monitoring of aerosol manufacturing and airborne pollutant particle concentrations. Significant advances have been made in characterization of agglomerates (physically –bonded particles) by employing fractal theory and relating agglomerate structure to its generation pattern through the fractal dimension, D_f. What might have been overlooked in characterization and simulations of fractal-like particles is that the above D_f values have been developed for agglomerates of *monodisperse* primary particles. For coagulating aerosols, however, this needs to be carefully examined as Brownian coagulation leads to polydisperse particles [2]. Furthermore, once coalescence or sintering starts between these primary particles, sinter necks are formed between them converting the agglomerates to aggregates [3]. Accounting for primary particle polydispersity is important as the characteristic sintering time depends strongly on primary particle size [4]. These properties may also affect their health impact [5], e.g. agglomerates may undergo restructuring & break-up [6] and release constituent primary particles.

The research during my PhD focused on the formation of aggregates (chemically- or sinter-bonded particles) by viscous flow sintering of amorphous materials (silica, polymers) [3] and grain boundary diffusion sintering of crystalline ceramics (titania, alumina) or metals (Ni, Fe, Ag etc.) [7] by multiparticle sintering simulations. A scaling law was discovered between average aggregate projected area and equivalent number of constituent primary particles during sintering: from fractal-like agglomerates to aggregates and eventually compact particles. This is a relation essentially independent of time, material properties and sintering mechanisms. So the surface area mean primary particle diameter was determined by (on-line) differential mobility analyzer (DMA) and aerosol particle mass (APM) analyzer measurements and this power law for aggregates. This primary particle diameter obtained during particle synthesis is in good agreement with ex-situ measurements like nitrogen adsorption and particle counts from microscopic images.

[1] P. Meakin, Fractal aggregates, *Adv. Colloid Interface Sci.* 28 (1988) 249.

[2] M.L. Eggersdorfer, S.E. Pratsinis, The structure of agglomerates consisting of polydisperse particles, *Aerosol Sci. Technol.* 46 (2012) 347.

[3] M.L. Eggersdorfer, D. Kadau, H.J. Herrmann, S.E. Pratsinis, Multiparticle sintering dynamics: from fractal-like aggregates to compact structures, *Langmuir* 27 (2011) 6358.

[4] M. Sander, R.H. West, M.S. Celnik, M. Kraft, A detailed model for the sintering of polydispersed nanoparticle agglomerates, *Aerosol Sci. Technol.* 43 (2009) 978.

[5] L.K. Limbach, P. Wick, P. Manser, R.N. Grass, A. Bruinink, W.J. Stark, Exposure of engineered nanoparticles to human lung epithelial cells: Influence of chemical composition and catalytic activity on oxidative stress, *Environ. Sci. Technol.* 41 (2007) 4158.

[6] M.L. Eggersdorfer, D. Kadau, H.J. Herrmann, S.E. Pratsinis, Fragmentation and restructuring of soft-agglomerates under shear, *J. Colloid Interface Sci.* 342 (2010) 261.

[7] M.L. Eggersdorfer, D. Kadau, H.J. Herrmann, S.E. Pratsinis, Aggregate morphology evolution by sintering: Number & diameter of primary particles, *J. Aerosol Sci.* 46 (2012) 7.

(3bs) Targeting Molecular Simulation Tools Toward Bioengineering Applications

Galen Collier, Chemical Engineering, Columbia University, New York, NY

The general motivation behind my research interests is a desire to find ways of harnessing the full capabilities of biomolecules as bioengineering tools. The majority of my research work has focused on the development and validation of protein simulation methods used to address fundamental challenges in bioengineering. My initial research focus was on nuclear magnetic resonance investigations of modified DNA structures for the purpose of elucidating the nature of their interactions with, and recognition by, various enzymes. This early experimental work was supplemented by molecular simulation, which has proved to be the most powerful and productive research tool currently available for nanoscale structural studies. Therefore, to quickly advance my studies of biomolecules as bioengineering tools, I transitioned to purely computational research efforts. This work has included studies of a variety of wild-type and engineered biomolecules used as, or in the presence of, bioengineered structures. Recently, my research has included a previously unexplored comparison of molecular dynamics force fields as used in the simulation of peptide and protein adsorption to material surfaces. Most recently, I have begun the development of new simulation tools that will enable the study of complex structural properties proteins and the mechanisms underlying protein function.

As an independent investigator, I expect to continue my work focused on the use of biomolecules as bioengineering tools with particular emphasis on the development and evaluation of purpose-built biomolecular structures. This work will be divided amongst the development of new computational research tools, studies of engineered biomolecules, and studies of the interactions between biomolecules and material surfaces. The nature of this work dictates regular collaboration with researchers focusing on experimental techniques, so my research will provide a variety of opportunities for collaboration. Through my research and teaching, I hope to engage students with diverse educational backgrounds in ways that will take advantage of their existing competencies and push them in the direction of developing new ones that will be needed soon in their research careers. I have a great deal of coursework development and teaching experience, as well as experience with mentoring graduate and undergraduate students.

During this poster session, in addition to welcoming the discussion of my qualifications as a new faculty candidate, I will present details of my work in developing simulation tools for the study of allosteric pathways in proteins. A brief overview of this project is included below.

Allosteric regulation of protein function is a mechanism by which a structural event occurring at one site within a protein's overall structure causes an effect at another site, and allostery plays a central role in a huge variety of mechanistic pathways in biological systems. The prediction and modulation of allosteric responses in proteins holds promise for applications in a diverse range of research areas, including drug design and materials engineering. The structural changes involved in protein allostery include deformations where local elastic moduli play a major role. Knowledge of the elastic modulus at each atom position or within specific regions in a protein can provide valuable information about the local mobility and stability under conformational deformations. However, most methods used for calculating local elastic moduli cannot be applied to proteins in a straightforward manner because they require second derivatives of the potentials that describe the interactions amongst a protein's atoms. To address this issue, I have begun the development of computational methods that enable the measurement of elastic moduli at different positions in a protein in order to build a 3D map of local mechanical properties that can be used to identify pathways of allostery in proteins and protein networks. Presented here are the results from a preliminary set of protein structural analyses completed using these methods to calculate elastic moduli in an atomistic fashion.

(3bt) Systems Biology of Cancer Signaling and Metabolism

Nicholas A. Graham, Medical and Molecular Pharmacology, University of California, Los Angeles, Los Angeles, CA

Biology has become a multiparameter, information-based science thanks to new measurement technologies (eg, genome-wide microarrays, mass spectrometry-based proteomics). This transformation has enabled an integrated, systems description of disease that bridges traditional biology and quantitative sciences. I am passionate about the power of quantitative systems biology to uncover novel biology and suggest new therapeutic routes, with a particular focus on cancer.

My graduate research at Caltech focused on crosstalk within biomolecular signaling networks and how these signaling networks interpret and process local cues to make decisions about fundamental cellular behaviors such as proliferation. These studies provided a foundation for my postdoctoral studies at the David Geffen School of Medicine at UCLA in the systems analysis of cancer, with particular focus on the links between signaling and metabolism.

Here, I will present two studies from my postdoctoral research: i) a systems analysis of crosstalk between glucose metabolism and tyrosine kinase signaling, uncovering a positive feedback loop by which metabolic perturbations lead to cell death, ii) development and application of a bioinformatic approach (self-organizing maps) to analyze multiparameter, single-cell data from a microfluidic device.

My future research will continue to synthesize my training in engineering and biological systems. I will focus on the integration of genomic, proteomic and metabolomic techniques to create systems models of disease. The future of biological research will require investigators who can synthesize and integrate quantitative biological data to generate novel biological insights, and my training in chemical engineering, biological sciences and bioinformatics has positioned me to become one of these interdisciplinary researchers.

PhD: California Institute of Technology (2007), Advisor Anand R. Asthagiri

Postdoc: University of California, Los Angeles, Advisor Thomas G. Graeber

(3bu) Influence of Microwave Band Irradiation On Catalytic Reforming Systems

Steven E. Edmund and Johannes W. Schwank, Chemical Engineering, University of Michigan, Ann Arbor, MI

Introduction:

Research and teaching in engineering have gone hand in hand during my time at the University of Michigan. I study the area of catalytic fuel reforming in the pursuit of alternatives to the traditional use of petroleum that has dominated the past century. The use of catalysts to create and move between different types of fuels holds great promise as a means to address our energy future. My research has provided many examples that I have been able to pull into the classroom, while by teaching I have achieved a greater depth of understanding in those fields. As an educator I seek to guide students into the deeper problem-solving and critical-thinking skills needed to solve challenging problems and ask insightful questions within the broader context of their field. Teaching has been a rewarding process as it continually morphs and changes with the students, who are themselves growing into more informed and educated people. I find an inherent delight in assisting someone to understand a new truth about the workings of the world.

Research:

Catalytic reforming of hydrocarbon fuels is frequently plagued by deactivation via carbon deposition and sulfur poisoning. Further deactivation may occur when catalysts are operated adiabatically without a sufficient exotherm to heat the catalyst and prevent oxidation of the active metal. The present study examines a means of dealing with hypothetical feed disturbance events and resulting deactivation from sulfur poisoning via irradiation of the catalyst in the microwave band (2.45GHz, λ=12.1cm). To probe interactions with electromagnetic radiation and to minimize internal field attenuation, a model system was chosen consisting of cordierite monolith supported catalysts, washcoated with a $Ni/Ce_{0.75}Zr_{0.25}O$

chemical synthesis applications. Sulfur was supplied to the reaction system in form of thiophene, with O/C and H_2O/C ratios chosen to represent scenarios where significant deactivation is expected.

For more information on my teaching and research please visit: http://www.stevenedmund.com/

(3bv) Design of Catalytic Materials with Targeted Electronic Properties

Hongliang Xin and Suljo Linic, Chemical Engineering, University of Michigan, Ann Arbor, MI

Catalysis plays an essential role in chemical conversion, energy production and pollution mitigation. Recent development of electronic structure theory leads to an unprecedented mechanistic understanding of molecular transformations at surfaces and the identification of improved catalysts from first-principles. However, the immense phase space of catalytic materials spanned by structural and compositional degrees of freedom precludes thorough screening, even with combinatorial quantum chemical calculations and high-throughput experiments. Development of predictive structure-reactivity relationships for rational design of catalytic materials with targeted electronic properties still remains a great challenge.

During my PhD studies, we have developed a general theoretical framework for the understanding of variations in the surface reactivity of catalytic materials upon a perturbation of their electronic properties. I will discuss our work on three projects: (1) *rapid screening of Pt multimetallic electrocatalysts for the oxygen reduction reaction in PEM fuel cells*; (2) *understanding of alkali promotion mechanisms for oxidation reactions on Ag surfaces*; (3) *coupling of phonons and energetic electrons for chemical reactions on optically excited nanostructures*. Each model system is characterized by one specific type of perturbation of the metal, introduced either by alloying with impurity elements, by doping of substrates with chemical promoters, or by exposing the surface of plasmonic nanostructures to visible photons. The fundamental knowledge-based developed in these studies can be readily applicable to other types of catalytic materials (e.g., oxides, sulfides and carbides) for many energy and environmental applications.

(3bw) Bio-Inspired Design of Adaptive, Dynamic, and Multi-Functional Materials and Architectures

Philseok Kim, School of Engineering and Applied Sciences, Harvard University, Cambridge, MA

Can we develop new materials and structures that can sense and respond to the environment to optimize, reconfigure, self-heal to maintain sustainability just like a living organism? My research theme is focused on "**bio-inspired design, synthesis, and characterization of adaptive and dynamic materials and architectures**" to develop energy-efficient adaptive building materials, nanostructured materials for electrochemical energy conversion and storage, and dynamic surfaces for interfacing biological systems and nanosystems. Keeping this big picture in mind, I have been always looking at the ***surface*** of a material or the ***interfaces*** of dissimilar materials. It would be impossible to discuss all the details of the projects I have been studying during this short meeting. However, I would be happy to discuss any of the following selected topics at your request at the AIChE meeting:

1. Hydrogel-Actuated Integrated Responsive Systems (HAIRS): Interfacing 'dynamic' materials with 'passive' structures to create adaptive materials

Taking the dynamic and adaptive nature of biological systems as a source of inspiration, I seek to develop sustainable, dynamic, and multi-functional materials that are able to adapt to the changing environment by sensing and responding to stimuli, reconfiguring their structures, and thereby generating new functions. One approach is to combine responsive materials, such as hydrogels, with various nano/microstructures in which the responsive materials harvest energy from the environment and transduce it into mechanical energy to reconfigure and reassemble the structures into a wide range of reversible surface patterns. These hybrid actuation systems generate a new class of adaptive materials that can respond to various environmental cues exhibiting reversibly switching functions important for applications in wetting, adhesive, color change, optics, anti-fouling, catalysis, capture and release, propulsion and transportation, and bio-nano interface. The versatility of this strategy is huge as any given stimulus-response pair can be put together in a variety of configurations to program how forces are synchronized and transferred between them, orchestrating responses that vary over a wide range of patterns on micro and macro scales. The design, synthesis, fabrication, and characterization of such new generation of multi-functional materials based on this strategy, hydrogel-actuated integrated response systems (HAIRS), along with potential applications as adaptive light redirecting and thermal gain regulating building envelopes (e.g. façades, glazings, and roofing systems) for energy-efficient buildings will be presented.

2. Bio-inspired Slippery Ice-repellent Coatings on Aluminum (SLIPS-Al): Wetting and adhesion of ice on a surface

Ice repellent coatings on aluminum can have significant impact on the global energy saving and improving safety in many infrastructures, transportation, and cooling systems. Recent efforts for developing ice-phobic surfaces have been mostly devoted to utilize lotus leaf-inspired superhydrophobic surfaces, yet these surfaces fail in high humidity conditions due to water condensation and frost formation and even lead to increased ice adhesion due to a large surface area. I have developed a radically different type of ice repellent coating material based on slippery, liquid-infused porous surfaces, SLIPS, that was previously developed in my group inspired by pitcher plant. SLIPS maintains an ultrasmooth and slippery liquid interface on a porous surface by infusing a water-immiscible liquid into it. I will discuss special advantages of SLIPS-coated aluminum as ice-repellent surfaces: extremely low contact angle hysteresis and ice adhesion strength, highly reduced sliding droplet sizes. I will also present results from icing/deicing experiments demonstrating the SLIPS coating on aluminum (SLIPS-Al) remains essentially 'ice-free' while conventional materials accumulate ice. These results indicate that SLIPS-Al is a promising candidate for developing robust anti-icing materials for broad applications even in high humidity, such as refrigeration, aviation, roofs, wires, outdoor signs, wind turbines, where ice can be easily removed by tilting, agitation (e.g. wind or vibrations), or even by a weak shear force.

3. Nanostructure Fabrication Using Conductive Polymers (STEPS and SCREEN): Electrodeposition of materials from solution onto a surface, Nucleation & Growth

Sophisticated 3D structures such as arrays of high-aspect-ratio (HAR) nano- and microstructures are of great interest for designing surfaces for applications in energy harvesting and storage, optics, and bio-nano interfaces. However, the difficulty of systematically and conveniently tuning the geometries of these structures significantly limits their design and optimization for specific applications. I have developed a low cost, high-throughput benchtop method that enables a HAR array to be reshaped with nanoscale precision by electrodeposition of conductive polymers. The method—named STEPS (structural transformation by electrodeposition on patterned substrates)—makes it possible to create patterns with proportionally increasing size of original features, to convert isolated HAR features into a closed-cell substrate with a continuous HAR wall, and to transform a simple parent two dimensional HAR array into new three-dimensional patterned structures with tapered, tilted, anisotropic, or overhanging geometries by controlling the deposition conditions. STEPS allowed for the fabrication of substrates with continuous or discrete gradients of nanostructure features, as well as libraries of various patterns, starting from a single master structure. By providing exemplary applications in plasmonics, bacterial patterning, and formation of mechanically reinforced structures, I will show that STEPS enables a wide range of studies of the effect of substrate topography on surface properties leading to optimization of the structures for a specific application. This work identifies solution-based deposition of conductive polymers as a new tool in nanofabrication and allows access to 3D architectures that were previously difficult to fabricate.

4. Mechano-responsive Optical Materials ("Stretch Me"): Mechanics of thin film under small deformation conditions

A stiff, thin membrane bonded to a compliant substrate undergoes surface buckling (i.e. wrinkles) when the membrane encounters a compressive load that exceeds the critical strain. The wrinkled patterns arising from these mechanically mismatched interfaces have been studied for mechanically tunable gratings, as well as a metrological tool for measuring the modulus of a thin film. Nevertheless, the dynamic changes in the optical transmittance of these wrinkled materials have not been studied in depth. I have investigated the mechano-responsive optical behavior of uni-axially pre-stretched and oxidized Sylgard 184 membranes and found that these materials undergo reversible transition between a transparent state and two opaque states. I will present detailed correlations between the optical transmittance spectra and the changes in the wrinkle patterns such as pitches, amplitudes, and orientations under given mechanical load, along with an analytical model to describe the observed mechanical behaviors and optical effects. The mechano-responsive and dynamically tunable optical properties of wrinkled elastomers offer numerous potential applications, including instantaneous privacy screens, reversible switching of microstructured patterns, data encryption, shadings and windows for buildings, and mechano-optical sensors, all from a low cost material and inexpensive processing method.

5. Academic Outreach: Teaching the concept of hydrophilicity and hydrophobicity of a surface

I have been closely working with local high school teachers for developing ideas for demonstrations that help K-12 and undergraduate students to understand important scientific concepts. I will discuss a recently developed demonstration entitled "Hydroglyphics" that visualizes the difference between hydrophilic and hydrophobic surfaces in a very exciting and approachable way. It involves placing a shadow mask on an optically clear hydrophobic plastic dish, corona treating the surface with a modified Tesla coil, removing the shadow mask, and visualizing otherwise invisible message or pattern by applying water, thus entitled as *hydroglyphics*. This demonstration was carried out in public at the Boston Museum of Science during the Nanodays event funded by National Science Foundation (March 31, 2012) and at the USAEF event funded by National Science Foundation (April 27-29, 2012).

(3bx) Lipid Phase Changes As a Mechanism of Microbial Pathogenesis

Angela C. Brown, Department of Pathology, University of Pennsylvania, School of Dental Medicine, Philadelphia, PA

Research Interests:

The goal of my research is to investigate the role of lipid phase behavior in microbial pathogenesis with a two-fold objective: (1) to understand the basis of the host-microbe relationship in disease pathogenesis and (2) to exploit these naturally evolved schemes to engineer novel therapeutic agents and drug-delivery devices.

Current Work:

Cell membranes are complex structures that are naturally impermeable to large molecules, and pathogens have therefore developed a range of schemes to penetrate the membrane during disease pathogenesis, many of which are highly specific at the cell, protein, and/or lipid level. Studying these interactions will provide insight into the mechanisms of disease progression, leading to the identification of potential therapeutic targets. In addition, we can use these naturally targeted mechanisms as a template to create more effective drug-delivery and therapeutic devices.

In this presentation, I will discuss some of the lipid-protein interactions involved in microbial pathogenesis, focusing on my current studies of the leukotoxin (LtxA) produced by the oral bacterium, *Aggregatibacter actinomycetemcomitans.* During my postdoctoral fellowship, I have applied by background in lipid phase behavior and colloidal science to study this organism, and have uncovered several previously unidentified mechanisms that may be conserved among other bacterial toxins. In particular, we have discovered the following LtxA-induced membrane and lipid changes:

1. LtxA damages the host cell bilayer without forming a pore, instead inducing the formation of a nonlamellar (inverted hexagonal) phase.

2. LtxA binds cholesterol, initiating a large-scale rearrangement of the toxin and its receptor into large, cholesterol-rich lipid rafts.

These newly defined mechanisms may be used by other bacterial toxins, including those produced by *Bordetella pertussis* (whooping cough), *Vibrio cholerae* (cholera), *Escherichia coli*, and *Streptococcus pyogenes* (streptococcal infections).

Future Directions:

The focus of my research group will be to study these interactions in more detail to determine how LtxA is able to exert these effects. In addition, we will extend our studies to investigate other pathogens, including viruses and fungi. The results of this work will provide information that will be valuable for the identification of potential therapeutic targets and the design biologically inspired therapeutic agents.

Potential Areas for Collaboration:

An understanding of lipid phase behavior and cell membrane functioning has applications beyond microbial pathogenesis, and could be applied in numerous fields including:

Drug Delivery - How do drugs and drug delivery devices penetrate the cell membrane?

Design of Antibiotics – How do antibiotics affect/damage bacterial cell membranes?

Biofilms and Quorum Sensing – How are bacterial signals passed through the membrane?

Funding:

NIH F32 Individual Postdoctoral Fellowship (Feb. 2010 – Jan 2013)

(3bz) Biomedical Applications of Single-Walled Carbon Nanotubes: Toward Design of Novel Optical Sensors

Bin Mu, Department of Chemical Engineering, Massachusetts Institute of Technology, MIT, Cambridge, MA

Semiconducting single-walled carbon nanotubes (SWNTs) are excellent candidates for optical sensors due to their stable fluorescence at near-infrared (nIR) wavelengths without photobleaching. It has been shown that SWNT emission is affected by changes in the local dielectric environment (solvatochromism), electron transfer (doping), or fluorescent quenching. The concepts can be utilized to create sensors for biologically important molecules such as glucose, nitric oxide, hydrogen peroxide, and proteins. Thus, design and synthesis of proper SWNTs wrapping reagents with specific recognition groups play a significant role in constructing a successful sensing interface. Over the past decade or so, polymers have been utilized for noncovalent dispersing of SWNTs in different media. Wrapping of SWNTs could be driven by chemical interactions between SWNTs and the functional groups comprising the polymers, which may include electrostatic interactions, π-π stacking, and hydrogen bonding. Understanding intermolecular interactions between wrapping polymers and SWNTs forms the basis for molecular recognition using fluorescent SWNTs and is essential for the proper design of polymer structures as synthetically selective receptors. My planned research in this area will include the fundamental and applied studies from both sides of experiments and simulation. Namely, to explore this sensing mechanism by designing more polymer-SWNT complexes and to develop novel sensing systems based on our understanding of interactions between SWNT and the polymer.

(3c) Sustainable Nanocomposites towards Electrochemical Energy Storage and Environmental Remediation

Jiahua Zhu, Integrated Composites Laboratory (ICL), Dan F Smith Department of Chemical Engineering, Lamar University, Beaumont, TX

Nanocomposites offer tremendous opportunities in the advancement of technologies. Facing the widely concerned challenges in energy shortage and storage as well as the environmental pollution control, sustainable nanocomposites towards electrochemical storage and environmental remediation have been developed to overcome the limited performance of current materials. Targeting a high capacitance composite material, an interfacial polymerization method has been developed to fabricate an integrated 3-dimensional (3D) structure comprising 2D graphite oxide (GO) sheet and 1D polyaniline nanofibers. This facile method can be used as a general method to synthesize similar hybrid structures which provides large surface area, high capacitance and satisfactory energy density. Hybrid materials synthesized from this method exhibits much higher energy density and power density than most of the carbon based composite materials. These findings open up a new window of GO-based composites for the capacitor applications with high energy and power densities.

Targeting the environmental remediation of nanocomposites, a comprehensive and sustainable waste-free coupled process of polymer nanocomposite (PNC) recycling is introduced to synthesize magnetic carbon nanocomposites (MCNCs) and to produce useful chemical radicals simultaneously. The embedded NPs serve as catalysts for the waste polymer decomposition. At the same time, the coke produced on the NP surface (poison the catalyst in conventional process) during heating serves as a carbon source for the in situ formation of a compact carbon shell structure and the volatile phase could be collected as a useful chemical resource or potential liquid fuel upon condensation. The MCNCs demonstrate an extremely fast Cr(VI) removal from the wastewater with an almost complete removal of Cr(VI) with extremely fast adsorption rate (< 10 min, conventional adsorbent takes hours even days). The large saturation magnetization of these novel magnetic carbon nanocomposites allows fast recycling of both the adsorbents and the adsorbed Cr(VI) from liquid suspension in a more energetically and economically sustainable way by simply applying a permanent magnet. The significantly reduced treatment time required to remove the Cr(VI) makes these MCNCs promising for efficient removal of the heavy metals from the wastewater.

SELECTED PEER-REVIEWED PUBLICATIONS (10 out of 35) & 2 BOOK CHAPTERS:

1. J. Zhu, S. Wei, H. Gu, R. Sowjanya, Q. Wang, Z. Luo, N. Haldolaarachchige, D. Young, Z. Guo, One-pot Synthesis of Magnetic Graphene Nanocomposites Decorated with Core@Double-shell Nanoparticles for Fast Chromium Removal, ***Environmental Science & Technology,*** 46 (2), 977-985 (2012)

2. J. Zhu, Z. Luo, S. Wu, N. Haldolarachchige, J. He, D. P. Young, S. Wei and Z. Guo, Magnetic Graphene Nanocomposites: Electron Conduction, Giant Magnetoresistance and Tunable Negative Permittivity, ***Journal of Materials Chemistry***, 22, 835-844 (2012).

3. J. Zhu, S. Wei, J. Ryu and Z. Guo, Strain Sensing Elastomer/Carbon Nanofibers "Metacomposites", ***Journal of Physical Chemistry C***, 115(27), 13215-13222 (2011)

4. J. Zhu, S. Wei, N. Haldolaarachchige, D. P. Young and Z. Guo, Electromagnetic Field Shielding Polyurethane Nanocomposites Reinforced with Core-Shell Fe-Silica Nanoparticles, ***Journal of Physical Chemistry C***, 115 (31), 15304-15310 (2011)

5. J. Zhu, S. Wei, Y. Li, L. Sun, N. Haldolaarachchige, D. Young, C. Southworth, A. Khasanov, Z. Luo and Z. Guo, Surfactant-free Synthesized Magnetic Polypropylene Nanocomposites: Rheological, Electrical, Magnetic and Thermal

Properties, ***Macromolecules***, 44(11), 4382-4391 (2011)

6. J. Zhu, S. Wei, Y. Li, S. Pallavkar, H. Lin, N. Haldolaarachchige, Z. Luo, D. P. Young and Z. Guo, Comprehensive and Sustainable Recycling of Polymer Nanocomposites, ***Journal of Materials Chemistry***, 21, 16239-16246 (2011)

7. J. Zhu, S. Wei, L. Zhang, Y. Mao, J. Ryu, A. B. Karki, D. P. Young and Z. Guo, Polyaniline-Tungsten Oxide Metacomposites with Tunable Electronic Properties, ***Journal of Materials Chemistry***, 21, 342-348 (2011).

8. J. Zhu, S. Wei, M. J. Alexander, T. D. Dang, T. C. Ho and Z. Guo, Enhanced Electrical Switching and Electrochromic Properties of Poly(p-phenylenebenzobisthiazole) Thin Films Embedded with Nano-WO_3, ***Advanced Functional Materials***, 18, 3076-3084 (2010).

9. J. Zhu, S. Wei, J. Ryu, M. Budhathoki, G. Liang and Z. Guo, In-Situ Stabilized Carbon Nanofibers (CNFs) Reinforced Epoxy Nanocomposites, ***Journal of Materials Chemistry***, 20, 4937-4948 (2010).

10. J. Zhu; S. Wei; M. J. Alexander; D. Cocke; T. C. Ho; Z. Guo, Electrical Conductivity Manipulation and Switching Phenomena of Poly(p-phenylenebenzobisthiazole) Thin Film by Doping Process, ***Journal of Materials Chemistry***, 20, 568-574 (2010).

11. "Nanostructured Ceramic Oxides: Preparation, Property Analysis and Applications" *Encyclopedia of Semiconductor Nanotechnology*, Edited by Ahmad Umar, in press (2010).

12. S. Wei, **J. Zhu,** P. Mavinakuli, Z. Guo, **Chapter 4:** **"Magnetic Polymer Nanocomposites: Fabrication, Processing, Property analysis and Applications"** *Multifunctional Polymer Nanocomposites*, CRC Press, Edited by Jinsong Leng and Alan K. T. Lau, (2010).

AWARDS/HONORS:

Chinese Government Award For Outstanding Self-Financed Students Abroad, The China Scholarship Council, Mar, 2012.

3rd Research Excellence Award in the *2011 Chinese-American Chemical Society Southwest Chapter (CACS-SW) Poster Competition, "Multifunctional Magnetic Graphene Nanocomposites for Environmental Remediation*", Houston, TX, Oct 29, 2011.

1st award for the poster "Effective Adsorption of Low Concentrated Arsenic(III) by Magnetic Nanoparticle Decorated Graphene Nanocomposite" presented in the *AIChE Annual Conference,* Environmental Division (ENV), Minneapolis, MN, Oct 16-21, *2011.*

3rd award for the poster "Multifunctional Magnetic Graphene Nanocomposites for Environmental Remediation" presented in the *AIChE Annual Conference,* Materials Engineering & Sciences Division (MESD), Minneapolis, MN, Oct 16-21, *2011.*

2nd award for the poster "Polypropylene Nanocomposites Reinforced with Different Carbon NanoStructures" presented in the *International Polyolefins Conference,* Houston, TX, Feb 27-Mar 2, 2011.

NSF Travel Scholarship, *2011 NSF-CMMI Research and Innovation Conference*, Atlanta, Georgia, January 4-7, 2011.

(3ca) Theory and Modeling of Artificial Molecular Machines in Biological Systems

Korosh Torabi, Department of Chemistry, Northwestern University, Evanston, IL

Living organisms offer numerous examples of molecular machines that convert different forms of energy into mechanical work in the form of directed movement or conformational change of other biological components. Within the past two decades, a considerable scientific research has been invested in understanding the biophysics of the molecular machines. In specific, artificial molecular machines that perform useful tasks on molecular scales or their cooperative assemblies capable of converting energy into mechanical work in scales much larger than individual molecules have presented a great promise in the fields of nanotechnology and alternative energy resources.

World of molecular machines is in many respects quite different than that of their macroscopic counterparts. Molecular machines operate in a domain where gravity and inertia are outplayed, by orders of magnitude, by Van der Waals, electrostatic and other types of molecular interaction and their far-from-equilibrium motion is always accompanied with considerable thermal fluctuations. To gain further understanding of the performance of the artificial molecular machines, we study the conformational change of several biological molecules under external mechanical force. We use theories of nonequilibrium thermodynamics along with novel molecular simulation techniques to investigate Helix-coil transition of peptides, deconstruction of coiled-coiled structures and unzipping of dsDNA under the action of artificial molecular machines. We are currently collaborating with Fraser Stoddart group at Northwestern University to implement our design principles in synthesis of molecular machines with mechanically interlocked molecules such as rotaxanes or cantenanes. We hope our theoretical studies guide the experiments and provide physical insight to the biophysics of the artificial molecular machines and their applications within biological systems.

(3cb) Next Generation Musculoskeletal Tissue Engineering

Bret D. Ulery, Institute of Regenerative Engineering and Department of Orthopaedic Surgery, University of Connecticut Health Center, Farmington, CT

As the field of tissue engineering moves forward and tackles greater challenges like organ regeneration, the need for novel, multi-dimensional strategies that facilitate complex and integrated tissue development becomes more necessary. Current tissue engineering solutions are comprised of one material with basic structure releasing a single bioactive factor cultured with one cell type and will likely be suboptimal for higher order systems. Instead composite scaffolds with complex micro- and nano-architecture releasing a number of bioactive factors to a mixture of incubated cell types must be designed to create artificial organs. This increase in complexity requires new research-based approaches where knowledge from a variety of scientific fields must be integrated to yield next generation technologies. Due to my diverse research background, I am uniquely qualified to design and develop novel solutions capable of facilitating complex tissue regeneration.

My research to date has focused on utilizing synthetic biomaterials to facilitate desired biological responses through drug delivery and tissue engineering strategies. I completed my PhD in Chemical Engineering with a minor in Immunobiology at Iowa State University under the guidance of Professor Balaji Narasimhan. In collaboration with Professor Michael Wannemuehler in the department of Veterinary Microbiology and Preventative Medicine, my research focused on the development of polyanhydride nanoparticle-based delivery vehicles for immunomodulatory applications. In specific, I investigated how polymer chemistry affected dendritic cell activation *in vitro* as well as enhanced and sustained vaccine-based immune responses *in vivo*. In September 2010 I began a postdoctoral fellowship at the Institute for Regenerative Engineering at the University of Connecticut Health Center working with Professor Cato Laurencin. My postdoctoral research focuses on designing novel biomaterial-based strategies capable of providing the necessary chemical, biological, and mechanical cues necessary to regenerate musculoskeletal tissues. Specifically, I have shown that ions and simple molecules can differentiate stem cells down desired lineages while simultaneously inducing these cells to produce their own growth and differentiation factors. A major focus of my research is the application of basic science and engineering principles to the development of new biomedical technologies which has so far yielded 7 research publications, 6 review articles and book chapters, and 2 patent applications.

During this poster session I will highlight examples from my graduate and postdoctoral research that demonstrate the capacity of synthetic biomaterials to initiate and sustain immune responses and direct musculoskeletal tissue development. Building on my past work, I will outline my plans to build a distinguished, independent research program at the interface of chemical engineering, biomedical engineering, materials science, and immunology. In specific, I will discuss how musculoskeletal tissue engineering provides opportunities to apply novel biomaterials strategies. The overarching goal of my research lab will be to exploit biomaterials to generate complex musculoskeletal tissue regeneration. This goal is comprised of three research thrusts: 1) improving vascularization in musculoskeletal tissue engineering, 2) facilitating immunomodulatory regenerative tissue engineering, and 3) utilizing combinatorial design and evaluation in tissue engineering.

Related References:

1) K.W.H. Lo*, **B.D. Ulery***, K.M. Ashe, H.M. Kan, and C.T. Laurencin. "Evaluating the feasibility of the small molecule phenamil as a novel osteoinductive factor for bone regenerative engineering." (*Submitted to Journal of Tissue Engineering and Regenerative Medicine*)

2) K.W.H. Lo*, **B.D. Ulery***, K.M. Ashe, and C.T. Laurencin. "Inducations and use of bone morphogenetic proteins in surgical repair of bone and spine." (*In Press*)

3) M.S. Peach, S.G. Kumbar, R. James, U.S. Toti, **B.D. Ulery**, D. Balasubramaniam, A.D. Mazzocca, M.B. McCarthy, N.L. Morozowich, H.R. Allcock, and C.T. Laurencin. "Design and optimization of polyphosphazene functionalized fiber matrices for soft tissue regeneration." Journal of Biomedical Nanotechnology. 8(1): 107-124. (**Feb 2012**)

4) **B.D. Ulery***, L.K. Petersen*, Y. Phanse, C.S. Kong, S.R. Broderick, D. Kumar, A. E. Ramer-Tait, B. Carillo-Conde, K. Rajan, B.H. Bellaire, M.J. Wannemuehler, D.W. Metzger, and B. Narasimhan. "Rational design of pathogen-mimicking amphiphilic materials as nanoadjuvants." Scientific Reports. 1(198) (**Dec 2011**)

5) L.K. Petersen, A.E. Ramer-Tait, S.R. Broderick, C.S. Kong, **B.D. Ulery**, K. Rajan, M.J. Wannemuehler, and B. Narasimhan. "Amphiphilic polyanhydride nanoparticle adjuvants activate innate immune responses in a pathogen-mimicking manner." Biomaterials. 32(28): 6815 - 6822. (**Oct 2011**)

6) **B.D. Ulery**, L.S. Nair, and C.T. Laurencin. "Biomedical applications of biodegradable polymers." Journal of Polymer Physics Part B: Polymer Physics. 49(12):832 - 864. (**Jun 2011**)

7) **B.D. Ulery*,** D. Kumar*, A.E. Ramer-Tait, D.W. Metzger, M.J. Wannemuehler, and B. Narasimhan. "Design of a protective single-dose intranasal nanoparticle-based vaccine platform for respiratory infectious diseases." PLoS One, 6(3): e17642. (**Mar 2011**)

8) **B.D. Ulery**, Y. Phanse, A. Sinha, M.J. Wannemuehler, B. Narasimhan, and B.H. Bellaire. "Polymer chemistry influences monocytic uptake of polyanhydride nanospheres." Pharmaceutical Research, 26(3): 683 - 690. (**Mar 2009**)

9) **B.D. Ulery**, K. Pustulka, Y. Phanse, B. Bellaire, and B. Narasimhan. "Amphiphilic polyanhydride chemistry affects monocytic association of nanospheres." Proceedings of the 37th Annual Biochemical Engineering Symposium, 37: 52 - 59. (**Jan 2009**)

** These Authors Contributed Equally*

(3cc) Photovoltaics and Catalysis for Photoelectrochemical Applications

Daniel V. Esposito, National Institute of Standards and Technology, Gaithersburg, MD

The energy available from sunlight striking Earth is orders of magnitude larger than the amount of energy currently consumed by mankind, yet our current energy system is dominated by CO_2–emitting, non-renewable fossil fuels. In order for solar technology to achieve widespread replacement of fossil fuels, major developments must be made in three interrelated areas: (i) solar conversion efficiency, (ii) materials and manufacturing costs, and (iii) energy storage. In my future research lab, I will explore solar and energy conversion technology that strives to address and overcome challenges in all three of these areas. A deep understanding of the trade-offs between each area is crucial for all levels of solar research, and I believe that there is a large and untapped potential for applying chemical engineering principles to evaluate these trade-offs and thereby optimize the overall worth of solar energy systems.

Central to my future research will be photoelectrochemical (PEC), solar thermal, and related hybrid technologies that are capable of directly converting sunlight into storable chemical and/or thermal energy. By generating energy that can be stored, PEC and solar thermal processes overcome the problem of solar intermittency, an inherent weakness of photovoltaic technology. My research will build off of my past and current research experiences at the University of Delaware and the National Institute of Standards and Technology (NIST). At the University of Delaware, my thesis research was focused on the development of hydrogen evolution catalysts and PEC device design, allowing me to gain broad expertise in the fields of electrochemistry, catalysis, and photovoltaics. At NIST, my current research efforts are centered on the development of well-defined metal-oxide-semiconductor (MOS) photoelectrodes for PEC applications. As I will discuss in my oral presentation at this year's meeting, these MOS structures hold great potential for efficient PEC energy conversion because they are well-suited for overcoming the fundamental efficiency/stability trade-off that so commonly hinders conventional photoelectrodes. The well-defined and ordered geometries of these MOS structures also make them ideal for micro-scale modeling and experimental efforts aimed at better understanding charge transfer phenomena in MOS-based photoelectrodes. In addition to presenting past, current, and future research efforts, I will also discuss my future teaching and outreach endeavors at this poster session.

Post-doctoral advisors: Dr. Thomas Moffat and Dr. Alec Talin, National Institute of Standards and Technology

Ph.D. advisors: Prof. Jingguang G. Chen and Prof. Robert Birkmire, University of Delaware

Selected publications

1. **D.V. Esposito,** S.T. Hunt, Y. Kimmel, and J.G. Chen, "A New Class of Electrocatalysts for Hydrogen Production from Water Electrolysis: Metal Monolayers Supported on Low-Cost Transition Metal Carbides". *Journal of the American Chemical Society*, vol. 134, 2012.
2. **D.V. Esposito** and J.G. Chen, "Monolayer Platinum Supported on Tungsten Carbides as Low-Cost Electrocatalysts: Opportunities and Limitations". *Energy & Environmental Science*, vol. 4, 3900-3912, 2011. **(Invited perspective article)**
3. **D.V. Esposito,** Y. Chang, J.G. Chen, R.W. Birkmire, and N. Gaillard, "Hydrogen Production from Photo-driven Electrolysis of Biomass-derived Oxygenates: A Case Study on Methanol using Pt-modified WO_3 Thin Film Electrodes". *International Journal of Hydrogen Energy*, vol. 36, 9632-9644, 2011.
4. **D.V. Esposito**, S.T. Hunt, A.L. Stottlemyer, K.D. Dobson, B.E. McCandless, R.W. Birkmire, and J.G. Chen, "Low-Cost Hydrogen Evolution Catalysts Based on Monolayer Platinum on Tungsten Monocarbide (WC) Substrates". *Angewandte Chemie International Edition*, vol. 49, 9859-9862, 2010. **(Cover article and press release)**
5. **D.V. Esposito**, O.Y. Goue, K.D. Dobson, B.E. McCandless, J.G. Chen, and R.W. Birkmire, "A New Photoelectrochemical Test Cell and Its Use for a Combined Two- and Three-Electrode Approach to Cell

Testing". *Review of Scientific Instruments*, vol. 80, 125107, 2009.

(3cd) Well Defined Nanomaterials Through Tunable and Smart Solvents

Steven R. Saunders[1], Christopher B. Roberts[2], Charles L. Liotta[1] and Charles A. Eckert[1], (1)School of Chemical and Biomolecular Engineering, Georgia Institute of Technology, Atlanta, GA, (2)Department of Chemical Engineering, Auburn University, Auburn, AL

The production of large quantities of monodisperse nanoparticles with unique properties remains a non-trivial task and is one hurdle to the wide use of nanoparticles in many applications. Nanomaterials have shown the potential to revolutionize catalysis, energy production, drug delivery, and composite materials. This potential is greatly increased if facile synthesis, processing, and modeling techniques are developed. Routes to produce large quantities of well-defined nanomaterials and leveraging their unique properties can be enabled through the use of smart solvent technology, sustainable engineering, and computational modeling.

My PhD work, performed at Auburn University under the supervision of Christopher B. Roberts, focused on the manipulation and prediction of the size distribution of metallic nanoparticles dispersed in various tunable solvent systems (i.e., CO_2-gas expanded liquids). Through a fine control in various system variables (including applied gas pressures, surface chemistry, and solvent conditions) I demonstrated a high degree of control over the size distribution of nanoparticles that can be dispersed or precipitated in a solvent. For example, a polydisperse (±2 nm) sample of dodecanethiol-stabilized gold nanoparticles dispersed in a mixture of hexane and acetone can be size-selectively fractionated into multiple monodisperse (±0.5nm) samples in a single processing stage without producing any waste. Also, I have developed a Total Interaction Energy Model that is capable of predicting the size distribution of nanoparticles that can be dispersed at various solvent conditions. My work at Auburn University demonstrated that tunable solvents and basic thermodynamics are one potential route to developing and understanding new nanoparticle processing techniques.

My postdoctoral work, performed at Georgia Tech under the supervision of Charles L. Liotta and Charles A. Eckert, focused on employing smart solvent technologies (e.g., reversible ionic liquids and organic-aqueous tunable solvents) and fundamental chemistry to improve and understand different processes (e.g., catalyst recovery/recycle and polymer degradation/stabilization). For example, we have developed a technique to synthesize metallic nanoparticles using a reversible ionic liquid (an ionic liquid capable of "switching" its ionic character on or off when exposed to the proper stimulus) with the potential to produce zero waste. Using various techniques (e.g., TEM, DLS, micro-structure analysis), we have developed a theory that describes the synthesis mechanism and have leveraged it to optimize the synthesis to provide monodispersed nanoparticles.

The work I have conducted during my PhD and postdoctoral fellowship, which will be presented in this poster, demonstrate that smart and tunable solvent technologies, sustainable engineering, and computational thermodynamic modeling enable novel methods of producing large quantities of monodispersed nanoparticle needed for many applications.

(3ce) Understanding Catalysis Through Organic-Inorganic Hybrid Catalytic Materials

Michael M. Nigra, Department of Chemical and Biomolecular Engineering, University of California Berkeley, Berkeley, CA

The design of metal cluster or nanoparticle catalysts that are both sinter-resistant (stable) and that have tunable electronic properties enables the understanding of structure-function relationships in heterogeneous and homogeneous catalysis. It also provides a method to tune the activity and selectivity of real-world catalysts by overcoming one of the major stumbling blocks for cluster catalysts – deactivation via sintering. I investigate the use of organic ligands, such as calixarenes, to control catalytically active noble metal sites on metal oxide supports and in homogeneous solution, utilizing an approach that is conceptually similar to that used by enzymes to control reactivity though organic side-chains acting as ligands. Noble metal catalysts have been synthesized with calixarene ligands bound to the metal nanoparticle yet still demonstrate reactant accessibility to the metal nanoparticle surface as measured by chemisorption probe experiments. Catalytic activity has been demonstrated for these using several different reaction systems such as reduction and oxidation reactions on these noble metal catalysts. Results provide a unique perspective into the structure-activity relationships present in catalysis.

(3cf) Electrochemistry of Oxygen with Protons and Lithium and Their Role in Fuel Cells and Lithium-Air Batteries

Venkatasubramanian Viswanathan, Chemical Engineering, Stanford University, Stanford, CA

With a growing energy demand, an energy economy based on cheap and sustainable energy sources such as solar and wind is likely necessary to power the planet in the coming years. However, the issue of the intermittent availability of these renewable energy sources needs energy efficient storage options. The coupling of renewable sources to energy efficient energy conversion devices would also be necessary to make efficient utilization of the renewable energy. Batteries and fuel cells stand at the forefront of energy storage and utilization options. I will discuss my research work in the areas of fuel cells and lithium-air batteries.

The electrification of road transportation has been severely limited by the limited capacity of the current batteries and there has been a recent push towards more energy dense batteries beyond Li-ion batteries. Of all the possibilites, Li-air is the most promising with a potential 5-10x gain in energy density. These batteries can be exploited in two configurations: (a) non-aqueuous and (b) aqueous. I will outline my work in both of these battery configurations.

There are many technical hurdles facing the non-aqueous Li-air batteries before they can become practical. Using a combined experimental and theoretical approach, we have identified two key fundamental issues involved in lithium-air batteries - (a) the origin of the sudden death involved in Li-air batteries and (b) the presence of carbonate impurities and their role in affecting the rechargability of these batteries. Using non-equilibrium transport calculations, we show that the origin of this 'sudden death' in flat electrode cells is due to the limitation imposed by electron transport to support the required electrochemistry. Using a simple metal-insulator-metal model, we show that discharge products beyond > 10 nm, have serious transport limitations and this leads to a bias rise across the film[1].

The fundamental issue during recharging the battery is associated with the presence of an impurity carbonate layer which has been recently identified experimentally. We develop a model incorporating details from the recent experiments and conclusively show that the origin of potential rise during recharge of the lithium-air batteries is due to a rising concentration of carbonate impurities at the Li_2O_2-electrolyte interface. This rise in the overpotential leads ultimately to the decomposition of the electrolyte and limits the rechargability of lithium-air batteries[2].

In the case of aqueous Li-air battery, although the inherent chemistry is more rechargable than the non-aqueous Li-air battery, fundamental issues associated with the discharge product and the need for precious metal-catalysts limit the practicality of this battery. We develop the first continuum-scale model for the aqueous Li-air battery identifying some of the key performance limitations of these batteries[3].

On the avenue of energy utilization, fuel cells are an attractive option. However, the electrochemistry associated with reduction of oxygen with protons has been severely limiting the performance of fuel cells. In this regard, we have develop a detailed first-principles based mechanism of oxygen reduction reaction on (111) and (100) facet of Pt which is the best elemental catalyst for oxygen reduction[4,5]. Using the detailed mechanism, we address the effect of particle size and support on the electrocatalytic activity of oxygen reduction on Pt nanoparticles using a combination of first-principles DFT calculations and Dynamic Monte Carlo simulations. We show that tuning the metal-support inteface could have a drastic effect on the electrocatalytic activity[6]. We have unified this analysis and shown that the activity trends on low-index facets of (111) and (100) follow a universal volcano[7].

This analysis has been used to understand effects of nanostructuring Pt monolayers on top of Rh(111). We show that a 3D Pt particle on top of Rh(111) has enhanced activity compared to a 2D monolayer of Pt on top of Rh(111)[8]. This design principle of nanostructing has been used to carry out a computational search for new candidate materials and one of them was identified which has been tested experimentally. This catalyst shows a higher specific activity than state-of-the-art TKK Pt/C catalyst.

Doctoral advisors: J. K. Nørskov and T. F. Jaramillo, Stanford University

As part of my PhD, I have been invovled in use and development of advanced computational methodologies in the exciting new area of 'engineering the new tools of scientific discovery'. I have gained expertise in first-principles based calculations for thermodynamics of reactions on surfaces, hybrid functionals for describing electronic structure of oxide materials, non-equilibrium transport calculations for calculating conductivity of various interfaces. I have developed Monte-Carlo based codes to examine complex geometric effects. I also have a firm grasp of contiuum scale modelling with a detailed system model developed for a realistic battery. A mix of engineering skills along with a deep fundamental understanding of physics of chemistry of surface processes and interface phenomenon puts me in a unique position to critically assess and develop the advanced energy-storage and conversion devices of the future.

Relevant publications:

[1] **V. Viswanathan**, K. Thygesen, J.S. Hummelshøj, J. K. Nørskov, G. Girishkumar, B.D. McCloskey, A. Luntz, J. Chem. Phys., 135, 214704 (2011).

[2] B. D. McCloskey, A. Speidel, R. Scheffler, D. C. Miller, **V. Viswanathan**, J. S. Hummelshøj, J. K. Nørskov and A. C. Luntz, J. Phys. Chem. Lett., (2012), 3, 997-1001.

[3] P. Albertus and **V. Viswanathan**, J. Electrochem. Soc. (in submission)

[4] **V. Viswanathan**, H. A. Hansen, J. Rossmeisl, T. Jaramillo, H. Pitsch,and J. K. Nørskov, J. Phys. Chem. C, (2012) 116 (7), 4698-4704.

[5] B.C. Han, **V. Viswanathan**, and H. Pitsch, J. Phys. Chem. C, (2012) 116 (10), 6174-6183.

[6] **V. Viswanathan** and F. Wang, Nanoscale, (2012) (accepted).

[7] **V. Viswanathan,** H. A. Hansen, J. Rossmeisl, and J. K. Nørskov, ACS Catalysis (2012) (submitted).

[8] D. Friebel, **V. Viswanathan**, D. J. Miller, A. Larsen, T. Anniyev, H. Ogasawara, J. K. Nørskov, and A. Nilsson, J. Am. Chem. Soc. (2012) (accepted).

(3cg) Synthetic and Spectroscopic Methods to Facilitate Design of Highly Selective Catalytic Sites

Nicholas Brunelli, School of Chemical & Biomolecular Engineeering, Georgia Institute of Technology, Atlanta, GA

Increasing energy costs and environmental concerns have increased the demand for highly selective and recyclable catalysts to eliminate costly purification steps and reduce waste and byproduct formation. The nature of selective sites can be probed with advanced spectroscopic methods to determine structure-function relationships. Investigating these relationships and creating synthetic methods to increase the number of highly selective sites are of paramount importance to achieving more selective and active catalysts.

My PhD research experience involved a plasma-based aerosol synthesis method for nanoparticles and characterization of nanomaterials. The limited ability of current technology to determine nanoparticle size *in situ* led me to design, build, simulate, and validate the nano-RDMA, a new instrument capable of classifying nanoparticle size in the one to ten nanometer size range. The instrument performance permitted *in situ*investigation of size effects on surface dehydrogenation of silicon nanoparticles, demonstrating a lower activation barrier for dehydrogenation from small clusters. Additionally, the capabilities of the instrument were tested through demonstrating the sputter production of atomic clusters (<1 nm in size) from an atmospheric plasma source. While the resolving power was substantial, the simulation results allowed design of a second generation instrument that has a factor of four improvement in resolution.

In addition to synthesizing and characterizing nanoparticles, my research involved designing, building, and characterizing an electrospray source that was used to incorporate platinum nanoparticles in membrane electrode assemblies (MEAs) for low catalyst loading, solid acid fuel cell applications. The electrospray source produced crystallites of the fuel cell membrane material on the scale of 100 nanometers – smaller than ever produced before – while simultaneously mixing platinum nanoparticles to create the MEA. The reduced size discrepancy between the catalyst particle (i.e. 10 nm) and the membrane material permitted low catalyst loading for the fuel cell.

My post-doctoral research examines the cooperativity between acid and organic base sites in bio-inspired cooperative catalysis. In enzymes, the catalytic active site is organized through a complex architecture to have (1) tuned interactions between acid and base groups, and (2) precise spatial organization. Through designed immobilization of acid and base sites on a rigid support, these beneficial aspects can be integrated into simpler materials, permitting the ability to design-in catalytic cooperativity at the nanoscale. Catalytic testing has shown that decreasing the acid-base interaction through using a weaker acid accelerates the rate of reaction. Additionally, the spatial organization of the active site proved important with longer more flexible linkers resulting in highly cooperative catalysts.

Integrating my background of instrumentation design, material synthesis, and catalysis, my future research will seek to control material functionality for applications in the fields of energy, catalysis, and advanced functional materials. Characterizing surface properties will elucidate properties that are responsible for catalyst selectivity, requiring new techniques and associated instrumentation to probe subtle yet significant changes that occur. The integrated and multidisciplinary probing of materials properties will lead to fundamental discoveries of structure function relationships that can be utilized to synthesize more selective and active catalytic materials.

Post-doctoral Advisor: Christopher W. Jones, Georgia Institute of Technology, School of Chemical & Biomolecular Engineering

Doctoral Thesis Committee: Professors Konstantinos P. Giapis, Richard C. Flagan, Sossina M. Haile, and J. L. Beauchamp, California Institute of Technology, Division of Chemistry and Chemical Engineering

Publications:

[1] N.A. Brunelli, R.C. Flagan, K.P. Giapis, *Aero. Sci. Tech.* **2009**, *43*, 53-59.

[2] A. Varga, N.A. Brunelli, M. Louie, K. P. Giapis, S.M. Haile, *J. Mater. Chem.* **2010**, *20*, 6309-6315.

[3] J. Jiang, M. Attoui, M. Heim, N.A. Brunelli, P. McMurry, G. Kasper, R.C. Flagan, K.P. Giapis, and G. Mouret, *Aero. Sci. Tech.* **2011**, *45*, 48-492.

[4] N. A. Brunelli, W. Long, K. Venkatasubbaiah, C. W. Jones, *Top. Catal.* **2012** (*in review*).

[5] J. Thompson, C. Blad, N. A. Brunelli, M. Leydon, R. Lively, C. W. Jones, S. Nair, *Chem. Mater.* **2012** (*in press*).

[6] N. A. Brunelli, K. Venkatasubbaiah, C. W. Jones, *Chem. Mater.* **2012** (*in review*).

[7] N. A. Brunelli, S. A. Didas, K. Venkatasubbaiah, C. W. Jones, (*in preparation*).

[8] N. A. Brunelli, E. L. Neiholdt, K. P. Giapis, R. C. Flagan, J. L. Beauchamp, (*in preparation*).

(3ch) Simultaneous Electronic and Ionic Conducting Block Copolymers for Lithium Battery Applications

Shrayesh N. Patel, Department of Chemical Engineering, University of California, Berkeley, Berkeley, CA

The primary use of polymeric materials for lithium batteries has focused on the electrolyte. Due to the flammability and reactivity of organic electrolytes used in traditional lithium-ion batteries, researchers have utilized solid polymer electrolytes, which are sufficiently ion conducting and stable during cycling. In addition, researchers have used nanostructured block copolymer electrolytes where one of the blocks allows for mechanical strength. However, researchers have not considered the use of nanostructured block copolymers for the cathode (positive) electrode of lithium batteries where both electronic and ionic charges are transported. For that reason, the use nanostructured block copolymer for lithium battery electrodes has been the primary focus of my PhD research. For reference, the traditional porous lithium battery cathode consists of a redox-active material, carbon black for electronic conduction, and non-conductive binder that holds the particles in place. In addition, the pores are backfilled filled with organic electrolyte for ionic conduction. Due to the poor electronic and ionic conduction, the active materials are often are used in the nanoparticle form, thus requiring the transport of charges to occur on the nanometer length scales. Therefore, the use of block copolymers that simultaneously conducts electronic/ionic charges and that self-assemble on the nanometer length scales is the ideal material. The modified cathode thus only consists of the redox-active material and the nanostructured block copolymers.

The primary focus of my PhD work involved the synthesis and characterization of the poly(3-hexylthiophene)-*block*-poly(ethylene oxide) (P3HT-*b*-PEO) as a nanostructured electrode binder where the P3HT-domains conduct electronic charges and PEO-domains conducts ionic charges. The results of ac impedance spectroscopy and dc measurements provided the first evidence of the simultaneous conduction of electronic and ionic charges in a block copolymer. In addition, my work was the first attempt to study the relationship between morphology and simultaneous electronic and ionic transport. Furthermore, the application of this material in a lithium battery with $LiFePO_4$ showed specific capacity reaching the theoretical limit and with minimal capacity fade, thus demonstrating the practical application of P3HT-*b*-PEO as conductive binder material. I believe the results of my work provides a foundation for substantial future work on nanostructured electronic/ionic conducting block copolymers for battery and other energy related devices. The scope of my PhD research fell in the fields of both polymer physics and electrochemistry. Knowledge in both fields will help develop my future academic research projects that both provide fundamental scientific knowledge and have a broad real world impact on energy applications.

(3ci) Single-Walled Carbon Nanotube Dynamics in Simple and Complex Media

Nikta Fakhri, Third Institute of Physics-Biophysics, University of Göttingen, Göttingen, Germany

Biological systems possess properties such as the capability to self-heal and adapt, which sets new horizons for materials engineering. The long-term goal of my research is to develop smart materials that incorporate dynamic and structural properties of complex biological systems while being robust and minimalistic in composition and design.

During my PhD work in Matteo Pasquali's lab, in the Chemical Engineering Department at Rice University, I employed visualization and analysis tools combined with theoretical modeling to study transport and diffusion of single-walled carbon nanotubes (SWNTs) in simple and complex soft condensed matter systems. Understanding the dynamics of SWNTs in such environments is crucial for establishing the potential of these nanostructures as both probes in biological systems and building blocks in soft materials.

During my postdoc in Christoph Schmidt's lab in University of Göttingen, Germany, I have been expanding my scope towards molecular and cellular biology and non-equilibrium soft matter. I am developing the application of SWNTs as probes with unique capabilities to study biological systems ranging from cytoskeleton networks via single cells to whole organisms (C-elegans).

(3cj) The Physical Genome: Force, Elasticity, and Transport in DNA Processing

Elena F. Koslover, Biophysics Program, Stanford University, Stanford, CA

The internal microenvironment of a living cell comprises an intricate choreography of molecules that must be transported from one location to another, elastic forces that must be overcome or harnessed into useful work, and interdependent chemical reactions whose rates must be carefully controlled. My research focuses on examining the basic physical phenomena that underly biological processes at the molecular and cellular level, with an emphasis on genome structure and dynamics. I develop multiscale models, employing both analytical and computational techniques, to approach questions of how the mechanical properties of DNA and its cellular environment impact the packaging and accessibility of the genome.

My thesis work demonstrates how the elasticity of the DNA molecule can give rise to tension-mediated cooperative binding between DNA-bending proteins, allowing them to sense each other across a wide range of length scales. I also investigate, using energy landscape optimization methods, how this elasticity plays into the packing of eukaryotic DNA in compact chromatin fibers. Furthermore, I calculate analytically how the statistical properties of DNA at long length scales are modulated by its interactions with proteins. This work gives rise to a generalized approach for coarse-grained modeling of polymer systems. Moving into the realm of dynamics, I use a combination of analytic reaction-diffusion models and simulations to study the target site search process of DNA-binding proteins under a variety of conditions relevant to in vitro and in vivo systems, uncovering a key role for confinement and a surprising robustness to DNA configuration. Finally, I examine the effect of force fluctuations on biomolecular kinetics, demonstrating how such fluctuations can qualitatively alter nucleosomal transcription by RNA polymerase, an essential process for eukaryotic gene expression.

My future research goals center on studying the fundamental physics underlying dynamic processes in the cell. Hijacking cellular machinery to engineer novel biomolecular circuits, to enhance production of pharmaceuticals, or to induce therapeutic effects, necessitates an ability to harness the kinetic properties of cell networks. The role of viscoelasticity, stochastic fluctuations, and dynamic sequestration in determining the interplay of complex biochemical processes presents a rich set of questions of direct relevance to cellular engineering. I intend to use my background in biophysical modeling as a springboard for addressing fundamental problems of how transport phenomena and spatial organization within the cell impact the structure and kinetics of biomolecular networks.

(3ck) Next-Generation Smart Polymeric Biomaterials

Murat Guvendiren, Bioengineering, University of Pennsylvania, Philadelphia, PA

Utilization of new methodologies to develop polymeric soft biomaterials, namely biohydrogels, that mimic native tissue environments is crucial for tissue engineering and biomedical applications. In this poster, I would like to highlight mainly three of my previous studies. First, I developed a novel procedure to develop hydrogel surface patterns with well-controlled shape, order and size [1], and utilized this platform to control stem cell shape and differentiation [2]. Then, I developed a hydrogel system to enable temporal stiffening, and studied short- and long-term stem cell response to dynamic mechanics [3]. Lastly, I developed a novel system with spatially and temporally displayed and reversible surface topography, and investigated the stem behavior (cellular and nuclear alignment) [4].

The main focus of my research will be development, characterization and processing of multi-functional and responsive polymeric biomaterials that display "dynamic" properties for biological applications, with a particular interest in tissue engineering and drug delivery. Although recent advances have produced unique biopolymers and materials-based strategies in tissue regeneration and repair, the majority of these materials display static properties (structural, chemical and mechanical) and lack the dynamic complexity of native tissue microenvironments. My overall goal is (*i*) to develop innovative material systems that display complex combinations of topographical, chemical and mechanical properties both in 2D and 3D towards tissue and stem cell engineering, (*ii*) to design and fabricate biomimetic hydrogels that mimic structure/property relationships and stimuli response in nature to towards developing smart biomaterials (adhesives, sealants, medical coatings, etc.), and (*iii*) to develop hydrogels that display tunable and user-defined gelation with spatial and temporal control of structural, chemical and mechanical properties towards next generation injectable biohydrogels. My approach will utilize biomimetic synthesis of polymeric materials (and their hybrids) coupled with micro/nano fabrication and advanced characterization techniques.

[1] M. Guvendiren, S. Yang, J.A. Burdick, Swelling-induced Surface Patterns in Hydrogels with Gradient Crosslinking Density, *Advanced Functional Materials*,19:3038-3045, 2009

[2] M. Guvendiren and J.A. Burdick, The Control of Stem Cell Morphology and Differentiation by Hydrogel Surface Wrinkles, *Biomaterials*, 31:6511-6518, 2010

[3] M. Guvendiren and J.A. Burdick, Stiffening Hydrogels to Probe Short- and Long-Term Cellular Responses to Dynamic Mechanics, *Nature Communications*, 3:792, 2012.

[4] M. Guvendiren and J.A. Burdick, Stem Cell Response to Spatially and Temporally Displayed and Reversible Surface Topography, in review.

(3cl) The Era of Self-Administration of Biopharmaceuticals

Jeong Woo Lee, Chemical and Biomolecular Engineering, Georgia Institute of Technology, Atlanta, GA

Since genetic engineering technologies opened the gate to the production of therapeutic biomolecules such as peptides, proteins, DNA, and vaccines, the number of biopharmaceuticals reaching the clinic has increased exponentially over the past two decades and still more biopharmaceuticals are expected to enter the market in the future. In contrast to low molecular weight chemical-based drugs, biopharmaceuticals cannot be given by the oral delivery route because they are degraded significantly in the stomach and intestine, not easily absorbed through the mucus membrane of the intestine, and also degraded in the liver after entry into the systemic circulation. For these reasons, biopharmaceuticals are administered mostly by injection using hypodermic needles to ensure high bioavailability of biopharmaceuticals; however, injections not only have low patient compliance due to fear and pain from the use of hypodermic needles, but also require medically trained personnel and specific protocols for safe use and disposal of used needles.

Two innovative methods of transdermal delivery of biopharmaceuticals were designed without using hypodermic needles for my doctoral research; dissolving microneedles patch (cited more than 90 times according to scholar.google.com in 4 years after publication) and microsecond skin ablation. These novel technologies are expected to empower patients to administer their medicine regardless of where they are. My research will emphasize pharmaceutical engineering by applying chemical engineering and interdisciplinary principles to the drug development and delivery ultimately for self-administration of biopharmaceuticals. My objectives as a faculty are to develop the collaborative research program for the delivery of biopharmaceuticals and help students understand the fundamentals of chemical engineering and apply their apprehension to the journey to explore new things.

POLYMER INTERFACES AND GRADIENTS AT WORK: BIOMATERIALS AND ENERGY MATERIALS

Julie N. L. Albert, North Carolina State University, Raleigh, NC

Introduction

Materials interfaces strongly influence transport and adhesive properties that dictate material or device performance in applications that span the biological, chemical, and physical sciences. For example, successful synthesis of materials for targeted drug delivery or evaluation of cell responses to chemical and mechanical stimuli rely on manipulation of interactions at the cell/material interface. Similarly, the efficiency of conductive membranes, such as those for organic photovoltaic devices, depends on the membrane/electrode interfacial properties as well as materials interfaces within the membrane for multi-component formulations. My research interests are centered on manipulating interfacial interactions to address challenges in two important areas of research: first, to understand and eliminate cancer metastasis, and second, to improve the efficiency of organic photovoltaic devices so that they can become a competitive alternative energy source. A key component of my approach to these problems is the use of gradient surfaces to study interactions in a high-throughput and combinatorial fashion. As detailed in the following section, my doctoral work on the self-assembly of block copolymer thin films (University of Delaware, advised by Prof. Thomas H. Epps, III) and my postdoctoral work on manipulating the mechanical and chemical properties of silicone elastomer networks for cell mobility studies (North Carolina State University, advised by Prof. Jan Genzer) have uniquely prepared me to address interfacial engineering challenges in both the biomedical engineering and alternative energy fields.

Previous Research

Doctoral Research

During my doctoral work, I designed, developed, and implemented new gradient methods for rapidly investigating the effects of substrate and free surface interfacial interactions on block copolymer thin film nanostructure orientation. Block copolymers are ideal candidates for the design of interfaces due to the diversity of their self-assembled nanostructures (e.g., spheres, cylinders, networks, lamellae). For example, they can be used to lithographically template organic and inorganic nanoscale features, to pattern peptides on surfaces for biological studies, or to fabricate membranes for nanofiltration or organic optoelectronics.[3] The methods I developed are illustrated schematically in Figure 1 and included: (1) generation of monolayer substrate surface energy/chemistry gradients via a controlled vapor deposition technique that retains the facile implementation of existing vapor diffusion methods while enhancing gradient tunability,[1] and (2) fabrication of a solvent-resistant microfluidic mixing device for gradient solvent vapor annealing of thin films that enables precise control over solvent vapor compositions and concentrations at the free surface of the films.[2]

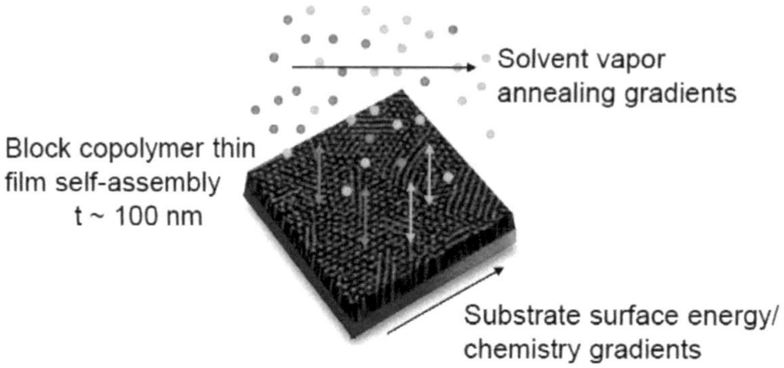

Figure 1: Schematic illustration of doctoral research using substrate surface gradients and solvent vapor annealing to affect the self-assembly of block copolymer thin films (~100 nm thickness).

Postdoctoral Research

Silicone elastomer networks (SENs) comprising poly(dimethylsiloxane) (PDMS) and/or poly(vinylmethylsiloxane) (PVMS) provide another interesting platform for interface manipulation due to their biocompatibility, tunable mechanical properties, and for PVMS, chemical modifiability via photoinitiated thiol-ene reactions to attach species to the pendent vinyl groups. With this platform for my postdoctoral work, I am developing a scheme that couples chain extension and networking reactions to achieve soft networks (E' ~ 10 kPa) that mimic the mechanical properties of soft biological tissues, and I am utilizing PVMS networks to direct the assembly of peptides on surfaces.

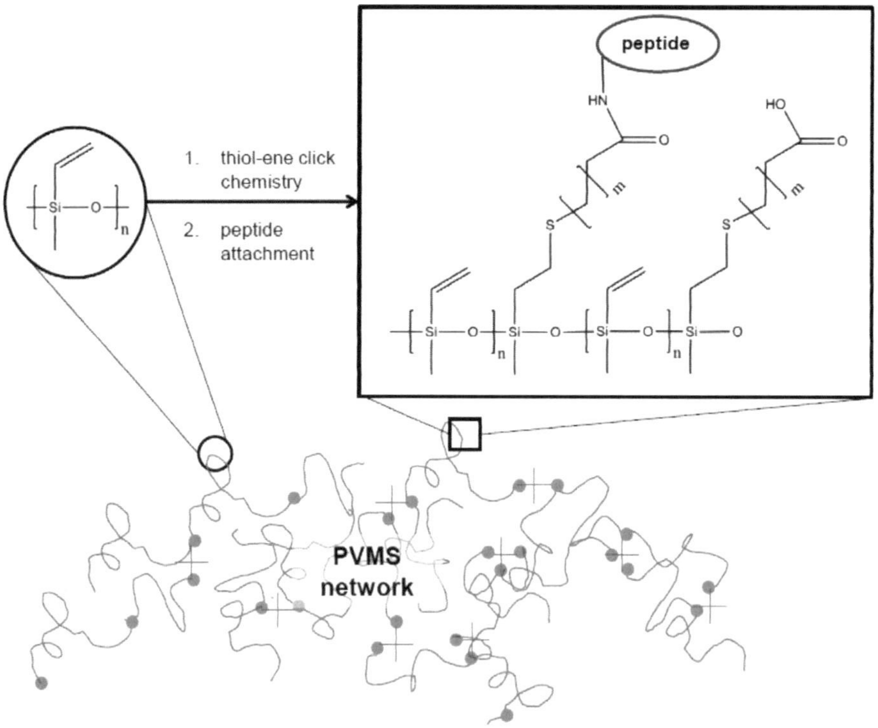

Figure 2: Schematic illustration of postdoctoral research on tuning PVMS network mechanical properties and modifying the surfaces of networks with peptide assemblies.

Proposed Research

My proposed research will draw upon my experience with block copolymer thin film self-assembly for templating or patterning interfaces, SENs for tuning the mechanical/adhesive component of interfacial interactions, and gradient methods for high-throughput study. I believe that by manipulating interfaces in soft matter systems, we can quickly and easily direct the self-assembly of novel biodegradable, biocompatible, and conducting polymers that can serve as materials in functional devices (e.g., in solar cells) or as platforms for creating new chemically and topographically patterned interfaces (e.g., for studying cell migration). Thus, I envision a research program that investigates the fundamental principles responsible for imparting materials with desired interfacial properties while simultaneously developing devices and platforms aimed at meeting societal and economic demands for more efficient processes, cleaner energy generation, and improved medical treatments.

Poly(vinylmethylsiloxane) (PVMS) containing copolymer systems will provide an excellent foundation for conducting this research. The mechanical properties of PVMS can be readily tailored from viscous liquid (natural state at room temperature due to the flexibility of the silicon-oxygen bond) to solid (via cross-linking or by blocking it with a glassy polymer). The potential for surface responsiveness and chemical modification of the pendent vinyl group via thiol-ene or vinyl-hydride chemistry to impart surface or bulk material functionality (e.g., photo-responsiveness, bioactivity) has also been demonstrated.[4,5] The combined benefits of block copolymer patterning and tunable PVMS mechanical properties make PVMS-containing copolymer systems an ideal research platform. For example, phase-separating PVMS blends could be used to understand cell responses to local mechanical stimuli for cancer metastasis studies, or the PVMS block of a copolymer could be doped with a photosensitive dye and coupled with a conducting block for organic photovoltaic devices. Employment of gradient methods would further enhance our ability to direct and understand the self-assembly of these copolymer systems. In the following paragraphs, I outline three projects that will establish PVMS as a platform for pursuing these research interests. These projects are illustrated schematically in Figure 3.

Project 1: Directing the Self-Assembly of Poly(siloxanes): From the Nanoscale to the Mesoscale

The first project will explore the fundamentals of block copolymer-homopolymer blend self-assembly under thin film confinement. These assemblies are expected to yield phase-separated morphologies that are highly influenced by surface interactions and which span multiple orders of magnitude (~10 nm to ~100 μm) depending on blend composition.[6] The majority of studies on block copolymer-homopolymer blends focus on one of two blend composition regimes: in the first, a homopolymer (A and/or B) swells the nanoscale phase-separated block copolymer (A-*b*-B) domains; in the second, the block copolymer (A-*b*-B) compatibilizes domains of macrophase-separated A and B homopolymers, much like a surfactant in an oil-water system. Few studies address the intermediate regime where one might expect to find interesting morphologies (such as the reported microemulsion phase) that bridge the gap between the nanoscale and the mesoscale, and even fewer have looked at off-symmetric compositions (i.e., volume fractions $f_A \neq f_B$).[7] The work from this project will contribute to these knowledge gaps, examine how thin film confinement changes the morphology as compared to the bulk, and explore ways to amplify or reverse surface segregation tendencies through substrate modification and solvent vapor annealing treatments. Gradient methods will be employed extensively for the latter studies.

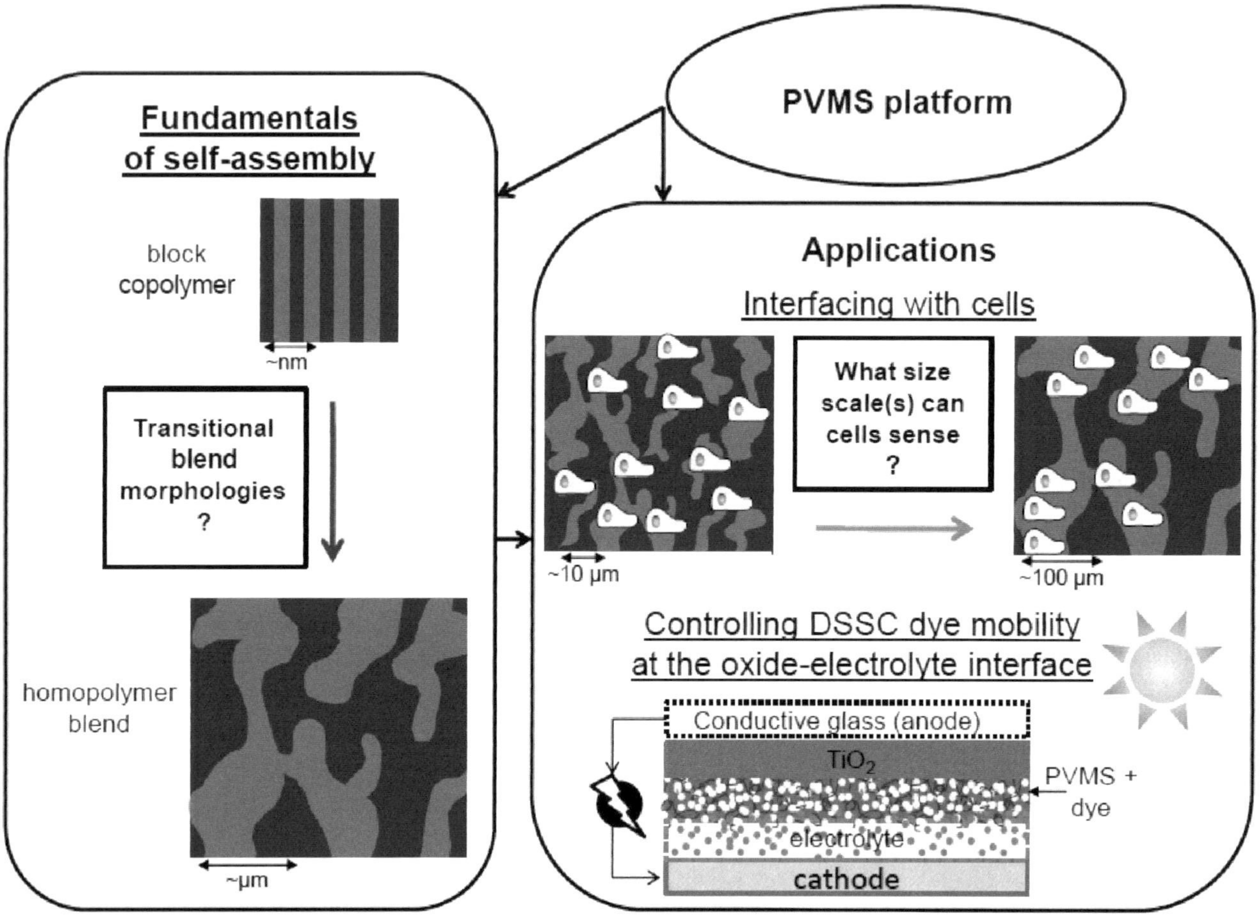

Figure 3: Schematic illustration of proposed research projects based on a PVMS platform: (1) Directing the self-assembly of block copolymer-homopolymer blends from the nanoscale (block copolymers) to the mesoscale (homopolymers) with block copolymer-homopolymer (A/A-*b*-B/B) blends. (2) Applying polymer blend surfaces to assess cell responses to chemical and mechanical stimuli involved in cancer metastasis. (3) Evaluating the role of dye mobility in dye-sensitized solar cell efficiency.

Project 2: Investigating Cancer Metastasis with Engineered Poly(siloxane) Blends

Surface patterns of PVMS homopolymer-copolymer blends will be used to assess the response of cancer cells to chemically and mechanically heterogeneous surfaces in order to elucidate the mechanisms behind cancer metastasis. The PVMS platform will provide tunability in surface chemical and mechanical properties and in the length scales over which they are presented. The envisioned blend system will consist of the block copolymer PVMS-*b*-PX, PVMS homopolymer, and PX homopolymer, where PX represents a complementary biocompatible block or homopolymer with a glass transition temperature below the curing temperature of PVMS networks (60-80 °C). Possible PX polymers include poly(lactic acid), poly(caprolactone), or poly(ethylene glycol). The final length scale of phase segregation will be controlled thermodynamically by the block copolymer content of the blend and kinetically by cross-linking the PVMS component to "lock-in" the structure at specific time points. Modification of the PVMS component with peptides via thiol-ene click chemistry and control over PVMS cross-link density will provide access to tunable chemical and mechanical contrast between blend phases. These surfaces will be used to investigate breast cancer metastasis by assessing the responses of healthy breast cells, non-invasive cancer cells, and invasive cells to segregated regions of chemical and mechanical stimuli. Future directions for this work will include the design of surfaces to distinguish between cell types and the fabrication of surfaces that can direct cell behavior by releasing cleavable moieties from the PVMS component for targeted delivery (e.g., chemotherapeutics, signaling molecules).

Proposal 3: Functional Poly(siloxanes) at the Interfaces of Dye-Sensitized Solar Cells

The aim of the third project is to elucidate the role of dye mobility in the efficiency of dye-sensitized solar cells (DSSCs). The dye absorbs light and interfaces directly with the semi-conductor oxide responsible for electron conduction and with an electrolyte that restores the lost electrons to the dye.[8] In this project, the dye molecules will be conjugated to PVMS pendent groups. The polymer will then be coated on or grafted to a titanium dioxide (TiO_2) semi-conductor. This architecture will enable more than a monolayer of dye to be present at the interface between the TiO_2 and the electrolyte solution and for dye mobility to be controlled by partially networking the PVMS chains through remaining side chain cross-links. Additionally, the flexibility of PVMS chains will maintain intimate contact between the dye and the TiO_2 for efficient electron conduction and may potentially improve contact between the dye and a solid-state electrolyte-replacement as well.

References

1. J. N. L. Albert, M. J. Baney, C. M. Stafford, J. Y. Kelly, T. H. Epps, III. "Generation of Monolayer Gradients in Surface Energy and Surface Chemistry for Block Copolymer Thin Film Studies," *ACS Nano*. 2009, *3*, 3977-3986.
2. J. N. L. Albert, T. D. Bogart, R. L. Lewis, K. L. Beers, M. J. Fasolka, J. B. Hutchison, B. D. Vogt, T. H. Epps, III. "Gradient Solvent Vapor Annealing of Block Copolymer Thin Films Using a Microfluidic Mixing Device," *Nano Letters*. 2011, *11*, 1351-1357.
3. J. N. L. Albert and T. H. Epps, III. "Self-assembly of Block Copolymer Thin Films," *Materials Today*. 2010, *13*(6), 24-33.
4. J. A. Crowe-Willoughby, J. Genzer. "Formation and Properties of Responsive Siloxane-Based Polymeric Surfaces with Tunable Reconstruction Kinetics," *Adv. Func. Mater.* 2009, *19*, 460-469.
5. S. Ahmed, H.-K. Yang, A. E. Ozcam, K. Efimenko, M. C. Weiger, J. Genzer, J. M. Haugh. "Poly(vinylmethylsiloxane) Elastomer Networks as Functional Materials for Cell Adhesion and Migration Studies," *Biomacromolecules*. 2011, *12*, 1265-1271.
6. F. Bates. "Polymer-Polymer Phase Behavior," *Science*. 1991, *251*, 898-905.
7. G. Liu, M. P. Stoykovich, S. Ji, K. O. Stuen, G. S. W. Craig, P. F. Nealey. "Phase Behavior and Dimensional Scaling of Symmetric Block Copolymer-Homopolymer Ternary Blends in Thin Films," *Macromolecules*. 2009, *42*, 3063-3072.
8. M. Grätzel. "Dye-sensitized Solar Cells," *J. Photochem. Photobiol. C: Photochem. Rev.* 2003, *4*, 145-153.

(3cn) Advanced Molecular Separations for Energy and Environmental Sciences

Tae-Hyun Bae, Chemistry, University of California, Berkeley, CA

Molecular separations are fundamental processes used in many energy and environmental sciences as well as chemical industries. Most separation processes widely used require multiple unit operations or consume large amount of energy to obtain high quality products. Hence, the development of new functional materials and processes capable of energy-efficient molecular separations is of interest in many fields.

In my first Ph.D. work at Seoul National University, as a polymer scientist, I have studied polymeric membranes and their processes for water and wastewater treatments. I have earned the second Ph.D. in chemical engineering at Georgia Institute of Technology after studying nanoporous materials such as zeolites and metal-organic frameworks for application in gas separation membranes. Currently, as a postdoctoral researcher in chemistry department at University of California, Berkely, I am working on chemistry of metal-organic frameworks for application in CO_2capture. Based on extensive experiences on nanoporous materials, polymer science and separation processes, I will be working on "advanced molecular separations for energy and environmental sciences" in the future.

I am interested in development of metal-organic framework membranes capable of carrier-mediated facilitated transport. Key requirement for successful fabrication of the facilitated transport membrane is to uniformly distribute carriers in membranes. In realizing such membrane, metal-organic frameworks have a great advantage over other materials due to their porous crystalline structures and tunable functionality, allowing uniform introduction of carriers to molecular paths in membranes. Development of nanopore membranes capable of sharp molecular discrimination is also my research interest.

Short bio

Postdoc: Jeffrey R. Long Research Group, Department of Chemistry, University of California, Berkeley

Ph.D. 2: School of Chemical & Biomolecular Engineering, Georgia Institute of Technology

Ph.D. 1: School of Biological Resources and Materials Engineering, Seoul National University, Korea

(3co) Molecular Modeling of Complex Chemical and Biological Processes for Human Health, Materials and Energy Applications

Diwakar Shukla, Bioengineering, Stanford University, Stanford , CA; Chemical Engineering, Massachusetts Institute of Technology, Cambridge, MA

Complex chemical and biological processes lie at the heart of some of the most challenging problems in science and engineering. The molecular-level description of these processes is required not only for understanding the physical phenomena but also for engineering novel solutions of the associated problem.

Multi-scale approaches are required for facing the challenging task of describing such processes that span several orders of magnitude in both length and time scales. These processes are being investigated by a spectrum of computational techniques ranging from quantum mechanics, molecular dynamics to multi-scale coarse graining and bridging between relevant scales. I have used computational methodologies such as world-wide distributed computing, rare event simulation techniques, GPU computing, network models etc. to study long time and length scale phenomena.

In this poster, I have discussed the applications of these techniques to problems related to the human health (such as protein folding/misfolding, aggregation, allosteric transitions in proteins, amyloid fibril formation), materials (crystallization of small molecules), and energy (transport of molecules in NiFe Hydrogenases, which are involved in reversible oxidation and production of Hydrogen).

More details can be found on my website: http://www.stanford.edu/~shukla

(3cp) The Design of Particulate Delivery Forms Via Single Drop Granule Formation Mechanisms

Heather N. Emady, School of Chemical Engineering, Purdue University, West Lafayette, IN

The overall goal of my future research program is to apply the engineering design approaches I have learned in my Ph.D. to industrially relevant particulate processing problems. My research interests lie in the broad realm of particle technology, with specific interests in single drop granule formation and the fundamentals of powder/liquid interactions.

Many industries, including pharmaceuticals, food, agricultural chemicals, and detergents utilize particulate processes to produce varying granular products. Wet granulation is one of these processes, where a liquid binder is added to a fine powder in order to get larger granules for improved particle properties. In typical granulation equipment, many of the granulation rate processes (wetting and nucleation, consolidation and growth, and breakage and attrition) occur simultaneously, making it notoriously difficult to control and predict product properties, such as size and shape.

Recently, a new granulation approach, regime separated granulation, has been proposed as a way to physically separate the different rate processes to get dramatically better control of the product granule attributes. The first and most important stage in any regime separated granulation process is nucleation, where new granules are formed by the addition of liquid to the powder bed. Drop controlled nucleation, where one drop forms one granule, is the most desirable operating regime. In order to apply regime separated granulation in practice, more knowledge is necessary regarding the mechanisms by which granules are formed from drop impact and penetration into a static powder bed.

In my doctoral work, I conducted single drop granule experiments with a syringe and a dish filled with powder to simulate drop controlled nucleation in static beds. From high speed camera videos of drop impact and penetration, I identified three different granule formation mechanisms: *Tunneling*, *Spreading*, and *Crater Formation*. Each mechanism produced distinctly different granule shapes. To quantify the conditions under which each mechanism will occur, I performed dimensional analysis and created a new regime map that plots the powder bed porosity (ε) against the modified granular Bond number (Bo_g^*), which is a ratio of the capillary force acting on a particle to the gravitational force acting on a particle. Subsequently, I derived a mechanistic model based on a force balance for the *Tunneling* mechanism and compared it to the regime map. This new knowledge about granule formation mechanisms opens up opportunities for designing processes and altering formulations to produce a given granule formation mechanism that leads to a particular granule shape and structure.

(3cq) Stochastic Modeling and Control of Neural and Small Length Scale Dynamical Systems

Gautam Kumar, Chemical Engineering, Lehigh University, Bethlehem, PA

In the last decade, dramatic advancements in experimental and computational techniques have flushed tremendous opportunities in the study of fundamental questions of science and engineering by taking the approach of stochastic modeling and control of dynamical systems. In parallel, emerging applications through these advances have ignited the development of new technologies by integrating advanced control strategies with stochastic dynamical models. One such example is the brain-machine interface. Over the course of my doctoral studies, my research interests have mainly been focused on developing optimal control strategies for applications in closed-loop neural prostheses and small length scale stochastic systems using a purely probabilistic framework. During this poster session, I will highlight examples from my graduate research which demonstrate my capability in advancing these subjects by integrating probabilistic tools with advanced optimal control policies. Building on these results, I will describe my future academic plans to commence a distinct, interdisciplinary and independent research program at the interface of engineering and computational neuroscience. I will explain how my interdisciplinary expertise in the theory of stochastic processes, neuroscience, optimization, nonlinear dynamical systems and process control uniquely positions me to capitalize on emerging opportunities at this area.

I am currently pursuing doctoral studies in Chemical Engineering at Lehigh University. The major theme of my studies is centered on developing control-theoretic framework for investigating the role of sensory feedback in closed-loop neural prostheses. A direct impact of such study can lead to a clinical deployment of neural prostheses in rehabilitation. In parallel, I have developed optimal stochastic control strategies for regulating behaviors of particles driven by Brownian motion. Emerging applications from this study include but are not limited to 1) understanding the dynamical behaviors of biological objects such as intra-cellular and DNA interactions in their natural environments for facilitating biomedical applications, 2) manipulation and regulation of nano and molecular structures for applications in drug delivery systems and laboratory on a chip technology.

(3cr) Simplifying the Complex Chemistry of Energy Conversion

Claude Franklin Goldsmith, Inorganic Chemistry, Fritz Haber Institute of the Max Planck Society, Berlin, Germany

Due to their high energy density and compatibility with existing infrastructure, hydrocarbon fuels will remain the principal energy storage medium for the foreseeable future. Although petroleum will continue to be the main fuel source, fuels derived from other materials – such as biomass, tar sands, and shale – will become major components. At the same time, increased efficiency and emissions regulations will impose new restrictions on engine and emissions control technology. Consequently, both the composition of and the requirements for next generation fuels are going to change dramatically in the coming decades. As the global demand for these fuels continues to increase, the development of sustainable technologies for the production and consumption of these fuels becomes increasingly urgent. In order to make these technologies commercially viable, we must advance our understanding of how the chemical bonds in these fuels are manipulated at the atomic level.

Short-contact time partial oxidation, flash pyrolysis, oxidative coupling, hydrothermal reforming, and Fischer-Tropsch can be used to synthesize fuels from diverse materials such as bio-mass, natural gas, and synthesis gas. Advanced conversion technologies such as catalytic combustion and solid oxide fuel cells could generate power with increased efficiency. Although the individual processes are quite different, each process involves temperatures and pressures that are high enough that the surface and gas-phase chemistry are coupled.

The challenge to developing these technologies is the staggering complexity of energy conversion. Alkane activation typically requires high temperatures and high pressures. Under these conditions, the heterogeneous chemistry on the catalyst surface is coupled with homogeneous chemistry in the gas phase. An accurate description of the chemical kinetics involves hundreds of species and thousands of elementary reactions, even for a species as light as propane. A detailed description of the coupling between homogeneous and heterogeneous chemistry requires quantitative descriptions for convective and diffusive fluxes in the gas phase and adsorption and desorption at the surface. This coupling is further influenced by the dynamic nature of the catalyst itself, which can undergo electronic and structural changes in response to the gas-phase chemical potential.

My research will help to elucidate the complex chemistry of energy conversion. I will combine experimental techniques with a suite of computational methods to create quantitative, predictive models for homogeneous/heterogeneous coupled systems. The benefit of detailed chemical kinetic mechanisms is two-fold. At a qualitatively level, they disentangle the layers of confusing phenomena, thereby providing engineers with insight into what is really going on at the atomic level. At a quantitative level, they make accurate predictions of future trends, thereby saving time and money. The techniques I am developing will provide fundamental insight into how fuels are manipulated at the atomic level, and they will accelerate technological breakthroughs in sustainable energy conversion.

(3cs) Design and Quantitative Characterization of Spatially-Patterned Collagen Biomaterials for Regenerative Medicine Applications

Steven R. Caliari, Chemical & Biomolecular Engineering, University of Illinois at Urbana-Champaign, Urbana, IL

Regenerating tissue structures such as orthopedic interfaces requires new biomaterial technologies that permit simultaneous control of microstructural, mechanical, and biochemical properties in a spatially-defined manner. My work has focused on the development of spatially-patterned dynamic materials to direct cell proliferation, phenotypic stability, and lineage-specific differentiation with broad applicability to a variety of regenerative medicine challenges.

The first part of my PhD developed and characterized collagen-glycosaminoglycan (CG) scaffolds for tendon tissue engineering. CG scaffolds are regulatory compliant extracellular matrix (ECM) analogs that have been applied to a variety of regenerative medicine challenges. We developed a directional solidification lyophilization approach to fabricate 3D scaffolds with highly aligned pores that mimic the native anisotropy of tendon and other aligned tissues [1]. Contact guidance cues provided by this microstructure support cell alignment within the 3D anisotropic network. We demonstrated that loss of the aligned contact guidance cues led to down-regulation of the tendon marker scleraxis (15-fold), indicating that this structure could aid long-term tenocyte phenotypic stability [2]. In order to improve the mechanical competence of these scaffolds, we created core-shell CG composites inspired by mechanically efficient core-shell composites in nature such as plant stems and porcupine quills that displayed high bioactivity and significantly (~ 36-fold) improved mechanical integrity [3].

We have also developed strategies for integrating multiple biomolecular signals into our scaffolds to provide a platform to simultaneously control multiple cell activities. We found single factor supplementation led to a dose-dependent trade-off between driving tenocyte proliferation versus maintenance of tenocyte phenotype. We then identified supplementation schemes using factor pairs (IGF-1, GDF-5) capable of rescuing tenocyte phenotype and gene expression profiles while simultaneously driving proliferation. We demonstrated a covalent immobilization strategy that allows efficient sequestration of bioactive levels of these factors, making this approach more amenable to *in vivo* translation. We also developed a benzophenone based direct, photolithographic approach to spatially pattern biomolecules within CG scaffolds and demonstrated creation of a wide range of patterns composed of multiple biomolecular species in a manner independent from scaffold fabrication steps [4].

This work became the foundation to build tendon-bone junction (TBJ) scaffolds with tunable, spatially-patterned microstructural, mechanical, and biochemical properties. TBJ scaffolds were fabricated by combining directional solidification [1] with a previously described liquid-phase co-synthesis method [5] to produce scaffolds displaying a distinct zonal structure with mineral content and geometric anisotropy mimicking the native TBJ. Mesenchymal stem cell (MSC)-seeded mineralized CG scaffolds displayed up-regulation of bone markers osteocalcin and bone sialoprotein as well as depressed expression of chondrogenic markers compared to non-mineralized scaffolds. Ongoing work is integrating mechanical stimulation with biomolecule-immobilized CG scaffolds to more efficiently drive multi-lineage MSC differentiation and long-term phenotypic maintenance across the graded scaffold.

I envision my own lab as an international leader in the design and quantitative characterization of 3D materials for applications spanning the interface of engineering and biology. We will leverage the technologies discussed here to design regenerative templates for tissues including bone, cartilage, muscle, and spine. These materials will integrate controlled presentation of therapeutic proteins, drugs, and genes. Additionally, quantitative tools to characterize scaffold permeability and pore size, shape, and fraction should be applicable to porous materials used in catalysis and a wide range of other engineering applications.

References:

1) Caliari SR, Harley BAC *Biomaterials*, 2011; 32(23): 5330-40.

2) Caliari SR, Weisgerber DW, Ramirez MA, Kelkhoff DO, Harley BAC *JMBBM*, 2012; in press.

3) Caliari SR, Ramirez MA, Harley BAC *Biomaterials*, 2011; 32(34): 8990-8.

4) Martin TA*, Caliari SR*, Harley BA, Bailey RC *Biomaterials* 2011; 32(16): 3949-57.

*equal contribution

5) Harley BAC, Lynn AK, et al. *J Biomed Mater Res A* 2010; 92(3): 1078-93.

(3ct) Enabling Technologies for High-Throughput Whole Tissue Analysis At Single-Cellular Resolution: From Model Organisms to Human Organs

Kwanghun Chung, Bioengineering, Stanford University, Stanford, CA

Understanding complex biological systems require integrated structural and molecular information. However, current techniques fall far short of providing complete system-relevant information with both global scope and molecular resolution. To address these limitations, I have developed two unique sets of technologies that enable rapid extraction of structural and molecular information at single-cellular resolution for broad biological systems. For small model organisms, such as C. elegans and Drosophila melanogaster embryos, I developed novel microfluidic technologies that allow precise and rapid manipulation of a large number of samples for imaging, screening, and laser ablation of single cells. These technologies have been used to study nervous system of C. elegans and embryogenesis of Drosophila melanogaster embryos. For large mammalian tissues, such as mouse brain and human tumor tissues, I have developed a novel technology that enables whole tissue imaging and molecular phenotyping without prohibitively inefficient mechanical sectioning and reconstruction. I have applied this technique to understanding function and dysfunction of mammalian brains. In conclusion, these enabling technologies promise to provide high-content system-wide structural and molecular information that may enable integrative understanding of a broad range of biological systems.

(3cu) The Thermodynamics and Chemistry of Atmospheric Organic Compounds

Scott A. Epstein, Department of Chemistry, University of California, Irvine, Irvine, CA

Ph.D. Carnegie Mellon University, Chemical Engineering. Advisor: Neil Donahue

Postdoctoral Work, University of California, Irvine, Chemistry. Advisor: Sergey Nizkorodov

The ubiquity of atmospheric particulate matter (PM) is a global problem that contributes to decreased life expectancies, respiratory effects, and may lead to premature cognitive decline. PM also effects visibility and influences climate by altering the radiative balance of the earth and changing cloud properties. The pervasiveness of particulate matter remains one of the most significant environmental problems; in the United States 74 million people live in counties that do not meet Environmental Protection Agency standards for PM less than 2.5 nm in diameter. PM contains a diverse mixture of chemical species, both inorganic and organic in nature. The inorganic fraction is relatively well-characterized; however, due to its complexity, the organic fraction (25 to 75% of the total particulate mass) remains to be adequately described.

The formation and evolution of atmospheric particulate matter is an ideal problem for chemical engineers. The organic fraction (organic aerosol, OA) consists of thousands of distinct chemical species in equilibrium with the gas phase. A large fraction of the species in OA have vapor pressures such that they will repartition between the particle and gas phase in response to changes in temperature or total condensed organic mass. I will present a semi-empirical relationship designed to correlate the heat of vaporization of an aerosol mixture with its distribution of volatilities[1]. This relationship has been used to help atmospheric modelers predict organic aerosol concentrations as a function of ambient temperature.

OA is either directly emitted by anthropogenic and biogenic sources or is formed by reactions of volatile organic precursors. The chemistry of these volatile organic precursors governs their eventual fate. If reactions with atmospheric oxidants such as O_3, NOx, and OH radical sufficiently reduce their volatility, the oxidation products will partition to the particle phase and contribute to the total PM mass. Ozonolysis oxidation reactions are complex, containing several intermediates and branching points. I will summarize experiments designed to elucidate the reaction barrier governing the reaction of the first intermediate, the *primary ozonide*, for a series of reactive alkenes[2, 3]. These experiments, performed at cryogenic temperatures under vacuum to isolate the short-lived intermediates, are the first measurements of these reaction barriers and enhance our understanding of ozonolysis.

In the presence of clouds or fog, gases or aerosol may partition into the aqueous phase and undergo cloud processing by dissolved oxidants or actinic radiation. The significance of aqueous photolysis is not well understood due to limited experimental measurements. I will show photolysis rate measurements of aqueous methyl peroxide[4] (the simplest organic peroxide) and aqueous glyceraldehyde[5] (an oxidized water-soluble photolabile compound), which are used to calculate atmospheric lifetimes. I will also present a theoretical framework that uses Henry's law constants and hydration equilibrium to determine the significance of aqueous photolysis for a large group of atmospherically relevant organic compounds[6]. This framework will help guide researchers in the selection of compounds for experimental measurements of aqueous rate constants.

My proposed research will focus on enhancing our understanding of the sources, composition, and evolution of particulate matter through laboratory studies and parameterizations designed to bridge the gap between experimentalists and chemical transport modelers.

REFERENCES

1. Epstein, S. A.; Riipinen, I.; Donahue, N. M., A Semiempirical Correlation between Enthalpy of Vaporization and Saturation Concentration for Organic Aerosol. *Environ. Sci. & Technol.* **2009,** *44*, (2), 743-748.

2. Epstein, S. A.; Donahue, N. M., The Kinetics of Tetramethylethene Ozonolysis: Decomposition of the Primary Ozonide and Subsequent Product Formation in the Condensed Phase. *J. Phys. Chem. A* **2008,** *112*, (51), 13535-13541.

3. Epstein, S. A.; Donahue, N. M., Ozonolysis of Cyclic Alkenes as Surrogates for Biogenic Terpenes: Primary Ozonide Formation and Decomposition. *J. Phys. Chem. A* **2010,** *114*, (28), 7509-7515.

4. Epstein, S. A.; Shemesh, D.; Tran, V. T.; Nizkorodov, S. A.; Gerber, R. B., Absorption Spectra and Photolysis of Methyl Peroxide in Liquid and Frozen Water. *J. Phys. Chem. A* **2011**.

5. Epstein, S. A.; Nizkorodov, S. A., Direct Aqueous Photolysis of Carbonyls in the Atmosphere. *In preparation.* **2012**.

6. Epstein, S. A.; Nizkorodov, S. A., A Comparison of the Chemical Sinks of Atmospheric Organics in the Gas and Aqueous Phase. *Atmos. Chem. Phys. Discuss.* **2012,** *12*, (4), 10015-10058.

(3cv) Rational Selection of Ionic Liquids for the Catalytic Conversion of Renewable Feedstock

Sameer Parvathikar, Andrew R. Tadd and Johannes W. Schwank, Chemical Engineering, University of Michigan, Ann Arbor, MI

Widespread adoption of biomass-derived fuels and chemicals has been hampered by low process efficiencies, inhibitive capital and operating costs, and other constraints inherent with current biomass conversion technologies. Ionic liquids (ILs), which are low melting salts, have unique properties like negligent vapor pressure, tunable physical and chemical properties, which make them great alternatives to conventional solvents. ILs, particularly, 1,3-dialkylimidazolium-based ones, have been demonstrated to be great solvents for reactions like the dissolution of cellulose, among others. This makes them promising alternatives to current biomass conversion technologies. While these ILs have proven to be quite capable of processing these large biomolecules, their widespread adoption has been limited due to their toxicity and inhibitive costs.

It is clear that there is a need for less expensive, and environmentally benign alternatives. The search for these alternative solvents currently, can be best described as Edisonian in nature. The objective of this work was to design solvent systems that result in increased yields of our desired product and be capable of dissolving and then converting higher fractions of macromolecules like cellulose and hemicellulose, by developing a deeper understanding of the interaction between the solute (biomolecules) and the solvent. A number of solvent properties like ability to donate or accept hydrogen bonds, acidity, etc., will be studied for both imidazolium-based ILs and lower-cost alternatives. Using probe reactions like the isomerization and dehydration of sugars, property-performance relationships will be established. The effect of using different catalysts to affect the reaction environment and thence product yields will also be examined. This work will provide insights into rationally selecting ILs to perform targeted chemistries.

My research and teaching activities, here at the University of Michigan, have complemented each other very well. Among other skills, teaching has taught me how to communicate to audiences with a wide variety of cognitive skills, offered opportunities to receive and implement constructive feedback, and hence develop personally and professionally. From teaching design of experiments and statistical significance of data in laboratory classes, to helping students develop problem-solving skills in theoretical classes, I have been able to use not just skills learned in my research, but offer concrete examples as well. As a teacher and an educator, my goal is to instill in students, a passion for learning, develop critical evaluation, insightful decision-making and risk-taking skills – all necessary tools for a successful engineer. I aim to create a caring, safe and creative environment for a student to grow into a confident and knowledgeable professional, and to help me, as a teacher, to grow with them.

(3cw) Modeling and Experiments of Nonspecific Interactions

Andrew D. White and Shaoyi Jiang, Chemical Engineering, University of Washington, Seattle, WA

Many of the difficulties of practical surface coatings and materials arise from the complexities of multiple interactions in biological environments. These interactions can be broadly broken into two categories: those which are important to function, high activity for a specific target, and nonspecific binding detrimental to function, low activity binding to many targets. These nonspecific interactions are central to many failures of drugs, sensors, and surface coatings in the harsh biological environments. Despite this fact, there is little modeling work on studying nonspecific interactions. Modeling all possible interactions within a such a system is difficult, yet there are ways to both indirectly study nonspecific binding and directly study many possible interactions using data-mining approaches. In this work, we describe recent work using bioinformatics, phenomenological modeling, molecular simulations, and experiments to study three types of nonspecific interactions. They are the weak nonspecific interactions between proteins in crowded environments, irreversible nonspecific protein adsorption onto surfaces, and nonspecific stabilization of proteins in molecular chaperones. These disparate systems have allowed us to generalize our conclusions into a small set of principles that other researchers should consider when designing specific activity in order to maintain resistance to nonspecific interactions.

(3cx) Advanced Photon Management for Solar Energy Conversion and Photocatalysis

Kevin M. McPeak, ETH- Zurich, Zurich, Switzerland

The late Nobel Laureate, Richard Smalley, declared "Energy is the single most important challenge facing humanity today." The use of large scale, renewable energy sources is necessary to meet the demands of both population growth and CO_2 reduction in the 21st century. Solar energy is the only clean, renewable energy source available that is capable of supplying the 13 terawatts that are necessary to power the planet. I will develop a research program which focuses on harnessing this abundant resource for energy conversion and photocatalytic devices. The utilization of advanced photon management techniques to enhance the performance of these devices will be a primary focus of my group.

During my PhD I studied the chemical bath deposition of nanomaterials for low cost photovoltaic devices. I developed a combinatorial approach for depositing oxide and chalcogenide nanomaterials which enabled rapid and unprecedented understanding of the effect of growth conditions on optical, electronic, and structural properties of the material. Furthermore, I pioneered the use of *in situ* x-ray absorption fine structure spectroscopy to better understand the fundamentals of chemical bath deposition.

Throughout my postdoctoral work at ETH Zurich I have developed novel, low cost techniques to fabricate large-area, complex, two dimensional metal grating structures for back reflectors in solar cell devices. I have used both finite element modeling and experimental results to verify the efficacy of these grating structures on improving solar cell efficiency.

As a faculty member, I will combine my bottom-up material synthesis, *in situ* characterization, and top-down fabrication experience to establish a leading program to enhance solar-to-fuel energy conversion and the photodisinfection of water. These application areas are fundamental to energy sustainability and public health in the world.

(3cy) Biomolecular Simulation Approaches for Proteins and Nucleic Acids with Novel Therapeutic and Biomedical Applications

Harish Vashisth, Chemistry and Biophysics, University of Michigan, Ann Arbor, MI

Engineering biomolecules for novel therapeutic applications requires developing a detailed mechanistic and functional understanding of potential drug targets. Cellular membranes often present a major physical barrier for the delivery of synthetic compounds that mimic their physiological counterparts such as small peptide hormones, growth factors, vitamins, and neurotransmitters. Binding of these molecular ligands to the integral membrane proteins largely mediates downstream cellular signaling events. My doctoral and postdoctoral work has spanned various classes of membrane proteins such as hormone receptors involved in diabetes and cancer, nutrient/toxin transporters, and G-protein coupled receptors that are targets of at least a third of marketed drugs. With the integration of biophysical theory, modeling, and atomistic simulation approaches on multiple length- and time-scales, I have explored the vast conformational repertoire of these biomolecules, which can be exploited for various pharmaceutical applications. Specifically, I am going to present part of my doctoral work in this poster where I studied the hormones and receptors of the insulin family, both of which are key players in the type-1 and type-2 diabetes at various levels, and also a homologous family of insulin-like growth factors that have been implicated in various cancers. Mechanistic principles that have emerged from this work have key implications for the design of novel insulin mimetic peptide agonists with tuned specificities for their cognate receptors. From my postdoctoral work, I will address key questions related to the coupling of conformational transitions in the nucleotide-binding motifs and the transmembrane domains of ATP-binding cassette (ABC) transporters that allow unidirectional transport of nutrients into, and toxins out of, cytoplasm. Lastly, I will also address the question of biogenesis of large ribonucleoprotein complexes known as the ribosomes (the protein-factory in all cells), which are targets of existing and novel antibacterial drugs.

(3cz) Design of Solid Acid Catalysts for Aqueous Phase Conversion of Lignocellulosic Biomass to Liquid Fuels and Fuel Precursors

Ronen Weingarten, Department of Chemical Engineering, University of Massachusetts Amherst, Amherst, MA

A worldwide initiative has been set forth to decrease our dependence on petroleum-based fuels and develop new renewable sources of energy. Particularly, the use of agricultural and forest residues and other lignocellulosic crops is a viable option compared to first generation biomass which demonstrate several drawbacks including competition with food crops and water. Lignocellulosic biomass is a sustainable source of liquid fuels which generates significantly less greenhouse gas emissions than do fossil fuels and may even be greenhouse gas neutral pending efficient production methods. The production of biofuels calls for the defunctionalization of highly oxygenated compounds which make up lignocellulosic biomass. The transformation of lignocellulosic biomass in aqueous media offers an environmentally friendly and economical route to produce targeted chemicals. The major challenge in the commercialization of lignocellulosic biofuels is the development of efficient technologies to achieve process sustainability and convert high volumes of biomass feedstock to displace crude oil as the primary source of fuels.

In part, further improvement of conversion and selectivity for aqueous phase processes are essential to overcome these challenges. Acid catalysts play fundamental roles in many applications focused on biomass upgrading and promising results have already been obtained with homogeneous acid catalysts. Nevertheless, due to their economic and environmental viability, it would be desirable to obtain solid acid catalysts that exhibit activities and selectivities at least comparable to their homogeneous counterparts. As such, the effect of water on the catalytic behavior of solid acid catalysts remains questionable, as exposure of the catalyst to a polar solvent such as water can potentially alter the intrinsic nature of the acid surface due to solvation effects. This reality in turn poses difficulties in determining the acidity of solid acids in aqueous media.

My research thus far in the field of aqueous phase biomass conversion has led me to significant findings pertaining to the chemistry and design of solid acid catalyst for aqueous phase conversion of carbohydrates. This consists of an integrated approach which includes: (1) understanding the fundamental chemistry and reaction pathways through kinetic studies with homogeneous acid catalysts and (2) investigating the structural properties of solid acid catalysts and their correlation with activity and selectivity. We developed a kinetic model for the dehydration of xylose to furfural in a biphasic reactor using a homogeneous catalyst.[1] Xylose serves as a model compound for the hemicellulose fraction of lignocellulosic biomass, and furfural is a valuable precursor for the production of liquid alkanes. We then used our model to describe the optimal reaction conditions for furfural production. Subsequently, we examined the role of Lewis and Brønsted sites of solid acid catalysts for the dehydration of xylose in aqueous media.[2] Our results have allowed us to deduce the optimal design of solid acid catalysts for this class of reactions.

In other studies we introduced a new process to produce levulinic acid from cellulose at high concentrations without the use of a homogeneous acid catalyst.[3] Levulinic acid is a versatile renewable platform molecule which can be used as the basis for the production of various high-volume chemicals and fuels. We also developed a kinetic model to predict the optimal reactor design and operating conditions for HMF (5-hydroxymethylfurfural) and levulinic acid production from glucose in a continuous reactor system.[4]

The objective of my research is to develop new catalytic materials and processes for the production of liquid fuels and fuel precursors from lignocellulosic biomass. Development of efficient technologies targeted for aqueous phase applications will play a significant role in the development of future biorefineries. This notion of a biorefinery, analogous to a petrochemical refinery, will consist of an integrated system of biomass conversion processes to produce transportation fuels, chemicals, heat and power. In particular, future work will focus on designing hydrothermally stable solid catalysts for a variety of aqueous phase reactions. Implementing advanced in-situ characterization techniques will be fundamental to elucidate the role of water on the intrinsic nature of the catalyst. Additional factors such as kinetic modeling and reactors engineering will also play fundamental roles in developing sustainable processes for the production of fuels and chemicals from biomass-derived feedstock.

References

1. Weingarten, R.; Cho, J.; Conner Jr, W. C.; Huber, G. W., Kinetics of Furfural Production by Dehydration of Xylose in a Biphasic Reactor with Microwave Heating. *Green Chemistry* **2010,** *12* (8), 1423-1429.

2. Weingarten, R.; Tompsett, G. A.; Conner Jr, W. C.; Huber, G. W., Design of Solid Acid Catalysts for Aqueous Phase Dehydration of Carbohydrates: The Role of Lewis and Brønsted Acid Sites. *Journal of Catalysis* **2011,** *279* (1), 174-182.

3. Weingarten, R.; Conner Jr, W. C.; Huber, G. W., Production of levulinic acid from cellulose by hydrothermal decomposition combined with aqueous phase dehydration with a solid acid catalyst. *Energy and Environmental Science* **2012**, DOI: 10.1039/c2ee21593d.

4. Weingarten, R.; Cho, J.; Xing, R.; Conner Jr, W. C.; Huber, G. W., Kinetics and Reaction Engineering of Levulinic Acid Production from Aqueous Glucose Solutions. *ChemSusChem* **2012**, DOI: 10.1002/cssc.201100717.

(3d) Nanotechnology for Biofuels

Leidy Peña[1], Keith L. Hohn[2] and Donghai Wang[1], (1)Biological and Agricultural Engineering, Kansas State University, Manhattan, KS, (2)Chemical Engineering, Kansa State University, Manhattan, KS

Uncertainty in the oil supply, and the environmental issues associated to the utilization of fossil fuels have motivated the interest to produce bio-ethanol as transportation fuel. Most of the ethanol produced in the US comes from the saccharification and fermentation of grains such as corn. Corn is a staple and its use a fuel generates food security concerns. In order to meet the government's goal of producing 35 million gallons of renewable fuel per year by 2035, about 20 million gallons of lignocellulosic ethanol should be produced per year. The production of ethanol cellulosic faces significant challenges. The structure and chemical composition of lignocellulosic biomasses makes these materials highly resistant to saccharification. Pretreatment of the biomass is required to make it susceptible to enzymatic hydrolysis. Mineral acids along with physical and thermal treatments can increase the digestibility of lignocellulosic materials to a high degree. The use of harsh chemicals demands for corrosion resistant materials, besides neutralization and waste disposal costs. Severe treatments can degrade the sugars to degradation products such as levulinic acid, hydroxyl-methyl-furfural, formic acid and furfural; these compounds can compromise the enzymatic hydrolysis and the fermentation of the carbohydrates because of their inhibitory character. Solid acids have effectively been utilized in commercial applications. Nanoparticles are the latest breakthrough in hetero-catalysis. Nanocatalysts have large surface areas for reaction, they can be recovered from reaction streams, and reused. Some research efforts have being made on the synthesis and characterization of acid-functionalized nanoparticles; however, effective acid nano-catalysts have yet to be developed. Acid-functionalized nanoparticles could be used to catalyze the pretreatment of lignocellulosic biomass; with the advantage that these catalysts could be recycled and reused.

(3da) Understanding the Relationship Between Nanostructure and Ion Transport in Membranes for Energy Applications

Megan L. Hoarfrost, Chemical Engineering, University of California, Berkeley, Berkeley, CA

Nanostructured membranes containing structural and ion-conducting domains are of great interest for a wide range of energy applications requiring high conductivity coupled with mechanical durability. Understanding the relationship between morphological and conductive properties of such materials is essential for designing optimal membranes. Here, I will discuss the relationship between nanostructure and ionic conductivity in mixtures of block copolymers with ionic liquids. The mixtures self-assemble into well-defined nanostructures, where the ionic liquid selectively resides in one microphase. Scaling relationships have been developed to describe the conductivity of block copolymer/ionic liquid membranes as a function of block copolymer composition, ionic liquid concentration, and temperature. Furthermore, confinement to block copolymer nandomains is shown to affect the conductive properties of the ionic liquid. For instance, the hydrogen bond network in the proton-conducting ionic liquid imidazolium bis(trifluoromethylsulfonyl)imide ([Im][TFSI]) is altered by confinement to lamellar block copolymer nanodomains, leading to high levels of a fast proton "hopping" transport mechanism. As a result, the activation energy for ion transport is lower than in comparable membranes that do not self-assemble. These findings mark important progress in understanding how membrane composition, structure and ion transport are interrelated in nanostructured membranes, and lay the foundation for my future research plans.

Optimizing ion transport in nanostructured membranes is essential for devices relevant for energy generation and storage as well as for water purification and biological processes. My future research will focus on understanding mesostructure-property relationships that enable materials design for these applications. I will combine my expertise in block copolymer synthesis and morphological characterization with expertise in nuclear magnetic resonance (NMR) microscopy techniques, which have traditionally been used to characterize pore structure and molecular diffusion in porous media, and have the advantage of simultaneously probing structure and transport. This is a unique and powerful toolset for investigating structure-property relationships in nanostructured, ion-conducting systems.

(3db) Finding the Rules That Determine Microbial Community Function

James Boedicker, Applied Physics, Caltech, Pasadena, CA

Controlling the functional outputs of microbial communities will be an essential tool for addressing many of today's scientific and technological challenges. Microbial communities are critical in processes as diverse as nitrogen fixation in plants roots and wastewater treatment, and there is increasing awareness of the role of such communities in human health. Recent interest in designing artificial communities of microbes for applications such as biofuel production or environmental remediation suggests that microbial communities are a central aspect to many emerging fields. In order to effectively address the scientific, health and technological challenges associated with microbial communities, we must first learn how these communities make decisions and regulate activity.

The goal of my research is to develop experimental and theoretical tools to predict and control the behavior of multispecies microbial communities. To that end, it will be essential both to understand how the local community composition influences community function and to develop predictive models of interspecies gene regulation. I discuss my progress towards these goals, including using microfluidic approaches to understand how microbial activity is modulated by the micron-scale spatial structure of microbial cultures and the combination of quantitative models of gene regulation with experiments to test the precise predictions of these models. My research will discover new functional communities, establish the input-output relationships of community function, determine strategies to control their activity, and enable us to apply microbial communities to address a broad range of biological and chemical challenges.

(3dc) Materials for 4D Biology: Spatial and Temporal Control of the Stem Cell Niche

Mark W. Tibbitt[1] and Kristi S. Anseth[2], (1)Department of Chemical and Biological Engineering, University of Colorado, Boulder, CO, (2)Department of Chemical and Biological Engineering, University of Colorado, Howard Hughes Medical Institute, Boulder, CO

Interactions between mammalian cells and the surrounding extracellular matrix (ECM) are critical in the regulation of stem cell differentiation, cell migration, tissue morphogenesis, and disease progression. Cell decisions are often made by integrating these complex and dynamic interactions in 4D: three-dimensional space and time. Synthetic polymer-based hydrogels have emerged as an effective cell culture platform to investigate the influence of these cell-matrix interactions on cell function in a systematic manner. Hydrogels are attractive mimics of the ECM on account of their high water content, tissue-like elasticity, and facile transport of nutrients, waste, and soluble factors. Furthermore, poly(ethylene glycol) (PEG) hydrogels can be formed under mild, cytocompatible conditions and are easily modified to present various biochemical and mechanical signals to cells during culture.

While cell-laden PEG hydrogel constructs have been employed for regenerative medicine and fundamental biological applications, there is still a lack of materials that allow the user to recapitulate the spatially heterogeneous and dynamic environment that cells interact with in vivo. My Ph.D. research has focused on the development, characterization, and application of photoresponsive PEG-based hydrogels that afford the user control of the mechanical and biochemical properties of the scaffold in both 3D space and time. This class of materials facilitates experiments that probe the relationship between spatially heterogeneous and dynamic signals from the ECM and cell function in a well-defined environment to study the fundamentals of the cell-material interactions, stem cell differentiation and plasticity, as well as the cellular response to local changes in soluble factor concentration.

Specifically, we have developed photoresponsive materials for the dynamic modulation of the ECM during 2D and 3D culture. These materials have been used to investigate stem cell response to dynamic changes in elasticity and the plasticity of stem cells during microenvironmental mechanotransduction. Further, we have utilized photoresponsive hydrogels to develop photodegradable microspheres for the targeted delivery of soluble factors to study tissue morphogenesis and chemotaxis. Thin films of photodegradable hydrogel have been employed for the selective capture and release of cells for diagnostic application. Click-based, enzymatically and photodegradable hydrogels have been used as advanced 3D culture platforms that promote cell function while permitting the analysis of cell behavior through the recovery of cultured cells.

(3dd) Controlling Reaction-Diffusion-Convection for Intelligent and Functional Chemical Systems

Siowling Soh, Chemistry and Chemical Biology, Harvard University, Cambridge, MA

Intelligent and functional systems detect cues from its environment, respond appropriately to these cues and perform their intended purposes. At present, development of these systems has relied mainly on electronic-based machineries, devices and robots for solving problems in practically all aspects of our lives. While this approach has largely been successful, the performance of these electronic systems is still far less intelligent than naturally-occurring systems, ranging from a single cell to any larger biological organisms. Despite their complexity, these systems from nature operate almost entirely out of chemical reactions coupled with diffusive and/or convective processes. Learning from these systems and engineering the appropriate reaction and transport processes can provide the solution to non-electronic-based intelligent systems with vast potential.

Examples of such intelligent and functional chemical systems constructed through controlling reaction-diffusion and/or convection include: (1a) Camphor boats which self-assemble and swarm collectively in a biased direction, (1b) oil droplets which solve mazes using the shortest possible route, (1c) programmable reaction-diffusion which allows fabrication of core-shell particles of desired geometries and (1d) core-shell porous materials which perform sequential reactions.

On the other hand, naturally-occurring systems, such as cells, also make use of reaction-diffusion and/or convective processes for intracellular transport and control (*). As such, I am also actively studying cells – in particular, cell mechanics and cell motility. Cells plated on micropatterns, such as (2a) on circular islands of different sizes allow a micromechanical model for different cell types to be constructed (cell shape is defined by cell membrane, cortical actin and supported by its nucleus) and (2b) on lines where movement of migrating non-metastatic, cancerous cells and metastatic, cancerous cells are found to be diffusive and super-diffusive respectively.

Through controlling reaction-diffusion-convection processes in artificial systems and studying natural systems, such as cells, hopefully, more advanced, intelligent and functional chemical systems can be realized.

(*) Soh, S.; Byrska, M.; Kandere-Grzybowska, K.; Grzybowski, B. A. Reaction-Diffusion Systems in Intracellular Molecular Transport and Control. *Angew. Chem.-Int. Edit.* **2010,** *49*, 4170-4198.

(3de) Understanding and Exploiting Protein Functional Dynamics to Combat Drug Resistance

Gregory Bowman, Chemistry and Molecular & Cell Biology, University of California, Berkeley, Berkeley, CA

Resistance to antibiotics and other drugs is increasing rapidly but pharmaceutical companies are investing less and less in combating this looming threat to human health precisely because resistance is likely to emerge before they can recoup their costs, much less make a profit. I propose to use a combination of theory, simulation, and experiment to understand the biophysical basis of drug resistance and develop new therapeutic strategies that will make combating resistance more tractable. I am uniquely suited for this work because of my strong background in understanding how a molecule's free energy landscape determines its structure, dynamics, and function and then using perturbations (like drugs) to manipulate these properties.

During my doctoral work at Stanford, I developed computational methods for extending the reach of atomistic models from hundreds of nanoseconds timescales to tens of milliseconds and applied them to develop a new theory of protein folding. The models I developed—called Markov state models (MSMs)—are discrete-time master equation models built from extensive molecular dynamics simulations using a combination of physics, Bayesian statistics, information theory, and network theory. They are essentially maps of the conformational space a molecule explores that capture both its thermodynamics and kinetics.

As a Miller Research Fellow at UC Berkeley, I am adapting/applying my methods to understand aspects of protein function like allostery, ligand binding, and protein-protein interactions. A major emphasis of my work is predicting how perturbations (like drugs and mutations) can alter a protein's function and then testing these predictions experimentally.

(3df) Multiscale Methods for Complex Systems

Brooks D. Rabideau, Mechanical Engineering, RWTH Aachen University, Aachen, Germany

Bio Postdoctoral Associate/*Transport in Ionic Liquid-Cellulose Mixtures*
PhD, Chemical Engineering, The University of Texas at Austin (2007)
BSc, Chemical Engineering, Northwestern University (2001

The conversion of waste biomass into tailor-made biofuels like that currently sought at RWTH has the potential to aid significantly in world's future energy demands through the use of clean, sustainable sources. Ionic liquids with their low-volatility and their surprisingly effective dissolution of cellulose are currently being evaluated as an environmentally-friendly replacement to traditional solvents used in the breakdown of biomass. Though a number of ionic liquids capable of this dissolution have been discovered, a myriad of possible cation/anion combinations exist. Interestingly, certain combinations of cation and anion have shown a very effective dissolution of cellulose, while other similar combinations do not. A number of guidelines exits as to what makes for a good solvent, such as the effective hydrogen-bond accepting ability of the anion, however many details, such as the exact role of the cation in this dissolution, remain unclear.

I'm using molelcular dynamics simulations with an explicit, all atom approach, to gain insight into the underlying mechanisms of dissolution. By focussing on small, crystalline bundles in various ionic liquids, we are able to observe the unzippering of individual strands from the main bundle. Through this study we have identified different underlying mechanisms for separate ionic liquids. In the future we hope to use advanced modeling techniques, such as kinetic Monte Carlo, to model the dissolution of much larger strands and compare these values with experimental results obtained by our collaborators. My future work will include more innovative multiscale methods and the extension of these techniques to understanding the underlying mechanisms of certain rheological phenomena.

Publications

B. D. Rabideau and A. E. Ismail. Molecular Dynamics Simulations of Dissolution Mechanisms of Cellulose in Ionic Liquids. (In preparation.).

A.N. Niazi, B. Rabideau and A.E. Ismail. Molecular dynamics simulations of ionic liquid/water mixtures: water concentration effects. (In preparation.)

B. D. Rabideau and A. E. Ismail. Spatially Ordered Local Environments around Cellulose-Derived Solutes in Water and 1-Butyl-3-methylimidazolium Chloride. (In preparation.).

R. Isele-Holder, B. D. Rabideau and A. E. Ismail. Definition and Computation of Intermolecular Contact in Liquids using Additively Weighted Voronoi Tessellation. J. Phys. Chem. A. (Accepted).

B. D. Rabideau, P. Moucheront, F. Bertrand, S. Rodts, Y. Mélinge, C. Lanos and P. Coussot. Internal Flow Characteristics of a Plastic Kaolin Suspension During Extrusion. J. Amer. Cer. Soc. 2012, 95, 494-501.

S. Rodts, J. Boujlel, B. Rabideau, G. Ovarlez, N. Roussel, P. Moucheront, C. Lanos, F. Bertrand and P. Coussot. Solid-Liquid Transition and Rejuvenation Similarities in Extrusion and Squeeze Flows of Thixotropic Materials Studied by NMR and MRI. Physical Review E. 2010, 81, 021402.

B. D. Rabideau, P. Moucheront, F. Bertrand, S. Rodts, N. Roussel, C. Lanos and P. Coussot. The Extrusion of a Model Yield Stress Fluid Imaged by MRI Velocimetry. Journal of Non-Newtonian Fluid Mechanics. 2010, 165, 394-408.

Q. Barral, G. Ovarlez, X. Chateau, J. Boujlel, B. Rabideau and P. Coussot. Adhesion of Yield Stress Fluids. Soft Matter. 2010, 6, 1343-1351.

B. D. Rabideau, C. Lanos and P. Coussot. An Investigation of Squeeze Flow as a Viable Technique for Determining the Yield Stress. Rheologica Acta 2009, 48, 517-526.

B. D. Rabideau and R. T. Bonnecaze. Simulation of the Hydrodynamically-Assisted Self-Organization of DNA-functionalized Colloidal Particles into 2D Arrays. Langmuir 2007, 23, 10000-10007.

B. D. Rabideau, L. E. Pell, B. A. Korgel and R. T. Bonnecaze. Observation of Long Range Orientational Order in Monolayers of Polydisperse Colloids, Langmuir 2007, 23, 1270-1274.

B. D. Rabideau and R. T. Bonnecaze. Computational Predictions of Stable 2D Arrays of Bidisperse Particles, Langmuir 2005, 21, 10856-10861.

(3dg) Structure/Property Relationships in Polymer Membranes for Water Purification and Power Generation

Geoffrey M. Geise, Deptartment of Chemical Engineering & The Texas Materials Institute, The University of Texas at Austin, Austin, TX

Global demand for clean water and renewable energy sources is increasing. These demands are inherently related because energy is required to purify water and, in most cases, purified water is required to produce energy. Polymer membranes are widely used in desalination applications, such as reverse osmosis (RO), nanofiltration (NF), forward osmosis (FO), and electrodialysis (ED), and they could be critical for many renewable power generation technologies, such as pressure-retarded osmosis (PRO), reverse electrodialysis (RED), and fuel cells. Improved membranes with tailored water and salt transport properties are required to extend and optimize these technologies. Structure/property relationships must be used to provide the fundamental framework for optimizing polymer materials for applications that require tailored water and salt transport properties.

Water and salt permeability, sorption, and diffusivity structure/property relationships are presented for a series of dense, nonporous polymers that include crosslinked poly(ethylene glycol diacrylate) hydrogels, sulfonated polysulfone random copolymers, and sulfonated styrenic pentablock copolymers. Salt transport through a polymer film depends, to a great extent, on whether or not the polymer is charged (e.g., sulfonated), and the effects of polymer charge on water transport and salt permeability, sorption, and diffusion are discussed. Additionally, the effects of polymer structure modifications, such as crosslinking and varying block copolymer architectures, are discussed to further demonstrate how water and salt transport properties can be tuned. Finally, water and salt transport data are modeled to further describe the permeability, sorption, and diffusion properties that govern water and salt transport.

(3dh) Towards the Rational Design of Materials; Effect of Ionizable Head Group Architecture On the Delivery Efficiency of Lipid-Based siRNA Nanoparticles

Christopher Alabi, Koch Institute for Integrative Cancer Research, Massachusetts Institute of Technology, Cambridge, MA

As a Postdoctoral fellow, my current research involves the design of non-invasive probes for tracking the assembly and stability of siRNA nanoparticles in both extracellular and intracellular environments. So far, I have designed two different siRNA-based fluorescent probes whose fluorescence emission changes in response to the assembly state of the nanoparticle. The first probe involves a redox-sensitive fluorescence-quenched probe that fluoresces only when the nanoparticle is disassembled in a reductive environment. The second is based on a FRET-labeled siRNA pair that fluoresces due to the proximity of the siRNA pair when the nanoparticle is intact. In both approaches, the delivery vehicle need not be labeled, thus making both techniques applicable to all types of delivery vehicles. This work was initiated based on my interest in understanding how the function of nanoparticles is related to their biophysical properties. In an effort to comprehend this relationship, I have synthesized a focused library of lipid-based nanoparticles and evaluated the effect of lipid structure and biophysical properties such as particle size, siRNA entrapment, pKa, lytic ability and extracellular stability on the in vitro and in vivo gene silencing performance.

The underlying theme of my research career so far is centered on the use synthetic design as a probe to investigate and answer fundamental mechanistic questions in an effort to improve the function of materials employed in both energy and biological research. As a graduate student, I explored both biological and energy-related research directions. My research in the latter involved the design and synthesis of novel cleavable polymerizable surfactants for the preparation of acid-modified mesoporous silicates towards their use as proton conductive membranes in hydrated fuel cells. Along this line of research, I also investigated the mechanism of proton transport in bitriazole-PEG composites towards their use in anhydrous membranes. With regards to biological research, I worked on the synthesis and evaluation of transferrin-targeted gold nanoparticles as imaging agents to elucidate the uptake mechanism and transport of targeted nanoparticles both *in vitro* and *in vivo*. The afore-mentioned research directions, both biological and energy-related, involved a careful combination of synthesis and mechanistic evaluation in order to understand and build a framework for future rational design.

(3di) Design and Characterization of Micro-Porous Hyaluronic Acid Hydrogels for in Vitro and in Vivo Non-Viral DNA Delivery

Talar Tokatlian and Tatiana Segura, Chemical and Biomolecular Engineering, University of California, Los Angeles, Los Angeles, CA

Natural wound healing and angiogenesis are a result of a cascade of bioactive signals being delivered at specified times in response to local biological cues. However, for chronic wounds which cannot heal naturally, external therapies are required. We are interested in engineering hydrogel scaffolds for cell-demanded release of non-viral DNA nanoparticles to more efficiently guide blood vessel formation in such tissues. Vascularization within tissue-engineered constructs still remains the primary cause of construct failure following implantation and while a myriad of approaches to enhance the vascularization of implants are being investigated none have completely solved the problem. We are currently investigating two hypotheses to enhance scaffold vascularization, both long-term mechanical support and DNA delivery. Our preliminary in vivo studies show that after subcutaneous implantation for 3 weeks enzymatically degradable hydrogels have cellular infiltration only at the periphery of the hydrogel, while hydrogels with micron sized interconnected pores (μ-pore) are extensively infiltrated. Significant positive staining for endothelial markers (PECAM) was also found for μ-pore implants and not for nano-pore implants, even in the absence of pro-angiogenic factors. We believe that an open pore structure will increase the rate of vascularization through enhanced cellular infiltration and that the added delivery of DNA encoding for angiogenic growth factors will result in long lasting angiogenic signals.

To test these hypotheses two approaches to make DNA-loaded enzymatically degradable μ-pore hydrogel scaffolds were developed. In the first approach, polystyrene nanoparticles, similar in size to DNA polyplexes, were immobilized to the hydrogel pore surface through protease sensitive peptide tethers. Enzymatically degradable tethers have been utilized for the immobilization and release of growth factors and small drugs, which are only liberated by cleavage caused by cell secreted proteases, such as matrix metalloproteinases (MMPs) or plasmins, during local tissue remodeling. These proteases are known to be up-regulated during wound healing, microenvironment remodeling, and in diseased states and can, therefore, serve as triggers for bioactive signal delivery. The goal was to use peptide sequences that have been shown to degrade at different rates through the action of MMPs to achieve temporally controlled nanoparticle internalization by cells that overexpress MMPs. Cellular internalization of the peptide-immobilized nanoparticles was shown to be a function of the peptide sensitivity to proteases, the number of tethers between the nanoparticle and the biomaterial and the MMP expression profile of the seeded cells. By immobilizing nanoparticles through protease sensitive peptide tethers, release was tailored specifically for an intended cellular target, which over-expresses such proteases.

Alternatively, in the second approach, μ-pore hyaluronic acid-MMP (HA-MMP) hydrogels were used to encapsulate a high concentration of DNA/poly(ethylene imine) polyplexes using a previously developed caged nanoparticle encapsulation (CnE) technique. Micro-pore hydrogels provide the additional advantages of being able to effectively seed cells in vitro post scaffold fabrication and allow for cell spreading and proliferation without requiring extensive degradation. Thus, release of encapsulated DNA polyplexes was assessed in the presence of mMSCs in hydrogels of various pore sizes (30, 60, and 100 micron). Steady release was observed starting by day 4 for up to 10 days for all investigated pore sizes. Likewise, transgene expression in seeded cells was sustained over this period, although significant differences between different pore sizes were not observed. Cell viability was also shown to remain high over time, even in the presence of high concentrations of DNA polyplexes. Combined these results suggest that DNA nanoparticle internalization and subsequent transgene expression could be controlled by both the protease expression profile of the seeded or infiltrating cells as well as the structural properties of the hydrogel.

Presently, we are using the knowledge acquired through these in vitro models to design and better predict scaffold-mediated gene delivery for local gene therapy in both subcutaneous implant and wound healing mouse models. We believe the proposed hydrogel systems have applications for controlled release of various DNA particles and other gene delivery vectors for in vivo tissue engineering and blood vessel formation.

(3dj) Mathematical Modeling of Biological Systems: Research At the Interface of Chemical Engineering and Biology

Jennifer Anne Pascal, Pathology, University of New Mexico, Albuquerque, NM

Within its Grand Challenges, The National Academy of Engineering poses the question, "How will engineering impact biological research?" Conducting research at the interface of engineering and biology by using mathematical modeling to describe and predict emergent behavior in biological systems has the potential to provide an answer to this question. Using my background in mathematical modeling, I seek to develop a research program that integrates chemical engineering and biology.

My doctoral research focused on using mathematical modeling to understand the effects of morphology on bioseparations in electrokinetics systems, such as nanocomposite gels. Microdomains of idealized geometries were used to describe the internal structure of the gels and the transport was modeled with continuum approaches. Using spatial averaging, optimal separation times of macromolecules, such as proteins, could be obtained for different morphologies. Through my coursework and doctoral research, I acquired fundamental knowledge in applied mathematics, transport phenomena and numerical analysis that I can apply to mathematically model a variety of systems. Because of my strong interest in biological applications, my postdoctoral work is highly interdisciplinary, at the interface of engineering and biology, focusing on modeling intracellular transport and drug delivery to tumors. Specifically, my current work strongly integrates mathematical models and experimental biological data. Through my doctoral and postdoctoral research, I have had the opportunity to interact with collaborators from a variety of disciplines, including mathematics, chemistry and pathology.

Because of my strong desire to be an effective teacher as well as researcher, my postdoctoral work also places an emphasis on education and teaching, I currently devote 25% of my time to improving my knowledge of pedagogical techniques and their implementation in the classroom, through participation in workshops, courses and conferences.

In this presentation, I will describe details of my previous and current research and how my foundational knowledge in mathematics, transport phenomena, and biology will be used to build my research program. I will also discuss my teaching philosophy and plan for facilitating courses.

(3dk) Integrating Process Systems Engineering with Microfluidic Device Development, Optimization and Control for Biomedical Applications

Mranal Jain, Cain Dept. of Chemical Engineering, Louisiana State University, Baton Rouge, LA

My post-doctoral/ doctoral research experience includes modeling of electrokinetic transport processes and design optimization of microfluidic devices. Microfluidic devices are portable devices capable of automating conventional analytical laboratory tests with short analysis time, low reagent/sample consumption. Such devices require fluid and particle manipulation to achieve various functionalities. My research focus is to gain fundamental understanding of transport phenomena at smaller length scales and apply it towards design of efficient microfluidic devices. My background in process systems engineering research (masters program) provides me with a unique perspective towards the microfluidics research problems. I believe that computational modeling at small scale would guide us towards better understanding of underlying physics and the performance of microfluidic devices can be improved further through application of optimization & control techniques.

In accordance with my background, my research interest lies at the interface of microfluidics and process systems engineering. I would like to contribute in following research areas:

- Computational transport phenomena in bio-microfluidic devices
- Multiphase microfluidics; Electrokinetic droplet generation
- Design optimization for microfluidic devices using surface and topology optimization
- Micromixing techniques and their performance analysis
- Classical and non-linear electrokinetics
- Application of process control techniques in microfluidics

In this poster, I will present summary of my research results. I will also discuss about various research challenges in above areas, their significance and possible approaches to tackle them.

(3dl) New Organic Semiconductors for Electronics, Optoelectronics, and Biomaterials

Eilaf Ahmed, Chemistry and Biochemistry, Massachusetts Institute of Technology, Cambridge, MA

Organic semiconductors have attracted significant interest due to their potential application in low cost plastic electronics, solar cells and biomaterials. The ability to engineer the electronic structures of organic semiconductors using molecular design and synthesis and understand structure-property relationships are essential for developing novel materials for the next generation high-performance organic electronics, and biomaterials. My graduate research was focused on the design and synthesis of new classes of *p*-/*n*-type organic semiconductors, understand factors that determine their self-assembly and morphology, elucidate structure-property relationship that govern their electronic and optoelectronic properties and explore their application in field-effect transistors, light-emitting diodes and solar cells. My postdoctoral work is focused on the design of highly fluorescent conjugated polymer nanoparticles for biosensor application, bioimaging and in vivo target drug delivery. The highly emissive NPs are developed as selective nanoprobes for DNA detection via a turn-on transduction mechanism. In addition, the NPs were also found to be effective for in vivo imaging and targeting tumor cells. As an independent investigator, my research interests broadly include: (i) the design of new organic materials, (ii) understanding self-assembly of organic nanostructures, and (iii) material applications ranging from biomaterials (biological imaging, drug targeted delivery, biosensors), to photovoltaic cells, and energy storage.

(3dm) High-Throughput Synthesis of Polymeric Nanoparticles Using 3D Flow Focusing in Parallel Microchannels

Jong-Min Lim[1,2,3], Pedro M. Valencia[4], Minsoung Rhee[1,2,3], Robert S. Langer[3,4], Omid C. Farokhzad[2] and Rohit Karnik[1], (1)Mechanical Engineering, Massachusetts Institute of Technology, Cambridge, MA, (2)Anesthesiology, Brigham and Women's Hospital - Harvard Medical School, Boston, MA, (3)Koch Institute for Cancer Research, Massachusetts Institute of Technology, Cambridge, MA, (4)Chemical Engineering, Massachusetts Institute of Technology, Cambridge, MA

Use of microfluidics for synthesis of nanoparticles has been of great interest due to controllability and reproducibility in their physicochemical properties. Recently, we reported microfluidic synthesis of poly(lactide-*co*-glycolide)-*b*-polyethyleneglycol (PLGA-PEG) nanoparticles for drug delivery using rapid nanoprecipitation by mixing polymer solution in acetonitrile with water in a 3D hydrodynamic flow focusing (3D HFF) geometry that isolates the precipitating polymers from channel walls, eliminating microchannel fouling. However, applications of the microfluidic synthesis were limited to *in vitro* studies due to low throughput. Here, we report multilayer microfluidic devices with 8 and 32 parallel channels for synthesis of nanoparticles using 3D HFF that enable sufficient throughput for *in-vivo* studies. In addition, sub-30 nm PLGA-PEG nanoparticles with various size distributions could be synthesized reproducibly using 3D HFF.

Master molds for the bottom and upper layers were patterned in SU-8 photoresist via conventional photolithography. Post arrays for interconnecting holes were patterned in SU-8 on top of the bottom layer with precise alignment using an UV aligner. A thin PDMS membrane was spin-coated on the master mold; spin speed was tuned to precisely control the membrane thickness to be slightly lower than the interconnect post height. The upper PDMS microchannel was prepared using conventional replica molding of PDMS in SU-8. The upper PDMS microchannel was aligned under a microscope and bonded on top of bottom PDMS membrane and peeled off. The multilayer PDMS microchannel was bonded to glass by oxygen plasma treatment to obtain the final device.

Because of the multilayer architecture, only 4 interconnecting tubes were required per microfluidic device. The PLGA-PEG in acetonitrile solution and pure acetonitrile could flow from the upper layer to the bottom layer through the interconnecting holes. The PLGA-PEG in acetonitrile solution was focused vertically by two distinct vertical acetonitrile sheath flows using the three sequential holes as confirmed in COMSOL simulations and confocal microscope images. It is noteworthy that stable vertical focusing could be achieved even at the high fraction of polymer flow in organic flow due to the optimal shapes of interconnecting holes.

At the junction between vertically focused PLGA-PEG in acetonitrile stream and lateral water sheath stream, the PLGA-PEG block copolymers self-assemble into the PLGA-PEG nanoparticles by nanoprecipitation. By using the multilayer PDMS microfluidic device, PLGA-PEG nanoparticles with various sizes could be synthesized with polymers of different molecular weights and concentrations. The nanoparticles synthesized from the microfluidic 3D HFF were smaller compared to the nanoparticles synthesized from the bulk synthesis method because the mixing time is shorter than the characteristic aggregation time scale in the microfluidic rapid mixing. Since the mixing time for hydrodynamic flow focusing is function of flow ratio of polymer stream to the total flow rate of water, the size of nanoparticles could be controlled in 3D HFF simply by changing the flow ratio of polymer stream to the total flow rate of water. Sub-30 nm PLGA-PEG nanoparticles with various size distributions could be synthesized when small molecular weight PLGA-PEG was used as polymeric precursor in the 3D HFF. Since the pharmacokinetics of nanoparticles highly depends on the physicochemical properties of nanoparticles, the reproducible and robust synthesis of PLGA-PEG nanoparticles by 3D HFF will be beneficial for *in-vivo* screening to find the optimal biodistribution and therapeutic efficacy.

This page intentionally left blank.

(3dn) Design Rules for Engineering Interfaces of Energy Materials

Kedarnath Kolluri, Materials Science, Los Alamos National Lab, Los Alamos, NM

My research goal is to obtain comprehensive understanding of transport phenomena and structural/phase transitions at internal interfaces and surfaces of inorganic materials. I seek to establish a research program that will develop interface structure-property relationships through computational investigations to enable design of nanostructured materials. My group will address the scientific grand challenge of designing materials with atom-by-atom control and the technological grand challenge of low-cost energy production, high-density energy storage, and no-loss energy distribution.

In the immediate future my research group will focus on the technological areas of

- accelerating ion transport in nanostructured ceramics for solid-oxide fuel cells
- storing/retrieving hydrogen easily in metals such as magnesium nanoparticles
- increasing realizable lithium storage capacity of nanostructured silicon
- trapping/removing radiation induced defects in nanostructured structural materials

We will strive to advance surface and interface science through

- comprehensive understanding of transport phenomena at buried interfaces in ceramics, metals, and earth-abundant materials such as silicon and magnesium
- comprehensive understanding of interface phase transitions
- development of atomistically-informed analytical and mesoscopic models that relate interface properties to its structure

In pursuing above mentioned scientific and technological goals, we will

- develop computational methods to study atomic structure and phenomena not accessible by existing ones such as transport by collective motion of atoms
- develop empirical potentials that describe accurately
    - charge distribution and transfer at ceramic-ceramic and metal-ceramic interfaces
    - chemical reactions such as hydrogenation of magnesium and lithiation of silicon
- participate in multi-investigator programs, especially with experimental investigators

(3do) Size Defined Catalysis: Tuning the Catalytic Properties by Selectively Designing Atomically-Precise Catalysts

Sarthak Gaur, Cain Department of Chemical Engineering, Louisiana State University, Baton Rouge, LA

Supported Au catalysts have been studied extensively for CO oxidation. These catalysts are known to catalyze this reaction even at sub-ambient temperatures. Numerous critical factors have been examined on these catalysts for CO oxidation including the size of Au, pretreatment, oxidation state of Au during the reaction, and the effects of support. However, the key factors that are believed to convert the "inert gold" into a highly active catalyst are (i) the size of Au particles, and (ii) the nature of support. While recent literature demonstrates catalytic activity of gold nanoparticles <2 nm, the next stage in fine tuning this catalysis process is to develop gold clusters prepared with atomic precision. Such atomically precise gold catalysts hitherto have not been investigated for CO oxidation.

To that end, we have recently demonstrated the preparation of atomically-tailored Au clusters consisting of 38-atoms (Au_{38}) [1] and also synthesized Au clusters with 25-atoms (Au_{25}) from a previously reported method [2]. We prepared atomically-precise Au_{38}/TiO_2 catalysts and have attempted various heat pretreatments to remove the ligand while retaining the size of the cluster. Furthermore, we examined these Au/ TiO_2 catalysts for catalyzing CO oxidation using *in situ* FTIR and synchrotron radiation-based EXAFS, and monitored the surface species while investigating the activity of these catalysts with time on stream. We noticed the presence of certain carbonates and bicarbonates (from DRIFTS) and thiolate species from EXAFS as residual sulfides (after a reductive pretreatment to activate the catalyst) which interact with Au apparently at the Au-support interface. Interestingly, we observed that these thiolates congregate at the Au-Ti interface and are quickly oxidized by O_2 under oxidative conditions to form sulfonyl (Au-SO_2R) linkages between the Ai and TiO_2 surface [3]. These linkages act as oxygen radical transfer points from activated O_2 from TiO_2 surface to the Au cluster. So despite being a poison, we have shown that these thiolates appear to act as radical transfer agents.

Another aspect of my research is to determine mechanistic pathways on Au_{38}/TiO_2 catalyst during CO oxidation by conducting real-time in-situ studies using synchrotron-radiation based X-ray absorption spectroscopy (XAS), operando IR (DRIFTS), and usage of isotopes. We have investigated CO oxidation on our catalysts using above techniques, and by carefully conducting planned experiments, we have also shown that CO oxidation takes place on our catalyst via a Mars van Kreven (redox) mechanism which involves the supply of lattice oxygen from TiO_2 support at the Au-TiO_2 interface. We support our experimental observations using theoretical methods, particularly using DFT, where we have studies CO oxidation on Au38 clusters and Au/TiO2 1-D nanorods.

(3dp) Engineering Polymeric Materials for Barriers, Hollow Fiber Membranes, and Hybrid Sorbents: A Path to a More Sustainable Future

Jong Suk Lee, Koros Group, Georgia Institute of Technology, Atlanta, GA

The understanding of gas/vapor transport in polymeric materials can guide advances for many applications: barrier material development, membrane based gas separation, and hybrid sorbent development. In fact, all of these applications are related to the same fundamental principle, the sorption-diffusion mechanism. The benefits to our society of doing research in these areas are enormous in terms of the environment as well as energy. As for barrier material development, the transport properties of poly(ethylene terephthalate) (PET), the current state of the art, can be significantly advanced by the combination of antiplasticization and crystallization to create the next generation of packaging. Furthermore, the effect of flavor molecules on the barrier properties of PET can be evaluated by using a new vapor/gas permeation system relying upon a highly accurate syringe pump. Membranes have also been utilized for separations involving gaseous feed streams and their industrial growth will be driven by their application to increasingly aggressive feed streams. The performance of defect-free Matrimid® asymmetric hollow fiber membranes for aggressive natural gas feed containing high pressure of CO2 and trace levels of toluene and n-heptane was studied. Lastly, a novel class of carbon capture system is being developed via hollow fiber-supported ionic liquid sorbents. Our emergent catch-and-release system has the potential to considerably cut the cost and energy associated with gas uptake and release. Also, it will provide a platform to utilize ionic liquid sorbents on an industry scale by integrating ionic liquid into hollow fibers. The scientific advances mentioned above will enable our society to be a leader in developing and deploying advanced technologies for more sustainable future.

(3dq) Algorithmic Exploration of "Building-Block" Chemistry

Christopher E. Wilmer, Chemical and Biological Engineering, Northwestern University, Evanston, IL

In the last decade a wealth of novel materials has been synthesized through what has come to be called "modular" or "building-block" chemistry. This new approach leverages the self-assembly of rigid molecular building blocks that only connect in very specific orientations and stoichiometries, which constrain the resulting assemblies into discrete geometries (e.g., polyhedra, tubes, spheres) or crystalline structures (e.g., metal-organic frameworks (MOFs)).

A remarkable feature of building-block chemistry is that the products are typically monodisperse, and their geometry and composition can be predicted *a priori*. Due to this predictability and the abundance of known building blocks (drawn from the vast organic chemistry literature), the field of MOF chemistry alone has reported thousands of novel structures in the past few years. While these numerous reports showcase the success of the building-block approach, they also belie the underlying combinatorial difficulty of finding the best material for a given application. There are millions of possible MOFs: trying to find the optimal material for a specific application by experimental trial-and-error is impractical and inefficient at best.

In my doctoral research at Northwestern University I developed algorithms that systematically enumerate all of the *hypothetical* MOFs that could be made from a given library of building blocks. Over 150,000 novel MOF structures were screened in a high throughput manner via molecular simulations. The scope of this computational screening effort was without precedent in the MOF field. For each MOF I obtained a range of material properties such as surface area, pore volume, pore size distribution, powder x-ray diffraction pattern, and gas adsorption. In addition to rapidly identifying MOFs that had higher gas storage performance than any previously known materials, I illuminated hitherto unidentified structure-property relationships that could only have been recognized by taking a global view of MOF structures. Working with synthetic chemists, we synthesized some of the highest performing MOFs predicted by my algorithms, and their measured gas adsorption agreed well with the simulations. Furthermore, by making our database of MOFs publicly available online, this research has already made a measurable, global impact in both academic and industrial labs, as well as stimulated significant commercial interest (>$1 million in venture capital). This disruptive approach to exploring the space of MOFs will significantly accelerate the discovery of better materials for applications such as hydrogen and natural gas storage.

More generally, the algorithmic investigation of materials self-assembled from building blocks can be usefully applied not only to MOFs but to discrete structures and other bulk material classes. Working closely with my current and anticipated experimental collaborator, Dr. Omar K. Farha, my future research will continue to interface closely with chemical synthesis and empirical measurements as I strive to develop *predictive* computational tools. My future research will also build on my anticipated post-doctoral research by incorporating modern high throughput quantum chemistry techniques (e.g., quantum Monte Carlo) to search for optimal catalysts and yield useful structure-property insights to create new catalysis design-rules. This algorithmic approach for exploring new chemical spaces, combined with the predictive power of quantum chemistry, will accelerate the pace of materials discovery and expand the scope of investigations in chemical separations, catalysis, and self-assembly.

PhD Advisor: Professor Randall Q. Snurr

(3dr) Molecular Simulations for Understanding Morphology At Interfaces

Naga Rajesh Tummala, Chad Risko and Jean-Luc Brédas, School of Chemistry and Biochemistry, Georgia Institute of Technology, Atlanta, GA

Understanding the morphology and packing of nano materials at interfaces and in the bulk often determines their applicability. In fact, many applications rely on preferential orientation and packing of nano-entities at the molecular level for efficient and reliable performance.[1,2]Computational tools such as molecular dynamics (MD), dissipative dynamics using coarse grained models, and quantum-mechanical calculations are being increasingly employed to investigate morphological and packing features, and to complement and guide experimental efforts.

My doctoral studies involved assessing the surfactant morphology at water-solid interfaces to aid in the description of admicellar polymerization. MD simulations of the surfactants at the water-nanotube interface were performed to discuss the effect of the curvature on the adsorbed surfactant morphology, which is important for the understanding of the dispersion of nanotubes. During my postdoctoral studies at University of Oklahoma, I investigated propagation of nanoparticles through porous media using the Lagrangian tracking model and the flow field in porous media was computed from a continuum Lattice-Boltzmann model.

At the Georgia Institute of Technology, I am performing simulations to gain molecular-scale understanding of the interfacial region in bulk heterojunction photovoltaic devices made of small-molecule donors and fullerene acceptors.[3] Random blends of benzothiodiazole-dithienopyrrole-benzothiadiazole (BTD-DTP-BTD),[4]a small molecule donor, and [6,6]-phenyl-C61-butyric acid methyl ester (PCBM), a commonly used acceptor, were created for this purpose and MD simulations were performed to calculate the solubility parameters and thermodynamics of these blends, and to provide a platform to perform coarse graining of these molecules. Coarse graining would potentially reveal the equilibrium molecular-scale morphologies at interfaces; then, semi-empirical quantum calculations on the snapshots obtained from MD simulations would allow to study the electronic couplings between these molecules, which in turn would be used to determine the optimal interfacial morphology.

(1) Kim, M.; Hohman, J. N.; Cao, Y.; Houk, K. N.; Ma, H.; Jen, A. K.-Y.; Weiss, P. S. *Science* **2011**, *331*, 1312.

(2) Zhang, H.; Hussain, I.; Brust, M.; Butler, M. F.; Rannard, S. P.; Cooper, A. I. *Nat Mater* **2005**, *4*, 787.

(3) Mishra, A.; Bäuerle, P. *Angewandte Chemie International Edition* **2012**, *51*, 2020.

(4) Polander, L. E.; Pandey, L.; Barlow, S.; Tiwari, S. P.; Risko, C.; Kippelen, B.; Brédas, J.-L.; Marder, S. R. *The Journal of Physical Chemistry C* **2011**, *115*, 23149.

(3ds) Smart Biomaterials

J. Dumas, Biomedical Engineering, Georgia Institute of Technology, Atlanta, GA

Smart Biomaterials

Advances in biomaterial research have greatly impacted biological research through new innovations in areas such as treatment. However, many grand challenges still exist as the boundaries between biology, chemistry, and engineering continue to merge. These challenges include the further improvement of biomaterial biocompatibility and the control of cell movement and differentiation *in vitro* and *in vivo* [1].

As a doctorate student in the chemical and biomolecular engineering department at Vanderbilt University, I developed a two-component allograft mineralized bone particle (AMBP)/polyurethane (PUR) system for the treatment of bone defects [2]. The biodegradable PUR phase, which consisted of lysine triisocyanate and polyester polyol, had tunable degradation rates and mechanical properties. The AMBP phase, which is bound to the PUR phase, was used as a filler to enhance the mechanical properties and provide a resorbable pathway for cellular infiltration. The adaptability of the AMBP/PUR composite system made it suitable for multiple applications including an implant or injectable platform [3]. Further, the AMBP/PUR composites served as a delivery system for biologics such as recombinant human morphogenic protein (rhBMP-2), accelerating new bone formation in critical size defects in New Zealand White (NZW) rabbits.

As an Institutional Research and Academic Career Development Award (IRACDA) postdoctoral fellow in the biomedical engineering department at Georgia Institute of Technology/Emory University, I am studying the role of proteases in cancer. I am developing a detection device based from multiplex cathepsin zymography, which uses a gelatin substrate in a SDS-PAGE gel to quantify enzyme activity. Cathepsin activity has shown to be upregulated in cancer tissue, and this difference is used to distinguish normal from tumor tissue [4]. Concurrently, I am utilizing hydrogel and polyurethane technology to create an in vitro bone metastasis model that will study key parameters such as protease (i.e., cathepsin) activity and cell invasion rate.

I intend to answer the grand challenges of biomaterials in my lab as a faculty member. My initial goals will include the fabrication of intelligent biomaterials for both treatment (e.g., tissue regeneration) and in vitro models capable of modeling disease pathology (e.g., cancer). I am excited about forging collaborations and expanding my knowledge to accomplish the challenges that await me in the future.

References.

1. Reichert, W.M., et al., *2010 Panel on the Biomaterials Grand Challenges.* Journal of Biomedical Materials Research Part A, 2011. **96A**(2): p. 275-287.

2. Guelcher, S.A., et al., *Synthesis, mechanical properties, biocompatibility, and biodegradation of polyurethane networks from lysine polyisocyanates.* Biomaterials, 2008. **29**(12): p. 1762-75.

3. Dumas, J.E., et al., *Synthesis, characterization, and remodeling of weight-bearing allograft bone/polyurethane composites in the rabbit.* Acta Biomater. **6**(7): p. 2394-406.

4. Chen, B. and M.O. Platt, *Multiplex Zymography Captures Stage-specific Activity Profiles of Cathepsins K, L, and S in Human Breast, Lung, and Cervical Cancer.* Journal of translational medicine, 2011. **9**: p. 109.

(3dt) Biofluids and Nanofluids Under Flow: Applications in Biomedical Engineering, Nanotechnology and Energy Harvesting

Amit Kumar, Chemical and Biological Engineering, University of Wisconsin-Madison, Madison, WI

My research interests focus on the dynamics of biofluids and nanofluids under flow. Two examples of such systems that I have extensively studied are suspensions of rigid non-spherical nanoparticles and blood. In this poster, I will focus on my current work on the flow induced segregation behavior in blood. Blood is a multicomponent mixture comprising mostly of red-blood-cells (RBCs) along with trace amounts of other components like leukocytes, platelets, and circulating tumor cells (in the case of cancer). Under physiological flow conditions both the leukocytes and the platelets segregate near the walls of the blood vessel, a phenomenon commonly known as margination, while the RBCs tend to migrate away from the walls. The key physical differences between RBCs, leukocytes, and platelets are their relative size and rigidity: the leukocytes are larger than RBCs, while the platelets are smaller; both are considerably stiffer than RBCs. However, how these differences in properties lead to the observed segregation behavior is poorly understood. Using detailed boundary integral simulations, we systematically delineate the effect of size and rigidity on the segregation behavior in particle mixtures and relate these to the observations of margination of leukocytes and platelets in blood flow. To gain a mechanistic understanding of these results, we introduce a novel hydrodynamic-Monte Carlo simulation technique, which incorporates two of the main ingredients of flow dynamics in confined suspensions: the wall induced migration and pair collisions. In particular, the model clarifies the important role played by heterogeneous collisions, i.e. collisions between two different species in a mixture, in the observed segregation behavior. The insights and tools presented in this poster will be helpful, e.g., in designing drug delivery particles for optimal vascular targeting and for designing microfluidic devices for separating/enriching trace components of blood. In this poster, I will also present several examples of future research directions, including the role of blood flow in cancer treatment, mechanical energy harvesting using dielectric droplets, programmable assembly of patterned nanoparticles under flow and separation of non-spherical nanoparticle mixtures in microfluidic devices.

(3du) New Predictive and Efficient Computational Tools for Studying Catalysis: From Transition Metal Surface Chemistry to Enzyme Engineering

Heather J. Kulik, Chemistry, Stanford University, Stanford, CA

Atomistic simulation provides valuable mechanistic insight into catalytic activity in systems ranging from the smallest molecular catalysts to complex surface phenomena in materials science and enzyme activity in the biological sciences. For this reason, rational, computational design is a promising pathway to the development of catalysts for applications in energy science, pharmaceuticals, and nanoscale electronics. Nevertheless, the effectiveness of a computational design strategy is dependent upon both the predictive accuracy and computational cost of the methods employed. Quantum mechanical methods, in particular, are impressively accurate: they directly incorporate the key physics needed to describe phenomena from bond rearrangement to charge transfer. However, computational cost constrains the most accurate QM methods to only a handful of atoms.

Density functional theory (DFT) is a widely-employed computational approach for determining the electronic structure of molecules and solids, but standard approximations in DFT cause it to suffer from a critical error known as self-interaction. This self-interaction error makes electrons over-delocalize and gives rise to erroneous descriptions of geometry and energetics, particularly for transition metals. I have developed and extended Hubbard-augmented DFT approaches that ameliorate self-interaction error and shown that these so-called DFT+U approaches reduce errors in the ordering of electronic states of small molecules and paradigmatic catalytic reactions by over an order of magnitude. Importantly, with little additional overhead, the U term is calculated self-consistently on each system considered and is thus not a fitting parameter. The efficient scaling of DFT+U and its extensions have permitted us to make new predictions that were later validated by experiment about systems ranging from self-assembled monolayers on transition-metal surfaces to the reactivity of metalloenzymes that are components of natural product biosynthesis.

In addition to accuracy, a critical concern is the system size that can be attacked by computational methods. As part of the first successful effort to dramatically speed up QM methods, I have developed and extended tailored algorithms that harness the massively parallel nature of graphical processing units (GPUs). These GPU-accelerated QM methods permit us to carry out unprecedented studies of proteins fully quantum mechanically, where less accurate force field methods were previously the only option. Using these new approaches I've quantified how quantum mechanical descriptions of proteins differ from those obtained with force fields and determined that QM methods excel in describing the disordered features of proteins that are likely to be key in mediating catalysis. I've also identified that quantum mechanical effects at enzyme active sites are much larger scale than usually considered through standard QM/MM approaches. A combination of both more accurate (DFT+U) and more efficient (GPU-accelerated) methods paves the way for predictive accuracy on large-scale, realistic systems relevant for applications in catalysis ranging from surface science to enzyme engineering.

(3dw) Understanding Charge Transport At Interfaces in Tough Solid Electrolytes to Enable Lithium Metal Batteries

Wyatt Tenhaeff, Materials Science & Technology Division, Oak Ridge National Laboratory, Oak Ridge, TN

Solid electrolytes offer the promise of safe, energy-dense secondary lithium ion and lithium metal batteries. To date, an obvious choice for a single solid electrolyte material with sufficient ionic conductivity, compatibility with lithium metal, and mechanical robustness does not exist. Mechanical properties are a key consideration given their role in suppressing lithium dendrite formation, which has plagued lithium metal anodes to date and ultimately must be solved to enable their safe commercial implementation. Fabrication of composite materials where both phases conduct Li cations is one approach to addressing these requirements.

Both phases being Li^+ conductive is critical and distinguishes these composites from several previous studies that demonstrated enhanced Li conductivities in nanocomposite solid polymer electrolytes where insulating nanoparticles, e.g. Al_2O_3 and SiO_2, comprise the minor phase. High loadings of the particulate phase are required for improved mechanical properties. With insulating fillers, the optimal weight loading is approximately 15%; above that, reduction of the conductive matrix phase overwhelms the enhancement from additional particles. For particulate phases that conduct Li^+, it is anticipated that much higher filler loadings can be utilized to improve mechanical properties while maintaining or enhancing Li^+ transport.

This poster will present my approach to understanding the transport of Li cations at interfaces between polymeric and inorganic solid electrolytes. Initially, charge transport was studied in laminated bilayers of thin films of polymer and inorganic electrolytes. A typical polymer electrolyte was poly(ethylene oxide) (PEO) or PEO-based copolymer mixed with a lithium salt, such as $LiClO_4$ or $LiCF_3SO_3$. Lithium phosphate oxynitride (Lipon) was employed as the stiff, inorganic electrolyte. It was discovered that the interfacial resistance was dominant in this system but could be controlled and essentially eliminated through fabrication techniques. The study was repeated for bulk electrolyte materials and similar conclusions were found. After understanding the charge transport at planar interfaces with well-defined areas, the study was extended to nanoparticulate composite solid electrolytes designed to optimize conductivities and mechanical properties. Initial results suggested that negligible charge transfer occurs through the higher conductivity particulate phase. Efforts to reduce the interfacial impedance in these nanocomposites will be presented.

Two additional research efforts concerning solid electrolytes will be discussed. The first involves the development of new flow battery designs that pair protected Li metal anodes with aqueous catholyte solutions. Laminating lithium metal between a water and air-impermeable solid electrolyte and current collector prevents the oxidation of Li by water, and the electrochemical cell can operate outside the standard electrochemical window of water. Low cost, renewable organic redox species with high aqueous solubilities and reduction potentials above 3V vs. Li^+/Li for energy dense catholytes have been discovered. In the second project, plating and stripping of Li metal is being characterized by neutron reflectometry. For long-term cycling of metallic anodes, electrodeposition of Li into perfectly uniform, fully dense, conformal layers over the entire electrically active area of the current collector is required. Subsequent reoxidation must maintain the uniformity of the layer such that local islands or grains with disparate mass transfer kinetics do not form and propagate throughout the layer, resulting in the mossy or dendritic morphologies. It is important to understand how the composition and operation of electrochemical cells influence the initial electrodeposition/reoxidation processes. Neutron reflectometry was used to characterize the density, composition, and roughness of ultrathin lithium films as they were deposited and reoxidized *in situ*. Reflectivity profiles of Ni and Lipon-coated Ni current collectors revealed changes in the sample structure for plated Li films as thin as 10 nm.

This page intentionally left blank.

(3dx) Designer Surfaces for the Study and Treatment of Human Injury and Disease

Anita Shukla, Bioengineering, Rice University, Houston, TX

In my future independent research program, I would like to focus on developing responsive and efficient biomaterials for the fundamental study and treatment of human injury and disease. During my Ph.D. in the Department of Chemical Engineering at the Massachusetts Institute of Technology, I developed expertise in the design of biomaterials for drug delivery. I focused on developing effective local delivery methods for multiple therapeutics over medically relevant timescales, enabling the treatment of cellular dysfunction while preventing drug toxicity. As an alternative to conventional systemic drug administration, I designed unique layer-by-layer assembled degradable polymer multilayer films as medical device coatings. Film architecture, assembly technique, and secondary interactions between film components were optimized to create active degradable films. These films were successfully engineered to deliver agents ranging in both chemical functionality and therapeutic properties, from small molecules to proteins, aimed at alleviating infection, biofilm formation, inflammation, and hemorrhage. Having developed biomaterials based approaches for disease treatment in my doctoral work, I realized that the design of such treatments is limited by an understanding of the cellular systems which are affected by them. Therefore, I decided to focus on acquiring expertise in designing materials to study fundamental cell behavior in my postdoctoral research.

As a postdoc in the Department of Bioengineering at Rice University, I am studying the effects of cytoskeletal tension on human mesenchymal stem cell differentiation. I am developing biologically inspired micropatterned protein surfaces with the aims of locally regulating cytoskeletal tension to direct the differentiation of human mesenchymal stem cells towards desired lineages. This work has tremendous potential to impact the design of biomaterials and tissue engineering scaffolds for regenerative medicine while simultaneously contributing to the fundamental understanding of stem cell mechanobiology. In my future research program, I will apply ideas from polymer self-assembly, cellular mechanobiology, and the study of molecular interactions to create novel responsive and efficient biomaterials.

(3dy) Advanced Polymeric and Bionanocomposite Solutions for Tissue Engineering and Drug Delivery Applications

Adam K. Ekenseair, Department of Bioengineering, Rice University, Houston, TX

Research Interests:

The rapidly evolving toolset available to polymer scientists is enabling the birth of a new generation of smart composite devices which can be tailored to suit the specific and demanding requirements of advanced biomaterials applications from efficient and site-specific drug delivery to the regeneration of complex tissue. With a strong background in polymer synthesis, characterization, and transport phenomena and development of polymeric therapies for tissue engineering applications, I am well-positioned to push the boundaries of scientific discovery at the interface of polymer science and biomedical research. Within this broad scope, early efforts will focus on three primary research thrusts:

I) Injectable *In Situ* Forming Hydrogels for Tissue Engineering and Drug Delivery

II) Bionanocomposite Polymers for Drug Delivery and Imaging

III) 3-Dimensional Bioprinted Composite Scaffolds for Tissue Engineering

Background:

My doctoral research focused on observing and comprehending the transport of small penetrant molecules in glassy polymers and elucidating the effects of structure and network parameters on the transport mechanism, with a particular emphasis on the occurrence of non-Fickian transport dynamics. Novel structure property relationships were developed for penetrant transport dynamics in the Case II transport regime, material-based mechanisms for controlling the nature of penetrant transport in glassy polymers were elucidated, and high-resolution X-ray computed tomography was adapted as a novel technique for quantitative, *in situ* imaging of penetrant transport processes.

In my post-doctoral studies, I have moved to the interface of polymer science and biomedical research, focusing on the design and application of injectable hydrogel-cell composite scaffolds for tissue regeneration in craniofacial defects. Specifically, we have developed and filed a provisional patent on a novel class of two-component, *in situ* dual-hardening hydrogels combining near-instantaneous temperature-induced gelation with epoxy-based crosslinking utilizing degradable polyamidoamine macromers. Such *in situ* dual-hardening, dimensionally stable, defect-filling, and degradable hydrogels with high gel water content are attractive substrates for tissue engineering and cellular delivery applications. In particular, the use of water-soluble and degradable polyamidoamine polyaddition-formed macromers offers tremendous synthetic flexibility and control over subsequent gel properties.

Post-doctoral Advisor: Professor Antonios G. Mikos, Departments of Bioengineering and Chemical and Biomolecular Engineering, *Rice University*

Ph.D. Advisor: Professor Nicholas A. Peppas, Departments of Chemical and Biomedical Engineering, *The University of Texas at Austin*

Publications:

1. **A.K. Ekenseair**, F.K. Kasper and A.G. Mikos. "Perspectives on the Interface of Drug Delivery and Tissue Engineering." *Adv. Drug Delivery Rev.* (in preparation).

2. **A.K. Ekenseair**, K.W.M. Boere, S.N. Tzouanas, T.N. Vo, F.K. Kasper and A.G. Mikos. "Structure-Property Evaluation of Thermally and Chemically Gelling Injectable Hydrogels for Tissue Engineering." *Biomacromolecules* (submitted).
3. **A.K. Ekenseair**, R.N. Seidel and N.A. Peppas. "Tuning the Transport Dynamics of Small Penetrant Molecules in Glassy Polymers." *Polymer* (submitted).
4. **A.K. Ekenseair**, K.W.M. Boere, S.N. Tzouanas, T.N. Vo, F.K. Kasper and A.G. Mikos. "Synthesis and Characterization of Thermally and Chemically Gelling Injectable Hydrogels for Tissue Engineering." *Biomacromolecules*, (in press), DOI: 10.1021/bm300429e.
5. **A.K. Ekenseair** and N.A. Peppas. "Network Structure and Methanol Transport Dynamics in Poly(methyl methacrylate)." *AIChE J.*, 58(5): 1600-1609 (2012).
6. **A.K. Ekenseair**, R.A. Ketcham and N.A. Peppas. "Visualization of Anomalous Penetrant Transport in Glassy Poly(methyl methacrylate) Utilizing High-Resolution X-ray Computed Tomography." *Polymer*, 53(3): 776-781 (2011).
7. **A.K. Ekenseair**, L. Duan, D.J. Carrier, D.I. Bransby and E.C. Clausen. "Extraction of Hyperoside and Quercitrin from Mimosa (Albizia julibrissin) Foliage." *Appl. Biochem. Biotechnol.*, 129-132 : 382-91 (2006).

(3dz) Unravelling Structure-Activity Relationships of Heterogeneous Catalysts

Ron C. Runnebaum, Department of Chemical and Biomolecular Engineering , University of California, Berkeley, Berkeley, CA

My PhD thesis, with Prof. Bruce C. Gates, developed a foundation for the fundamental understanding of the catalytic chemistry for deoxygenation of oxygen-containing hydrocarbons. Our goal was to determine such a detailed reaction network for compounds representative of lignin-derived bio-oils and an important functional group characterizing compounds in them. Data characterizing these reactions determined an approximate reaction network for each reactant and generalization of the results to reactant types and reaction classes, and conversion data sufficient to approximately quantify kinetics of the principal reactions in the networks. The data provide understanding to allow resolution of the roles of the separate catalyst functionalities (e.g. solid acid and metal) and provide a basis for prediction of reactivities of whole bio-oils.

One postdoctoral research project, with Prof. Alexander Katz and Dr. Stacey Zones, focuses on the synthesis and characterization (including catalysis) of delaminated zeolites, which retain a high-degree of crystallinity and have a large external surface area available for reaction of bulky reactants and products. Another research project with Prof. Alexander Katz investigates the effects of organic ligands on catalysis by heterogeneous metal-cluster catalysts; catalytic characterization of these grafted metallocalixarenes by using probe reactions enable elucidation of structure-reactivity relationships for these materials.

Future work will build upon my expertise in both catalyst synthesis and heterogeneous catalysis to develop fundamental understanding for the formation of useful fuels and chemicals and to provide a basis for understanding more complicated systems.

(3e) Regenerative Medicine: From Tissue Engineering to Organ Engineering

Yeonhee Kim, Surgery, Massachusetts General Hospital / Harvard Medical School, Boston, MA

The only definitive treatment for end-stage liver disease is orthotopic liver transplantation. Although successful, this procedure is limited by the severe shortage of donor organs with over 27,000 patients dying annually due to lack of the available donor organs, hence, alternative methods to supply viable cells and tissues to patients is urgent. A promising, innovative approach is tissue engineering in which the most appropriate cell populations and biological scaffolds are incorporated and assembled to mimic *in vivo* structure and physiology of intact organ. Recently, we developed a successful liver decellularization protocol to create transplantable grafts from decellularized whole livers. Decellularized liver matrix (DLM) scaffolds retain a fully intact structure and composition of native organ extracellular matrix (ECM) as well as vascular network, and can be repopulated with hepatocytes (the cells that carry out most of the liver's primary functions), demonstrating function *in vitro* and *in vivo*after transplantation. Although we have shown successful transplantation of the recellularized grafts in the short term (8 h), long-term survival of recipients was limited by occurrence of thrombosis due to contact of blood with the naked ECM scaffold upon transplantation. Therefore, a DLM graft that is thromboresistant is necessary to extend the transplantation time and evaluate therapeutic benefits of the implant in animal models of liver failure.

The long-term goal is to engineer a transplantable liver graft for curing and treating relevant liver diseases. The objective is to generate a thromboresistant and functional liver graft by forming a heparin-coated surface in the DLM via layer-by-layer (LbL) self-assembly technique and repopulating with primary hepatocytes and microvascular endothelial cells. The central hypothesis is that the heparin immobilization of the DLM scaffold will extend survival of the graft recipients by blocking thrombosis *in vivo* and allow for functional evaluation of the grafts in animal models of liver failure in the long-term. The rationale is that heparin anticoagulant surface modification of the decellularized matrix will greatly aid in preventing thrombosis during blood perfusion while allowing for repopulation with cells *in vitro*; these findings will lead to create a desirable tissue-engineered liver that enables to support long-term recipient survival after transplantation.

(3ea) Interdisciplinary Approach to the Design, Synthesis, and Evaluation of Inhalable Therapeutics

Timothy Brenza, Chemical and Biological Engineering, Iowa State University, Ames, IA

Infections and diseases of the respiratory system contribute to morbidity and mortality in both the developing and the developed world. The therapeutic advantages of inhalable therapeutic treatment of respiratory diseases include the application of the therapeutic to the site of infection which enables a high localized concentration of therapeutic with low systemic exposure. However, the successful development of inhalable therapeutics requires the ability to successfully formulate the therapeutic into an inhalable form, the successful deposition of the material into specific respiratory regions, and controlled interactions of the material with extracellular and cellular environments. The solution to these challenges requires the expertise and interaction of multiple disciplines including: material science, engineering, molecular biology, immunology, pharmaceutics, and physiology.

My graduate and post-doctoral experiences in chemical engineering and pharmaceutics have uniquely prepared me to tackle these challenges. For drug delivery vehicle design, I have synthesized biodegradable polymers and used these to encapsulate antibiotics and proteins into particles on both the micron and nano scales. I have also used oligosaccharides from the surface of non-typeable *Haemophilus influenzae* and polyethylene glycol to modify the surfaces of polystyrene nanospheres to provide cell targeting and mucosal avoidance, respectively. The biodistribution, extracellular and cellular interactions of particles were evaluated utilizing *in vivo* murine and *in vitro* models.

To supplement the *in vitro* experience, translational studies into particle aerosolization and aggregation of dry powder formulations have provided an understanding of deposition and distribution of inhalable therapeutic formulations. Combining this knowledge with my experience with human respiratory physiological measurements provides me with the tools required to address the challenge of designing inhalable therapeutic delivery systems for treatments of respiratory diseases.

(3eb) Engineering Functionality Into Layer by Layer Assembled Nanocomposites

Christine M. Andres and Nicholas A. Kotov, Chemical Engineering, University of Michigan, Ann Arbor, MI

Hybrid nanocomposites create functional materials with enhanced and diversified properties as compared to their bulk counterparts. However, actual application of the highly desired properties of nanomaterials into functional devices requires radically new approaches to fabrication which achieve multiple combinations of diverse materials and complex multi-scale 3D architectures in a simple, low cost and versatile approach. Typical methods for preparing materials with 3D microscale features are often restricted by limited material variability, lack of control over the nanophase and challenging transitions to macroscale materials. Layer-by-layer assembly (LBL), the sequential adsorption of monolayers attracted by chemical forces, is an excellent technique for the fabrication of an extensive variety of highly controlled nanocomposites with exceptional properties unavailable by other means. Here we present a toolbox of new approaches to introduce controlled multiscale architecture to LBL materials for the simple transition into advanced technologies. As the ability to precisely tune material properties is critical for successful development of novel and improved functionalities, the developed techniques based on inkjet printing, photolithography, and colloidal crystal templates provide systematic control of chemical and structural features from the molecular scale through the macroscale. We will demonstrate successful elimination of intermediate rinsing steps for the generation of direct-write patterning of LBL materials, the creation of self-supporting channels, microcontainers and actuators from LBL materials, self-folding and shape memory strategies and finally 3D porous macroscale structures from LBL materials with controlled negative thermal expansion properties.

(3ec) Solar Grade Silicon Production in a Fluidized Bed Reactor

Juan Du, Department of Chemical Engineering, Carnegie Mellon University, Pittsburgh, PA

Solar Grade Silicon Production in a Fluidized Bed Reactor

Juan Du

Carnegie Mellon University, Pittsburgh, Pennsylvania, USA

Poly-silicon, the feedstock for semiconductor and photovoltaic industries, has witnessed an explosive demand due to the expansion of the photovoltaic (PV) industry and limited recovery of rejected silicon from the semiconductor industry. Siemens process is the classic method to produce electronic grade silicon, the price of which for semiconductor industry is not as critical as in the solar industry. Fluidized bed reactors have excellent heat and mass transfer characteristics and can be utilized for solar grade silicon production to overcome the energy waste problem in Siemens process. The energy consumption is reduced because the decomposition operates at a lower temperature and cooling devices are not required. Moreover fluidized beds have higher throughput rate and operate continuously to reduce further capital and operating cost. FBR process has been commercialized by several companies and promises to deliver solar grade silicon at reasonable and stable price.

We present a multi-scale modeling approach for solar grade silicon production in fluidized bed reactor by thermal decomposition of Silane. Silane decomposes to form hydrogen and silicon. The seed particles grow due to the heterogeneous chemical vapor deposition and scavenge of silicon powder produced in homogeneous gas phase reactor. The dynamics of gas and particle phases are modeled in computational fluid dynamics (CFD) model and reaction model. The CFD module shows the hydrodynamics. The volume fraction of solid phase is imported into the reaction model. The reaction model describes the emulsion phase and bubble phase of FBR. The mass exchange and heat transfer between two phases are developed by establishing mass and heat balance for each phase. The output of the reaction module, i.e. total deposition rate, is passed to population balance which describes particle growth process. A novel discretization scheme is presented to obtain discretized version of the population balance. The particle size distribution calculated by discretized population balance is used as input of CFD module. The complex interplay between gas phase and particle phase is captured by integrating those modules. The multi-scale model is validated by experimental data provided by pilot plants in industry.Based on the multi-scale model, an inventory control system is designed to control particle size distribution and hence to regulate the average size of the product by manipulating the product withdrawal rate and seed addition rate.

(3ed) Mechanics of Electrochemical Energy Storage Materials

Vijay Sethuraman, School of Engineering, Brown University, Providence, RI

Recently, much emphasis has been placed on reducing the use of fossil fuels and the emission of greenhouse gases. To address these two interrelated issues, newer governmental regulations demand a sequential increase in energy-efficiency standards from auto manufacturers over the next two decades. Electrochemical energy-storage and energy-conversion have been identified as critical enabling technologies by the National Academies for meeting such regulatory demands. For example, automakers have already begun manufacturing hybrid, plug-in hybrid and electric vehicles with lithium-ion batteries in an effort to address such regulatory demands to increase the overall fuel-efficiency standards. However, the state-of-the-art energy-storage devices used in hybrid (*e.g.,* Toyota *Prius*), plug-in hybrid-electric (e.g., Chevrolet *Volt*) and electric (*e.g.,*Nissan Leaf) vehicles do not meet the durability and performance requirements at a cost comparable to conventional vehicles. Furthermore, the widespread use of such electrochemical devices expected in the near future, especially in the automotive industry, requires the training of students in fundamental as well as applied aspects of electrochemistry.

In light of these requirements, my current research program concentrates on functional materials and characterization techniques for electrochemical energy-storage and energy-conversion devices (*e.g.,*batteries, super-capacitors, and fuel cells) with an emphasis on understanding failure mechanisms. My interest in this area stems from my graduate and postdoctoral work on electrode components for lithium-ion batteries and polymer electrolyte fuel cells. I will continue to study electrochemical energy-storage and energy-conversion because this area has vast societal and economic importance and also presents compelling scientific and engineering challenges.

My current and past research efforts focus on understanding the properties (with importance to performance and failure characteristics) of electrochemical energy-conversion and energy-storage materials – *e.g.,* materials used in batteries and fuel cells. My approach towards understanding a candidate material consists of (1) extracting thermodynamic, kinetic and transport parameters, as well as mechanical properties by performing relevant experimentation, (2) testing theoretical predictions using the estimated parameters, (3) bridging material properties to practical performance, (4) obtaining insights for material-synthesis as well as system-evaluation efforts, and finally (5) quantifying the ability of a candidate chemistry to meet its performance goals set by the US Department of Energy. A central theme of my research program is to apply the engineering insights gained from this approach to the development of high-performance, low-cost and durable lithium-ion batteries and high-temperature fuel cells. My overarching research goal is to develop a comprehensive set of research tools and expertise to enable the engineering of high-performance materials for energy applications.

For recent publications, please visit http://www.nanopolitan.com

(3ee) Modeling and Simulation of Interphase DNA and the Programmable Self-Assembly of DNA-Coated Nanoparticles

Jonathan D. Halverson, Center for Functional Nanomaterials, Brookhaven National Laboratory, Upton, NY

Despite the discovery of the double helix structure of DNA in 1953, the spatial arrangement of the genome remains poorly understood on length scales above that of nucleosomes or a few tens of nanometers. Progress on the genome folding problem has been made in recent years due to experimental methods with the HiC method being the most notable. While it has been known for decades that chromosomes remain in distinct territories during the interphase of the cell cycle, a key finding from these latests experiments is that on the scale of 0.7-10 million base pairs the chromatin fiber within a chromosome is organized in an unentangled, hierarchal structure called a fractal globule. A melt of unknotted nonconcatenated ring polymers can serve as a simple model for such a system largely because rings are unentangled and they lack free ends so reptation is suppressed. We have conducted molecular dynamics simulations using a bead-spring model to study the statics, dynamics and rheology of ring polymer melts. For comparison linear chains are also studied. Topological interactions cause the rings to be partially collapsed and partially segregated. Additionally, the contact probability of monomers within a single ring is found to follow a power law consistent with a fractal globule, as is also the case for human chromosomes. Proposed research involves investigating the model further by studying the diffusive motion of a particle in a ring polymer melt and by examining the mechanism of stress relaxation of the rings due to a rapid change in density.

Today DNA can be synthesized in the laboratory with any desired sequence. This capability provides a route to the programmable self-assembly of nanoparticles. Nanoparticles have novel optical, electronic, magnetic and chemical properties due to their small size. Envisaged applications such as nanoplasmonics require the nanoparticles to be precisely arranged into structures with a specific overall shape. Nanoparticles coated with ssDNA can form effective bonds with other particles through the hybridization of complementary strands. In principle, an arbitrary nanostructure can be encoded by using a large number strands. Using a coarse-grained model we have conducted computer simulations to study the self-assembly of pre-programmed finite clusters from an initially homogeneous nanoparticle solution. We find that the constituent nanoparticles can be designed such that the clusters are produced in near perfect yield and free from errors. In addition, by controlling the kinetic pathway the clusters form rapidly.

While the experimentalist is able to vary the particle size and shape it is still a challenge to controllably decorate nanoparticles with multiple strand types. However, even with only a small number of strand types a wide variety of structures have been produced and the resulting structures are often not easily understood. Proposed research involves developing models for these systems and conducting simulations to explain the dynamics of self-assembly and the phase behavior.

Research carried out in part at the Center for Functional Nanomaterials, Brookhaven National Laboratory, which is supported by the U.S. Department of Energy, Office of Basic Energy Sciences, under Contract No. DE-AC02-98CH10886.

(3ef) Poly(ionic liquid) Block Copolymers for CO2 Capture

Brian Adzima, NETL, Pittsburgh, PA; Department of Chemistry, Carnegie Mellon University, Pittsburgh, PA

Ionic liquids possess low vapor pressures, high ionic conductivities, wide electrochemical stability ranges, broad thermal stabilities, and tunable solubilities enabling their use in a wide suite of devices such as lithium batteries, fuel cells, solar cells, light emitting diodes, and gas separation membranes.[1] Attempts to synthesis improved materials by polymerizing ionic liquid molecules have however produced mixed results. For example, in gas separation membranes a trade off exists between the mechanical integrity gained by polymerizing an ionic liquid and the permeability that is lost in the process.[2] Fortunately, phase separated polymer membranes formed by diblock copolymers may offer a suitable comprise between mechanical integrity and desired properties for all of the applications listed above. Already membranes formed from conventional diblock copolymers with poor CO_2 solubility have have shown improved separation results for CO_2/N_2 and CH_4/N_2 separations, with the increases in selectivity attributed to the orientation of cylindrical block orthogonal to the surface.[1]

At NETL I have developed techniques for preparing diblock copolymers incorporating monomers with improved CO_2 solubility, such as ionic liquids, using a combination controlled radical polymerization techniques and click chemistry.[3] The core synthesis centers on the polymerization of c-vinyl triazole monomers[4] using free radical and reversible addition-fragmentation chain-transfer (RAFT) polymerizations.[5] Quaternization with amines, followed by ion-exchange reactions with $LiTf_2N$ produces a poly(ionic liquid). Approaches for producing block copolymers from these pILs will be discussed, as will strategies for synthesizing, and subsequently crosslinking block copolymers to form membranes. Alternative approaches such as directly polymerizing of c-vinyl triazolium salts, and the incorporation of ionic liquid groups unto polymer backbones using the thiol–ene reaction will also be discussed.

This experience, combined with expertise using photo and click chemistry gained during my Ph. D. will support my goal as an assistant professor of starting a research group focusing on the development of materials with improved properties for a myriad of end applications.

[1] D. Mecerreyes, *Prog. Poly. Sci.* **2011**, *36*, 1629.

[2] J. E. Bara, D. E. Camper, D. L. Gin, R. D. Noble, *Acc. Chem. Res.* **2009**, *43*, 152.

[3] H. C. Kolb, M. G. Finn, K. B. Sharpless, *Angew. Chem. Int. Edit.* **2001**, *40*, 2004.

[4] H. Nulwala, K. Takizawa, A. Odukale, A. Khan, R. J. Thibault, B. R. Taft, B. H. Lipshutz, C. J. Hawker, *Macromolecules* **2009**, *42*, 6068.

[5] G. Moad, E. Rizzardo, S. H. Thang, *Aust. J. Chem.* **2005**, *58*, 379.

(3eg) Functional Materials: Molecules, Polymers, Particles, and Fibers

Sangyeul Hwang, Chemical Engineering Department, University of Michigan, Ann Arbor, MI

My research is situated in between the biological and material sciences, where a variety of functional materials from small molecules to micron-sized polymeric fibers have been prepared. Such diverse materials have different functions depending on their sizes. Therefore, it is categorized into four different groups; molecules, polymers, particles, and fibers.

For small molecules, organic synthesis has been utilized to discover and develop new drugs including enzyme inhibitors (Thrombin[1] and HIV protease inhibitor[2]), antibiotics (Quinolone, Anti-Methicillin-Resistant Staphylococcus Aureus (MRSA), and Oral Cephalosporin), antivirus (Anti-Hepatitis B Virus (HBV)), and natural compounds. Such an extensive organic synthesis has offered a fruitful tool in my research, thus became a key platform for the other studies.

As a part of polymeric functional materials, organometallic polymers as novel nitric oxide (NO) generating biomaterials[3] and new amperometric sensors[4]based on these materials have been developed. The NO generating polymers can catalyze the spontaneous decomposition of endogenous S-nitrosothiols found in blood (e.g., S-nitrosoalbumin, S-nitrosoglutathione, and S-nitrosocysteine) to produce NO locally at the polymer/blood interface for prolonged periods of time. Because NO has a variety of potent biological activities, including anti-platelet adhesion/activation, anti-proliferation of smooth muscle cell, anti-microbial, as well as wound healing, NOGPs have a variety of potential biomedical applications including stents, vascular grafts, indwelling sensors, catheters, and extracorporeal circuits in order to improve their biocompatibility.

The nano/micron sized anisotropic particles have been prepared by electrohydrodynamic (EHD) co-jetting. The prepared particles are comprised of two distinctive compartments within individual particles. Various combinations of different polymers, inorganic molecules (magnetite, titanium dioxide, gold nanoparticles, carbon nanotube, and silica), and dyes (optical and/or fluorescence) have been selectively embedded in each compartment,[5] potentially useful for drug delivery, catalyst, and particle swimmer. Additionally, magnetic Janus particles have been studied for controllable self-assemblies that are highly ordered in two-dimensional arrangements including a hexagonal close packing, and applied into a form of magnetically switchable surface.[6]

A novel approach to manufacture multiblock magnetic nano/microwires is established via spatiotemporally programmable magnetic self assemblies and concomitant particle interconnections.[7]The other functional fibers fabricated by the EHD co-jetting are studied for de novo two-way shape memory polymers and anisotropic polymeric niches for stem cell differentiation.

References

1. Kim, S. S.; **Hwang, S.**; Kim, Y. K.; Yun, M. K.; Oh, Y. S., "*Rational Design of Selective Thrombin Inhibitors*" *Bioorg. Med. Chem. Lett.*, **1997**, *7*, 769-774.
2. Lee, K. W.; **Hwang, S.**; Kim, C. R.; Nam, D. H.; Chang, J. H.; Choi, S. C.; Choi, B. S.; Choi, H.-w.; Lee, K. K.; So, B.; Cho, S. W.; Shin, H., *"The Chemical Development of LB71350" Organic Process Research & Development* **2003**, *7*, 839-845.
3. **Hwang, S.**; Cha, W.; Meyerhoff, M. E. "*Polymethacrylates with Covalently Linked Copper(II)-Cyclen Complex for In Situ Generation of Nitric Oxide from Nitrosothiols in Blood*" *Angew. Chem. Int. Ed.* **2006**, *45*, 2745-2748.
4. **Hwag, S.**; Cha, W.; Meyerhoff, M. E. "*Amperometric Nitrosothiol Sensor Using Immobilized Organoditelluride Species as Selective Catalyst*" *Electroanalysis*, **2008**, *20*, 270-279.
5. **Hwang, S.**; Roh, K.-H.; Lim, D. W.; Wang, G.; Uher, C.; Lahann, J. "*Anisotropic Hybrid Particles Based on Electrohydrodynamic Co-Jetting of Nanoparticle Suspensions*" *Phys. Chem. Chem. Phys.* **2010**, 12, 11894-

11899.
6. **Hwang, S.**; Nguyen, T. D.; Yoon, J.; Bhaskar, S.; Hitt, J.; Chickering, D.; Bernstein, H.; Glotzer, S. C.; Lahann, J. "*Magentophoretic Surface Switching*" **2012**, *In Submission*.
7. **Hwang, S.**; Park, T.-H.; Lahann, J. "*Spatiotemporally programmable Self-Assemblies of Magnetic Particles*" **2012**, *In Submission*.

(3eh) Molecular Engineering Chemical Imaging Probes for Super Resolution Fluorescence Microscopy

Younghoon Kim, Department of Chemical and Biomolecular Engineering, University of Illinois at Urbana-Champaign, Urbana, IL

My research area spans from protein engineering to single molecule fluorescence microscopy extending application to protein formulation and synthetic biology. Through my PhD and post-doc training, I gained experience in the interdisciplinary environment based on chemical engineering education. My current post-doc projects will be presented.

1) New dye was developed for super resolution microscopy, stochastic optical reconstruction microscopy (STORM). Recent advances in imaging and single molecule fluorescence microscopy (SMFM) have enabled direct observation of biological processes at the molecular level. Super-resolution imaging techniques such as photoactivated localization microscopy (PALM) and STORM have been developed using photoswitchable fluorescent proteins and organic dyes. However, there is a strong need for development of bright and photostable nanoscale probes for biological imaging as pointed out from inventors of STORM and PALM. In this work, we suggest dye-conjugated dendrimers as nanoscale imaging probes for high-resolution fluorescence microscopy, which overcome the diffraction limited optical resolution. Collective effects in multichromophoric systems generated controlled blinking and enhanced localization precision compared to single organic dyes such as Cy5 or Alexa flour 657. With this new generation of fluorescent dye, we were able to achieve a 10 nm spatial resolution for reconstructed images from *d*STORM, which surpasses the spatial resolution of single organic probes (25 nm).

2) We demonstrate labeling of transcription factor proteins by incorporation of non-natural amino acids containing bioorthogonal chemical functionalites (azide or alkyne) followed by copper-free click chemistry. The cyclooctyne promotion [3 + 2] dipolar cycloaddition with azides, "copper-free click chemistry" provides rapid and site-specific labeling of transcription factors (TFs). Following expression, purification and labeling, transcription factor proteins retain their sequence-specific DNA-binding activity under native conditions, whereas the activity of TFs labeled by copper-mediated click chemistry was ablated. Copper-free click chemistry exhibited high reaction yields, even at low temperatures or in the presence of high salt concentrations, which are generally unfavorable reaction conditions for common methods involving hydrolysis-mediated protein modification. Residue-specific replacement of methionine by azidohomoalanine was used to achieve mono- or multi-functionalization of engineered or "synthetic" transcription factor proteins.

(3ei) Single Molecule Studies of Heterogeneous Catalysts

Xiaojiao Sun, Chemical Engineering, Kansas State University, Manhattan, KS, Keith L. Hohn, Chemical Engineering Department, Kansas State University, Manhattan, KS and Daniel A. Higgins, Chemistry Department, Kansas State University, Manhattan, KS

With the development of space and time resolution and detector sensitivity of microscopic equipment, it is possible to study a single site on heterogeneous catalysts since the concentration of probe molecules that responding to various properties of catalytic sites are down to nanomolar scale. The fluorescence data collected from the well-separated dye molecules can be interpreted to the properties of the interacting sites on the solid matrix. The improvement of fluorophores, which are originated in biochemistry and biochemical engineering, make the study of using single molecule spectroscopic (SMS) method to characterize solid catalysts promising, especially the ones that are sensitive to acid sites. The material under study is mesoporous silica synthesized by sol-gel method. Compared with other solid catalysts, this material can be modified with a wide variety of functional groups and the pore structure can be strictly ordered, providing a simple and well-defined system to study. With the SMS method and appropriate dye molecules, the active sites can be located and characterized by the fluorescence behaviors of the probes, and this resolution is difficult to achieve for conventional methods. In-situ characterization is also possible by using SMS method, and the reaction condition can be realistic without the requirement of high vacuum or conductive surface. The current work is to study the aligned mesoporous silica functionalized with sulfuric groups. A coumarin-based dye, which is low pH sensitive, is used to map the acid sites and obtain the acidity distribution on the solid matrix. To confirm the result, some conventional methods will also be used to verify the data and the data interpretation by SMS method. This work will indicate the potential of applying single molecule spectroscopic method for the characterization of heterogeneous catalysts.

(3ej) Assessment of Fouling in Native and Surface-Modified Water Filtration Membranes

Daniel J. Miller, Donald R. Paul and Benny D. Freeman, Chemical Engineering, University of Texas at Austin, Austin, TX

Fouling is a major obstacle to implementation of polymeric membranes in water purification applications, causing deterioration of membrane performance and, consequently, increases in operational expenses. Surface modification is one way of alleviating membrane fouling. Most porous membranes are made of hydrophobic polymers, promoting the adhesion of foulants via strong hydrophobic-hydrophobic interactions. Hydrophilization of the membrane surface tends to reduce fouling because a layer of water molecules becomes tightly bound to the hydrophilic surface, acting as a buffer against foulant ingress. Dopamine was recently found to non-selectively deposit from buffered alkaline solution to form "polydopamine" on virtually any surface, rendering the surface very hydrophilic after only a few minutes of contact with dopamine solution. The chemical structure of polydopamine, which was previously ambiguous, was investigated by many techniques, including solid state NMR, FTIR, UV/vis and powder X-ray spectroscopies. Polydopamine was also used as a platform for the molecular conjugation of other anti-fouling materials, such as poly(ethylene glycol), to the membrane surface.

This work has focused on the evaluation of fouling behavior in native and surface-modified commercial membranes. Water filtration membranes can be operated in two ways: constant permeate flux with variable transmembrane pressure (TMP) or constant transmembrane pressure with variable permeate flux. Membrane fouling may be measured using either of these two methodologies. A constant permeate flux crossflow system was carefully constructed and used to challenge native and surface-modified membranes with an oil/water emulsion. The critical flux—the flux below which minimal fouling occurs—was determined by a flux-stepping technique during constant flux crossflow filtration. The membrane performance during fouling was evaluated at fluxes higher than and lower than the critical flux. Membrane resistance evolution during fouling is compared for constant-flux and constant-TMP operation. The efficacy of surface modifications in reducing biofouling was also studied. Many evaluations of anti-biofouling behavior rely on short-term tests of protein or bacterial adhesion. The results of such experiments were compared to those obtained for comparatively long-term operation under hydrodynamic conditions relevant to industrial module operation.

(3ek) Nano-Electochemical Systems for Energy Conversion

Carlos Hangarter, Metallurgy Division, NIST, Gaithersburg, MD

Nanomaterials have been rigorously pursued in the energy challenge and now comprise a significant effort towards high performance fuel cells, solar fuel production and solar cells. Nanoengineering is an attractive route to efficiently utilize materials and enhance functionality due to confinement effects and interfacial phenomenon. The diminutive sizes and large surface to volume ratios of these structures can augment chemical sensitivity and surface reactions or influence charge transport and photonic coupling.

Electrochemistry, which is central to life sustaining energy systems, is also a major theme among sustainable energy storage and conversion technologies. Manifesting as redox reactions, electrocatalysis and photoelectrochemistry, the high efficiency of electrochemical reactions and capacity to directly interface high energy density fuels continues to drive these efforts.

Many energy systems are inherently electrochemical in nature and nanostructuring provides a format to efficiently utilize material, manipulating their features on a length scale to control photon interactions and chemical reaction pathways. My research is at the junction of electrochemistry and nanomaterials and aims to address the challenges of sustainable energy technologies.

(3el) Multifunctional Protein-Based Materials for the Synthesis and Organization of Nanomaterials

Alia P. Schoen, Materials Science & Engineering, Stanford University, Stanford, CA

Recent advances in materials synthesis have revealed unique properties that occur at the nanoscale; however, the long-range organization of nanomaterials into the hierarchical structures required for applications remains an unmet challenge. In striking contrast, such hierarchical organization is an integral feature of natural bioinorganic materials, where the concerted action of proteins exquisitely orchestrates structural organization. One of the key challenges in protein-based synthesis of inorganic materials is developing robust and versatile strategies for a chosen protein scaffold to interact with inorganic components in a specific and controlled manner. In my PhD thesis research, I have developed a new biomimetic strategy to modify protein scaffolds based on site-specific molecular recognition. This versatile strategy, termed Template Engineering Through Epitope Recognition (TEThER), is based on the design of bi-functional TEThER peptides that mimic the activity of natural adaptor proteins to site-specifically assemble with a single native protein scaffold for a variety of functions.

This project focused on understanding the unique self-assembling protein clathrin and developing the TEThER strategy to modify self-assembled clathrin scaffolds for the synthesis of inorganic nanomaterials. The self-assembly of clathrin was studied through a combination of dynamic light scattering and cryo transmission electron microscopy which showed it to be a highly dynamic process that can progress through different kinetic pathways to arrive at ordered three-dimensional structures. By designing a family of TEThER peptides, I have demonstrated the versatility of the novel site-specific method TEThER strategy to bridge the interface between self-assembled clathrin and two classes of inorganic materials, metal oxides and noble metals, to synthesize titanium dioxide, cobalt oxide, and gold nanoparticles. Crystallographic characterization of nanoparticle morphology and phase using transmission electron microscopy gives insight into the nucleation and growth mechanism for each of these reactions. Peptide sequences with affinity for various inorganics can be combined with site-specific recognition sequences for other clathrin scaffold binding sites in a mix-and-match manner to enable the nucleation and growth of a wide variety of materials and nanoscale patterns. Additionally, since this strategy is based on molecular recognition that occurs at protein binding interfaces, it is broadly applicable to other protein systems.

In my research program I aim to develop multifunctional protein-based materials for the synthesis and long-range organization of nanomaterials through self-assembly and the design and use of multi-component, modular systems. Such a strategy offers a promising new way forward for the manipulation and organization of materials at the nanoscale. Through continued development of strategies to interface a variety of native and engineered biomolecules with inorganic materials, the bottom-up synthesis of designed hierarchically-ordered biomimetic nanomaterials for applications such as energy conversion and storage and bone tissue regeneration will be possible. My interdisciplinary background uniquely positions me to accomplish this goal by bringing together insight from diverse fields, including materials science, protein engineering, and chemistry.

(3em) High Performance Conducting Polymer Based Nanomaterials for Energy Storage Devices

Nasim Hyder, Chemical Engineering, Massachusetts Institute of Technology, Cambridge, MA

Next generations electrochemical energy storage devices require high-power, high-energy density, simultaneously. To achieve this goal, it is essential to identify novel electrode materials based on nanoparticles, nanotubes or nanofibers with decreased transport length of ions/electrons and high surface area. Energy storage devices (capacitor, battery) based on electroactive polymers and carbon nanomaterials recently gained significant interest to supply high-power, high-energy, simultaneously, for hybrid electric vehicles, plug-in hybrid electric vehicles. The development of conjugated polymeric nanofiber with controlled aspect ratio and morphology exhibits attractive potential in terms of improved electronic conductivity and porosity that are competitive with state-of-the-art RuO_2.

Conducting polyaniline nanofibers and multiwalled carbon nanotubes with their exotic properties can significantly increase power (W/kg) and energy density (Wh/kg) in addition, to their longer life cycles and charging-discharging efficiency. We assembled free-standing multilayered thick (10 to 60 microns) electrodes of conducting polyaniline (PANi) and multiwalled carbon nanotubes (MWNTs) that are robust, binder-free using rapid spray and vacuum-assisted layer by layer techniques. The electrode microstructure reveals porous nanostructure with high surface area; Preliminary results for these highly conductive electrodes (4.4 S/cm) show high pseudocapacitance of 210 F/cm^3 and are stable over 1000 cycles. I will present our findings on the optimization of the free standing (along with thin film) asymmetric electrode morphology and electrochemical performance in aqueous and organic electrolytes including the ongoing research in this area.

(3en) Advancing Genome Scale Models

Patrick F. Suthers, Department of Chemical Engineering, The Pennsylvania State University, University Park, PA

The development of genome scale models (GSM) for a variety of microbial, eykaryotic and multi-cellelar organisms is proceeding with an accelerating pace. Automated reconstruction tools such as ModelSEED, Pathway Tools, and others are increasingly being utilized to quickly arrive at draft compilations of metabolic models followed by detailed curation to improve upon modeling components crucial for answering the queries relevant to the specific organism and study. Typically, simulations are carried out for an idealized growth substrate typically containing a single carbon substrate (e.g., glucose). Regulation is customarily handled by shutting off specific reactions in the presence of a specific nutrient environment. For example, in the iAF1260 *E. coli*model, there are specific reaction restrictions for anaerobic, aerobic, and aerobic with glucose growth conditions. In this poster, a second generation model for Costridium acetobutylicum is presented.

We make extensive use of MetRxn (http://metrxn.che.psu.edu), a web-based knowledgebase that includes standardized metabolite and reaction descriptions and integrates information from 8 databases and 44 metabolic models into a single unified data set. In addition to metabolic biotransformations, we describe the inclusion of regulatory information. The establishment of detailed gene-protein-reaction associations enables us to properly localize regulatory interactions at the transcriptional, protein or metabolite level and rely on the gene-protein-reaction (GPR) connections to propagate their action through metabolism. We make use of gene and protein expression data and a variety of growth conditions.

(3eo) Separation of Macromolecules by Photonic Crystal Defects Chromatography (PCDC)

Nicolas Alvarez, Chemical Engineering, Carnegie Mellon University, Pittsburgh, PA

Size Exclusion Chromatography (SEC), also known as Gel Filtration Chromatography, is currently one of the most prevalent separation and characterization techniques for synthetic and natural polymers, including proteins and carbon nanotubes. The basic separation mechanism is an entropic exclusion related to the size of the molecules compared to pore size. The application of SEC for the separation of complex mixtures is severely impaired by its poor separation resolution, large reagent consumption and intrinsic difficulties for the integration into automated multistep analysis. The underlying problem of SEC and other gel-based conventional separation techniques is a lack of engineering control of the molecular separation process in terms of both pore geometry and surface interaction. Further development of artificial sieving structures with tailored surface functionalities (e.g. by modifying surface hydrophobicity and surface charge density) has been considered. However, since an engineering control of the interaction potential is still missing, these methods are subject to undesired surface adhesion and clogging by macromolecules.

Recently, the barrier of optical nanomanipulation has been surpassed by utilizing the field amplification and localization in photonic crystal (PC) cavities. PCs are composite dielectric structures possessing a periodic spatial variation in their refractive index on the order of the wavelength of light. Introducing defects into the periodic lattice of a PC structure induces formation of localized photonic modes within the otherwise forbidden photonic band gap. This leads to a significant enhancement of the optical gradient trapping force. Such devices are capable of trapping nanoparticles with radii down to 10 nm with a low input power of 5 mW in aqueous media. PCDC utilizes differences in the interaction of analyte molecules in a flow (mobile phase) with localized optical electric fields (OEFs) (representing the stationary phase). Macromolecules interact with the optical fields depending on their intrinsic properties such as size, shape and refractive index. The lateral concentration distribution depends on the interaction strength between the solute molecules and the gradient of the localized field. The stronger the field gradient the stronger the interaction strength of a given solute molecule and the more nonuniform the concentration profile normal to the surface of the PC structure. Separation is achieved by confining the distribution of solute molecules to different lateral positions in a non-uniform flow (i.e. Poiseuille flow). This separation mechanism is known as field flow fractionation (FFF). The lateral position of a given solute's average concentration determines its residence time in the channel.

In this presentation, the separation mechanism for OEF-FFF is outlined and the relevant dimensionless parameters are analyzed considering a discreet number of OEFs confined to a microchannel whereby a plug of dispersed solutes is undergoing Poiseuille flow. There are three relevant dimensionless parameters that determine the selectivity, resolution, and dispersion or quality of separation: the dimensionless field strength, the normalized field decay length, and the Peclet number. The first two dimensionless groups determine the selectivity and resolution of separation, while the Peclet number is responsible for overall dispersion and quality of separation. A detailed numerical analysis of these dimensionless groups considering experimentally feasible axially non-uniform OEFs is discussed and the results are compared to an analytical analysis of a simpler axially uniform OEF. An experimental design is outlined and preliminary experimental results are correlated to the theoretical results.

(3ep) Tunable Drug-Encapsulated Ultrasound Contrast Agents

Yoonjee Park[1], Tuan Pham[1], Carl Beigie[1], Robin Cleveland[2], Jon O. Nagy[3] and Joyce Y. Wong[1], (1)Biomedical Engineering, Boston University, Boston, MA, (2)University of Oxford, Oxford, United Kingdom, (3)NanoValent Pharmaceuticals, Bozeman, MT

Ultrasound contrast agents using micron-sized bubbles (less than 8 μm in diameter) have potential to be drug delivery vehicles in addition to targeted molecular imaging. However, challenge remains in terms of producing monodisperse microbubbles stable against gas dissolution for improving blood circulation times. In the case of drug delivery, it is crucial to control destruction to release encapsulated contents *only* to the targeted area with minimal prior passive leakage of drug. By using a photopolymerizable diacetylene lipid as a component in a microbubble shell matrix, the surface properties such as elasticity or resistivity against gas dissolution can be tuned,[2] which therefore enables imaging at one frequency and drug release in the area of interest at a different frequency. Drug-encapsulated monodisperse bubbles are created using a double emulsion microfluidic technique to increase capacity of drug encapsulated.

Polymerizable lipid mixtures, consisting of 10 mol% of ethylene glycol diacetylene lipids (h-PEG_1PCDA), 15 mol% of PEG-diacetylene lipids (m-PEG_{5000}-PCDA) and L-α-phosphatidylcholine, hydro soy PC were used. The microfluidic flow focusing device designs were adapted from Hettiarachchi et al.[3,4] Paclitaxel in triacetin (0.1mg/ml) was used as a test drug and 1 mol% of the total paclitaxel was fluorescently labeled oregon green 488-paclitaxel for a FITC filter.

The size distribution of the polymerized shell microbubbles (PSM) produced by microfluidic flow focusing technique and commercially available microbubbles (Vevo MicroMarker, VMM) was observed to be 2.2 μm with 8% polydispersity, and 2.3 μm with 36% polydispersity, respectively. The PSM showed a significantly slower decrease in intensity of gray-scale ultrasound image brightness than VMM or nonpolymerizable soy PC shell microbubbles (NSM). The half-life (t_H the time for the ultrasound signal to drop 6 dB in brightness) for the 25%DA was t_H = 5 min, and for both VMM and NSM, $t_H < 1$ min. The brightness of the 25%DA decreased by 14 dB after 15 min, whereas the VMM brightness decreased by 36 dB, suggesting rapid microbubble destruction. In addition, the bubbles polymerized to different extents (t_{polym}) showed variable destruction rates at different ultrasound power levels. This suggests that polymerization can not only provide passive bubble longevity but tunable rupture capability. The double emulsion microfluidic device produced the paclitaxel encapsulated microbubbles. The FITC-filtered image shows the location of paclitaxel within the shell. These results indicate that the dissolution of microbubbles in the bloodstream or under ultrasound stimulation is tunable by varying the fraction of polymerizable lipid or polymerization time. Therefore, these tunable microbubbles have the potential to be customized ultrasound contrast agents for targeted molecular imaging and therapeutic treatment applications.

References: [1] Park et al. *Langmuir* **2012**, *28,* 3766-3772. [2] Hettiarachchi, K.et al. *Lap Chip* **2007**, *7*, 463-8. [3] Hettiarachchi, K.et al. *Biotechnol. Prog.* **2009**, *25*, 938-945.

(3eq) Acid Functionalized Magnetite Nanoparticles for Carbohydrate Hydrolysis

Myles A. Ikenberry, Chemical Engineering, Kansas State University, Manhattan, KS

Acid catalyzed hydrolysis of lignocellulosic materials is a well-studied and promising method of producing glucose and other precursors to ethanol and industrially useful chemicals. However, this method of hydrolyzing biomass is limited by the costliness of downstream acid separation and disposal. A potential solution to this problem is the use of solid acid catalysts, which can be separated more easily and reused. Magnetite nanoparticles have an advantage over other solid acid catalysts in that they can be separated by magnetic extraction, rather than requiring centrifugation or filtration. Additionally, the inherently high surface area and concentration of acid groups at the outer surface of the material gives the functionalized particles the potential to outperform the activity of other solid acid catalysts. Despite the potential to improve the economics of renewable energy production from biomass, acid functionalized magnetite nanoparticle applications remain limited by difficulties such as achieving economically viable synthesis methods for enhanced control of particle size, agglomeration, and acid stability. Some recent research in these areas is discussed, in addition to a novel, inexpensive synthesis technique whose functionalized particles have been tested for the hydrolysis of some carbohydrates.

(3f) Feedback Controlled Colloidal Self-Assembly

Jaime J. Juárez, Materials Science and Engineering, University of Illinois at Urbana-Champaign, Urbana, IL

The ability to assemble nano- and micro- components into ordered configurations provides the basis for fabricating tunable materials to manipulate electromagnetic energy (e.g., meta-materials, photonic bad gaps) for existing (e.g., photovoltaics, reconfigurable antennae) and emerging technologies (e.g., optical computing, sub-diffraction limited imaging, invisibility cloaking). The limiting factor in assembling microscopic components into ordered materials is the inability to actively remove defects (e.g., grain boundaries, point defects) that form during assembly. This work demonstrates a conceptually new approach to guide the dynamic evolution of a colloidal crystal self-assembly process by using feedback control to quantitatively and reversibly guide the dynamic evolution between disordered fluid and crystalline configurations. Models for self-assembly are used to inform feedback control laws that close the loop between real-time sensing (via order parameters) and actuation (via tunable electrical potentials). Feedback control tunes the ordering internal components and performs active reconfiguration in colloidal crystals of a targeted size.

(3g) Tuning the Molecular Packing of Organic Semiconductors for High Performance Using Metastable Crystallization

Gaurav Giri and Zhenan Bao, Chemical Engineering, Stanford University, Stanford, CA

Solution deposition of organic semiconductors (OSC) is a leading contender for producing large-area, inexpensive, and flexible organic electronics. Our work involves using a new solution processing technique called solution shearing in order to cause OSC crystal lattice strain and thus modify the molecular packing of a wide range of OSCs through processing alone. We quantitatively measure the molecular ordering resulting from different deposition conditions and show effective tuning of molecular packing that yields higher charge carrier performance without changing the underlying chemical structure. We investigate the stability of the novel molecular packing structure and also measure the relationship between the morphological features and its impact on charge carrier transport. We also develop an in-situ technique that uses X-ray diffraction to study how the strained crystals grow as a function of solution shearing parameters. This in-situ technique utilizes high speed optical and X-ray diffraction imaging (~100 fps) in order to study OSC crystal growth. My proposed research will expand on the concept of metastable crystallization of OSC and other organic molecules in order to reliably generate new polymorphs and strained crystals. Through the use of confined and controlled solvent evaporation, novel solute packing motifs can be achieved. X-ray diffraction and differential scanning calorimetry (DSC), among other techniques, can be used to screen for these crystal polymorphs.

(3i) Multiphase Flow Phenomena in Chemical and Biological Systems

Travis W. Walker, Chemical Engineering, Stanford University, Stanford, CA

The addition of multiple phases to flow systems drastically increases the complexity of the flow physics. These complexities reveal themselves on the macroscopic and microscopic length scales and can involve solids as well as immiscible and miscible fluids. Interfacial interactions between miscible systems are further complicated by the presence of gradients in the chemical potential that vary in space and time. These gradients distinguish miscible systems from immiscible systems, which approximate these gradients as discontinuities. Additional complexities such as the presence of polymers, surfactants, colloids, and particulates to flow systems create complex fluids or soft materials that respond nonlinearly to stress. A vast number of manufacturing practices involve multphase systems that are highly structured and rheologically complex.

My most recent research activities have focussed on two problems involving flowing, miscible liquids: "rinsing flows," or flows where one liquid is ablated by a second fluid, and the interaction of miscible drops and jets.

The work on rinsing flows was motivated by the desire to remove colloidal, particulate contaminants in the processing of advanced microdevices. Specifically, we investigated the discovery that polymeric liquids could be turned into soft adhesives that effectively eliminate particles without damage to delicate surfaces.

Most recently, I have been investigating the interaction of droplets and jets of a liquid that is miscible with its surroundings. This basic flow problem is central to numerous industrial and natural processes, where mixing of cleansing liquids and creating biocompatible implants for drug delivery are two prominent examples.

The research program that I aim to develop will combine experiment and theory to understand the flow physics of miscible, often non-Newtonian, liquids. A problem of particular interest is mucous transport in the human body, specifically investigating how rheological changes in mucus affect cilia transport in patients with cystic fibrosis.

(3j) Dynamic Response of Associating Polymers: From Blood Clotting to Kinetically-Driven Assembly

Charles Sing, Department of Materials Science and Engineering, Massachusetts Institute of Technology, Cambridge, MA

Recent developments in the fields of supramolecular chemistry and biophysics have demonstrated that there are a rich array of both naturally-occurring and synthetic interactions that rely on complex, long-lived, but ultimately reversible associations. In the case of biological systems, these long-lived associations represent motifs composed of anywhere between one and even hundreds of amino acids depending on the physiological situation; disulfide bonds, integrins, hydrophobic interactions, and any number of more specialized interactions are ubiquitous. I have elucidated the role of such binding behaviors in the specific case of the multimeric protein von Willebrand Factor (vWF), a blood clotting protein that functions as a targeted adhesive at the site of a wounded blood vessel. Despite its complex structure on sub-quaternary levels, the quaternary structure of vWF conveniently lends itself to coarse-grained simulation models that I use to illustrate concepts described through analytical theory. vWF's counterintuitive behavior has remained largely unexplained until my PhD work, which describes how long-lived binding states couple to shear flows to drive an otherwise inert polymer globule into an adsorbed and elongated state despite unfavorable hydrodynamic forces.

Throughout the course of the aforementioned investigation, I developed the simulation and theoretical tools to study polymer systems with long-lived, reversible associations. Due to the coarse-grained nature of these simulations, the principles I have elucidated are largely applicable to designed synthetic systems based on supramolecular chemistry ideas. A host of counterintuitive and kinetically-driven behaviors were observed in the course of understanding single-chain self-associating systems, and it is already clear that these concepts will provide routes to novel, stimuli-responsive, and transient structures. In my future plans, I anticipate using these ideas as the platform upon which I can build designed materials that can form structure in either the bulk or the single-chain limit through these temporal constraints.

My poster will outline the most crucial aspects of my PhD work on von Willebrand Factor, along with an outline of my anticipated projects involving designed, kinetically-driven systems as it relates to both my previous PhD experience and my anticipated Postdoctoral work.

(3k) Heterogeneous Catalysis and Sustainable Energy Production: From Fundamentals to Applications

Bingjun Xu, Chemical Engineering, Caltech, Pasadena, CA

The interplay of fundamental mechanistic investigation, application-oriented engineering, and novel materials synthesis plays a critical role in energy-related research. My previous work on heterogeneous catalysis and thermal water splitting uniquely positions me to exploit the synergy of these fields to tackle the challenges of sustainable fuel production and developing green chemical processes in my future research. In the poster session, I will present some of my previous and current research related to these topics.

My Ph.D. thesis established the mechanistic framework for oxidative coupling reactions on the single crystal Au(111) surface under ultrahigh vacuum conditions. Gold-based catalysts have shown remarkable activity towards a number of important reductions, though the mechanistic steps remain controversial. Elucidating the intrinsic activity of metallic gold is a critical step in fully understanding gold catalysis. Reaction pathways and rate-limiting steps for oxidative coupling reactions on metallic gold surface have been elucidated by a combination of surface-sensitive spectroscopies and density functional theory calculations. Oxidative coupling reactions carried out on nanoporous gold catalysts under ambient conditions faithfully followed the proposed mechanism, demonstrating the applicability of the mechanism.

My postdoctoral research focuses on developing a low-temperature, thermal water splitting cycle aiming to take advantage of waste heat from nuclear power plants. A manganese-based thermochemical cycle has been proposed, with the highest operating temperature (850°C) close to that of the now piloted sulfur-iodine cycle. Furthermore, the Mn-based cycle is founded on earth-abundant elements without involving any toxic or corrosive chemicals. In addition, I am working to synthesize zeolites with chiral structures via rational design of structure-directing agents.

Integrating my background in heterogeneous catalysis, clean energy, and materials synthesis, my future research aims to address such important challenges as the production of fuel from carbon neutral/negative sources and the development of green chemical processes.

Ph.D. Advisors: Prof. Cynthia M. Friend and Prof. Robert J. Madix

Postdoc Advisor: Prof. Mark E. Davis

(3l) Nanoconfined Organic Molecules and Polymers: Fundamentals and Device Applications

Dun-Yen Kang, School of Chemical & Biomolecular Engineering, Georgia Institute of Technology, Atlanta, GA; School of Chemistry and Biochemistry, Georgia Institute of Technology, Atlanta, GA

Recent work on the optical, electronic, and photonic applications of 1D-nanostructured organic molecules and polymers strongly highlights the enhancement in their properties over 2D thin films of the same materials. Nanoporous materials could provide nanoconfinement for organic molecules or polymers with controllable channel dimensions, orientation, and interface chemistry. On the other hand, scalable techniques for fabricating thin films and membranes of nanoporous materials have been developed. Hence, the creation, scientific investigation, and application of photon/electron/phonon-conductive organic guest molecules or polymers in 1D nanoconfined geometries within nanoporous thin films could revolutionize the functionality of these guest entities while possessing excellent processibility covering a wide range of length scales (10^{-6}-10^{2} m). A host of new device platforms for "green" energy, photoelectricity, biomimetic, and biomedical devices can be created via the "marriage" of functional organic molecules/polymers and 1D nanoporous membranes.

The realization of these exciting possibilities requires a new level of understanding of the structure and properties of these complex hybrid nanostructures, as well as the capability to engineer and optimize their performance. My interdisciplinary research program will therefore be dedicated to the science and engineering of nanostructured host-guest devices comprising functional molecules and polymers confined in 1D-nanoporous thin films. My proposed novel device platform, and my formulation of reliable fabrication-structure-property relations in such devices, will together allow us to pursue technological applications that include: (1) thermoelectric generators containing thermoelectric organics for harvesting waste heat from portable electrical devices, vehicles, or power plants; (2) artificial antenna systems with dye-molecules mimicking green plants for storage of photonic energy in an enthalpic form; and (3) biosensors with conducting functional polymers for cancer clinic testing.

I have acquired a strong research background in advanced nanoporous materials synthesis and characterization, surface chemistry and functionalization in nanoconfined environments, study of molecular transport phenomena in nanostructured materials, and fabrication of nanotubular thin films and membranes. This interdisciplinary expertise will strongly support the development of the proposed research program.

Publications:

• Yucelen, G. I.; **Kang, D.-Y.**; Guerrero, R. C.; Wright, E. R.; Beckham, H. W.; Nair, S., Shaping Nanotubes at the Molecular Scale from Precursors of Controlled Curvature, *Nano Letter,* **2012,** *12* (2), 827-832

• **Kang, D.-Y.**; Tong, H. M.; Zang, J.; Sholl. D.S.; Jones, C. W.; Nair, N., Single-Walled Aluminosilicate Nanotube / Poly (vinyl alcohol) Nanocomposite Membranes, *ACS Appl. Mater. Interfaces,* **2012,** *4* (2), 965-972

• **Kang, D.-Y.**; Jones, C. W.; Nair, N., Modeling Mass Transport in Composite Membranes with Tubular Fillers, *J. Membr. Sci.,* **2011,** *381* (1), 50-63.

• **Kang, D.-Y.**; Zang, J.; Jones, C. W.; Nair, S., Single-Walled Aluminosilicate Nanotubes with Organic-Modified Interiors. *J. Phys. Chem. C* **2011,** *115* (15), 7676-7685.

• **Kang, D.-Y.**; Zang, J.; Wright, E. R.; McCanna, A. L.; Jones, C. W.; Nair, S., Dehydration, Dehydroxylation, and Rehydroxylation of Single-Walled Aluminosilicate Nanotubes. *ACS Nano* **2010,** *4* (8), 4897-4907.

(3m) Biopreservation: From Single Cells to Organs

O. Berk Usta, Surgery, Massachusetts General Hospital / Harvard Medical School, Boston, MA

I have a strong background on theoretical and computational modeling of complex and multiphase flows, and elastic solids using mesoscopic approaches such as lattice-Boltzmann and lattice-spring methodologies. I have worked with the pioneers in this field and have investigated questions of both fundamental and biological relevance ranging from colloidal self diffusion, polymer migration in microchannels, separation of microcapsules, viscoelastic droplet deformation to self propulsion and communication of microcapsules. During the lastt few years I have added an experimental component to my research toolbox working at the Center for Engineering in Medicine with Dr. Martin L. Yarmush and Dr. Mehmet Toner. We are currently investigating several questions in the broad field of biopreservation, highthroughput single cell analysis, microfluidic separations and drug delivery in efforts to solve problems related to liver and burn injuries.

During this session I will primarily discuss biopreservation schemes such as cryopreservation and supercooling to preserve liver, its cells and tissues where our groups has made great strides recently. I will also introduce the use of microfabricated microwell arrays in the quest to understand the effect of cell heterogeneity on successful preservation; this platform combined with powerful statistical analysis is a potent tool for describing many hidden relations when one looks only at average population responses.

(3n) Structure and Dynamics of Block Copolymer Based Soft Materials

Sangwoo Lee, Department of Chemical Engineering and Materials Science, University of Minnesota, Minneapolis, MN; Chemical Eng. and Materials Sci., University of Minnesota, Minneapolis, MN

Block copolymers, made by joining two or more linear polymer blocks, have been studied over a half century, and these compounds are still being intensively explored. We recently discovered an equilibrium phase of packed spheres in poly(1,4-isoprepe-*b*-DL-lactide) (IL) diblock and poly(styrene-*b*-1,4-isoprene-*b*-styrene-*b*-ethylene oxide) (SISO) tetrablock copolymers, known as Frank-Kasper σ-phase using small angle X-ray scattering (SAXS) and transmission electron microscopy (TEM) experiments. The σ-phase contains 30 microphase-separated spheres in a large tetragonal unit cell, nucleates and grows from the body-centered cubic (BCC) phase similar to its occurrence in metal alloys. It is also known as a crystal approximant to dodecagonal quasicrystals, indirectly predicted by the Landau theory in diblock copolymer melts.

The mean field treatment on block copolymer thermodynamics near the order-disorder transition has given valuable understandings on this subject, but it assumes an infinitely large degree of polymerization (N) with an infinitesimally small Flory-Huggins interaction parameter (χ). All experimentally accessible block copolymer compounds have N and χ far from these limits, and the fluctuation effect near ODT from finite N has not been fully elucidated. We have investigated the thermodynamics of poly(1,4-isoprene-*b*-DL-lactide) diblock copolymers near the ODT with small N and relatively large χusing SAXS, dynamic mechanical spectroscopy (DMS), and differential scanning calorimetry (DSC) techniques. These studies provide fresh insights into the role of fluctuations on the state of segregation and ordering kinetics in symmetric and asymmetric diblock copolymers, prompting new questions regarding the application of mean-field and fluctuation corrected theories to block copolymer melts with finite molecular weights.

Block copolymer surfactancy using polyalkyl-based non-ionic surfactants have been investigated over several decades, these macromolecular surfactant compounds form mostly non-equilibrium states in water and/or oil due to the extremely low critical micelle concentrations (CMC's), which are at least several orders lower than the CMC's of traditional non-ionic surfactants. We investigated the non-ergodic effects by examining two ternary mixtures, each containing the same components, poly(1,2-butadiene-b-ethyelene oxide) (OB), water, and 1,5-cyclooctadiene, but the mixtures were prepared using different protocols. Both systems split into three layers, and the middle layer was evaluated using cryo-scanning electron microscopy (cryo-SEM) and SAXS. Qualitatively different microstructures were found in each system. This difference demonstrates that three-component block copolymer/water/oil mixtures are much more significantly affected by non-equilibrium effects than the analogous block copolymer based two component systems.

Along with above research conclusions, future research plans will be discussed.

(3o) Nanocatalysis From First Principles: From Reaction Mechanisms to New Materials

Jeffrey A. Herron, Chemical & Biological Engineering, University of Wisconsin-Madison, Madison, WI and Manos Mavrikakis, Department of Chemical and Biological Engineering, University of Wisconsin - Madison, Madison, WI

The atomic-scale design of heterogeneous catalysts from fundamental principles has the potential to yield new materials with improved catalytic properties. Using state of the art Density Functional Theory (DFT) methods, we elucidate the detailed reaction mechanisms of selected catalytic reactions on a number of model, monometallic surfaces. Surface reactivity trends from these systematic studies allow for the identification of key reactivity descriptors, which are then utilized for designing bimetallic and ternary alloy catalysts with improved catalytic properties for specifically targeted chemical reactions. As the atomic-scale composition and architecture of catalytic nanoparticles plays a vital role in determining their catalytic properties, these fundamental studies provide invaluable insight into the structure sensitivity of reactions, which can serve as a guide to catalyst synthesis. Here, we discuss the application of these techniques towards a number of industrially important reactions. Topics include hydrogen production and purification, as well as electrocatalysis for fuel cell applications.

(3p) New Biocatalysts for the Production of Renewable Chemicals: Bioinformatics and Molecular Simulation Studies On Fatty Acid Synthesis Enzymes

David C. Cantu, Chemical and Biological Engineering, Iowa State University, Ames, IA

Being capable of designing enzymes to produce compounds not found in nature, or existing ones in greater quantities, is a desired goal in bioengineering. Computational molecular simulation and bioinformatics techniques can be used to study biological systems towards that end, with applications in fuels and specialty chemicals. Further, these techniques may also be directed toward designing new drug molecules and protein therapeutics. Work done for my doctoral dissertation is presented in this poster. Fatty acid and polyketide synthesis enzymes are the focus of our research as part of the NSF Engineering Research Center for Biorenewable Chemicals (CBiRC). Working with my major professor, Peter Reilly, my specific project is to find or design enzymes capable of catalyzing desired reactions. Toward that goal, we have classified enzymes into families through sequence analysis, and we have built a database containing known sequences and structures (www.enzyme.cbirc.iastate.edu). In a collaborative effort, thioesterases were studied experimentally based on our statistical and phylogenetic analyses. We recently conducted normal vibrational mode analysis and predicted three-dimensional structures of acyl carrier proteins. We are currently working on mixed quantum mechanics and molecular mechanics (QM/MM) simulations to find thioesterase reaction mechanisms and transition states.

(3q) Building Hierarchical Catalysts with Ultra-Small Units

Xueyi Zhang and Michael Tsapatsis, Chemical Engineering and Materials Science, University of Minnesota, Minneapolis, MN

Zeolites are aluminosilicates with ordered micropores, where the size of the micropores (0.5-2 nm) are similar to industrial molecules. It has long been recognized that in the processing of certain petrochemical products, fine chemicals, and biomass, the ability to combine mesoporosity with the zeolitic microporosity of conventional zeolites could provide new opportunities to fine tune reaction rates, selectivity and resistance to deactivation. Hierarchical zeolties are a new class of zeolites with precisely defined and often ordered mesoporosity. In addition to hierarchical pore architecture, ultra-small (down to the size of a unit cell) building units can lead to catalysts with higher specific surface area, vastly extending the degree of manipulation of their surface catalytic properties.

My research involves the synthesis of hierarchical zeolite catalysts with diffusion lengths at various scales. Hard and soft templates were involved in the preparation of hierarchical zeolites, such as 3DOm-i zeolites [1] and supported lamellar zeolites [2], where the diffusion length can be tuned from 20 nm to 1.6 nm. In addition to the synthesis of these materials, characterization techniques (such as electron microscopy, X-ray diffraction and scattering) suitable for the small length scales are also involved in my research. These techniques, along with state-of-the-art modeling methods, enable comprehensive understanding of the structure and properties of these materials [3]. Repetitive branching strategy originally used in the preparation of semiconductor tetrapods was also applied to zeolites for the first time, enabling the direct-synthesis of self-pillared pentasil (MFI/MEL) zeolite nanosheets with 2 nm-thick MFI lamellae, where the diffusion length of zeolite catalysts was further reduced to 1 nm [4]. Preliminary results show that this self-pillared material enables higher reaction rates for bulky molecules compared to those of other hierarchical and conventional MFI zeolites [4].

In my future research, the repetitive branching strategy is expected to be applied to other catalysts. For example, other intergrown zeolite structures can be prepared in a self-pillared pattern to allow catalytic reactions of bulky molecules over Brønsted and Lewis acid sites. Moreover, metal and metal oxides that show twinning growth pattern can be prepared for hydrogenation reactions, photocatalysis and base catalysis.

[1] P.-S. Lee, X. Zhang, J.A. Stoeger, A. Malek, W. Fan, S. Kumar, W.C. Yoo, S. Al Hashimi, R.L. Penn, A. Stein, M. Tsapatsis, Sub-40 nm zeolite suspensions via disassembly of three-dimensionally ordered mesoporous-imprinted silicalite-1, J. Am. Chem. Soc., 133 (2011) 493-502.

[2] X. Zhang, M. Tsapatsis, Mesoporous silica nanoparticles from a clear sol and their transformation to lamellar silicalite-1 particles and films, Microporous and Mesoporous Materials, 138 (2011) 239-242.

[3] K. Varoon, X. Zhang, B. Elyassi, D.D. Brewer, M. Gettel, S. Kumar, J.A. Lee, S. Maheshwari, A. Mittal, C.-Y. Sung, M. Cococcioni, L.F. Francis, A.V. McCormick, K.A. Mkhoyan, M. Tsapatsis, Dispersible exfoliated zeolite nanosheets and their application as a selective membrane, Science, 333 (2011) 72-75.

[4] X. Zhang, D. Liu, D. Xu, S. Asahina, K. A. Cychosz, K. V. Agrawal, Y. Al Wahedi, A. Bhan, S. Al Hashimi, O. Terasaki, M. Thommes, M. Tsapatsis. Synthesis of self-pillared zeolite nanosheets by repetitive branching. To be presented at AIChE Annual Meeting, Pittsburgh, PA (2012).

(3r) Directing Molecular Assembly At Interfaces for Pharmaceutical, Electronic and Energy Applications

Ying Diao[1,2], Allan S. Myerson[3], T. Alan Hatton[3], Bernhardt L. Trout[3], Stefan Mannsfeld[2] and Zhenan Bao[1],
(1)Chemical Engineering, Stanford University, Stanford, CA, (2)SLAC National Accelerator Laboratory, Menlo Park, CA, (3)Chemical Engineering, Massachusetts Institute of Technology, Cambridge, MA

Molecular self-assembly processes, such as nucleation, crystal growth, aggregation, nano-/micro- phase separation, have a profound impact on the solid-state properties of materials. For instances, difference in crystal packing (polymorphism) influences bioavailability, stability and processibility of pharmaceutical compounds; morphology, molecular packing and orientation of organic semiconductors are critical in determining their charge transport characteristics; controlling the nanoscopic phase separation is key to achieving high-efficiency organic solar cells.

In most practical circumstances, these molecular assembly events occur at interfaces, especially solid-liquid interfaces during solution processing. By engineering the nano-topology, microstructure and chemical patterns of surfaces and interfaces, my research has established new methodologies and demonstrated unprecedented control over nucleation kinetics, polymorphism and crystal morphology. Furthermore, I investigated the role of nanoconfinement on nucleation and employed it as a powerful tool for directing the crystal packing of organic crystals at interfaces. With the new physical insight obtained and the novel approaches developed, I aspire to carry this research further towards 1) developing new technology platforms for fabricating next-generation pharmaceutical and energy materials, 2) deciphering the mechanism of molecular assembly processes at interfaces.

For further information, please visit my website at http://www.stanford.edu/~diao

(3s) Characterization of Novel Inhalable Dry Powder Particles Containing Paclitaxel and the Development of Three-Dimensional Multicellular Spheroids for the Treatment and Study of Lung Cancer

Samantha A. Meenach, Pharmaceutical Sciences - Drug Development Division, University of Kentucky, Lexington, KY; Chemical and Materials Engineering, University of Kentucky, Lexington, KY

Despite the significant advances in the treatment of lung cancer, it is a disease which still signifies poor prognosis and challenges in implementation of treatment. Targeted pulmonary inhalation drug delivery offers many advantages for lung cancer patients in comparison to conventional systemic chemotherapy. These include: the potential to deliver local therapeutically effective concentrations of drug directly to the lung, minimized side effects due to limited systemic delivery, and ease of use for the patient. Inhalable dry powder formulations of nanoparticles and microparticles containing a chemotherapeutic are advantageous in their ability to deliver drug deep in the lung via optimally sized particles, higher local drug dose delivery, and long-term storage capability. In this work, novel advanced spray-dried inhalable PEGylated phospholipid microparticle/nanoparticle powders containing the chemotherapeutic paclitaxel were successfully designed and produced via dilute organic solution spray drying under various conditions. Fixed ratios of dipalmatoylphosphatidylcholine (DPPC) and dipalmatoylphosphatidylethanolamine poly(ethylene glycol) (DPPE-PEG) were mixed with various ratios of paclitaxel in a dilute methanol solution. Upon optimization of the spray drying conditions (e.g. pump rate), the physicochemical characterization of the particles was completed. Scanning electron microscopy (SEM) images showed the spherical particle morphology of the inhalable particles. The size of the particles was statistically analyzed using these images SigmaScan software and determined to be 600 nm – 1.2 μm in diameter, which is optimal for deep lung alveolar penetration. Differential scanning calorimetry (DSC) and x-ray powder diffraction (XRPD) were performed to analyze solid-state transitions and long-range molecular order, respectively, and allowed for the confirmation of the presence of phospholipid bilayers and/or paclitaxel and their physical states. The amount of paclitaxel loaded into the particles was analyzed via high performance liquid chromatography (HPLC) and their aerosol performance was evaluated using a Next Generation Impactor (NGI) to determine the fine particle fraction and aerodynamic diameter. These results demonstrate this novel therapeutic platform as one capable of effectively delivering paclitaxel directly to the lung for the treatment of lung cancer.

As for all cancer but particularly for lung cancer, applying *in vitro* outcomes to *in vivo* applications has limitations because conventional two-dimensional (2D) cell culture does not recreate a physiologically representative model for cells. We investigated a three-dimensional (3D) cell culture technique to model lung tumors *in vitro*. A 3D lung cancer model was created by applying collagen (a semi-non-adhesive material) to a cell culture Transwell, which allows for nutrient transfer through the collagen. Two lung cancer cells lines (H358, a bronchioalveolar carcinoma and A549, a lung adenocarcinoma) were seeded on the collagen. The non-adhesive collagen caused the cells to be more inclined to attach to one another rather than its surface allowing for the formation of multicellular spheroids (MCS). To better mimic the environment for lung cancer specifically, an air-interface culture (AIC) condition was created. For AIC conditions, the cell media on the apical side of the Transwell was removed and the basolateral side with media provided effective nutrient transport to the MCS while still exposing the cells to air. The AIC model yielded viable MCS at sizes similar to MCS formed in LCC conditions (100 to 150 μm in diameter). With the optimized 3D model, LCC cells were exposed to paclitaxel in media, and paclitaxel-loaded dry powder aerosol particles described above were delivered to AIC cells through direct application with an insufflator. Using viability analysis, it was shown that both applications of paclitaxel showed variance in efficacy when comparing 2D and 3D culture conditions (where the IC_{50} values for paclitaxel were higher for 3D compared to 2D). Transepithelial electrical resistance of A549 cell monolayers was evaluated before and after particle delivery to illustrate that the particle application does not affect the permeability of the cells, which infers that this form of drug therapy will not affect the permeability of lung tissue. Overall, a much more representative *in vitro* model has been developed that is expected to be an improved predictor of efficacy of alternative drug delivery methods such as direct pulmonary delivery for lung cancer, which could lead to more efficient drug therapies for lung cancer patients.

(3t) Application and Mechanism Understanding of Nano-Structured Catalyst in Biofuel Production

Cun Wen, Department of Chemical Engineering, University of South Carolina, Columbia, SC

APPLICATION AND MECHANISM UNDERSTANDING OF NANO-STRUCTURED CATALYSTS IN BIOFUEL PRODUCTION

Cun Wen, University of South Carolina, Columbia, SC

Department of Energy has set the goal of biofuel production to 36 billion gallons by 2022, while the present production rated at 1.1 billion gallons at 2011. The capacity needs more than 3200% increase in total or 40% increase in annual for the next 10 years. One of the limiting factors in biofuel production is the low throughput of the batch reaction mode. Though it has been long known that heterogeneous processes can accelerate the biofuel production, the actual progress is hindered by poor understanding of reaction mechanisms and requires sophisticated material design and synthesis. The reaction mechanism research is intricate with numerous potential reaction pathways, similar to the possible routes to overcome Himalayas. Our strategy to tackle this issue is to combine the mechanism research with kinetics studies, which serves as loyal contour-mapper and tells us the most feasible way. This strategy has been proven to be effective.[1-9] Herein, I bring in nano-structured heterogeneous catalysts with our strength in advanced nanomaterial synthesis and fundamental reaction-mechanism studies to expedite the biofuel production.

To upgrade biomass to biofuels, the aromatic rings and the C-O bonds need to be broken down and C-C/C-H bonds need to be build up, on which stage heterogeneous catalysts can show off. Firstly, 2-methyfuran acid catalysis reaction is chosen as a model reaction to exemplify our idea of heterogenerizing biofuel production. The 2-methylfuran is a mode molecular representing biomass feed stocks, which usually need acid catalysts for deoxygenating because of oxygen as heteroatom and need chain growth for generating biofuel because of low carbon numbers (<C6). Usually concentrated sulfuric acid is used as acid catalyst to serve the purpose, but it is unfavorable for large scale industrial production. Our results show that the zeolite catalyst can replace the concentrated sulfuric acid with even better production selectivity in terms of cetane number. Furthermore, the cetane number of the product can be increased by decrease the acid strength of the zeolites. To understand the relationship between acid strength and product cetane number, the reaction mechanism on the zeolite catalyst is probed by *in situ* techniques, such as FTIR. The reaction mechanism study reveals that high acid strength increases the adsorption energy, and thus prolongs the resident time of the intermediates on the zeolite. Because longer resident time facilitates the C-C chain growth, the branched products with poor cetane number are favored on zeolites with high acid strength.

Secondly, the products from 2-methfuran acid catalysis reaction need further hydrogenation to upgrade to final biodiesel, and Co is reported to be active for the hydrogenation reaction, whose activity can be further promoted by Cu doping. Thus, bimetallic core-shell nanoparticle Cu@Co with alloy of CoCu on the shell (for high activity) and Cu as the core (for low catalyst cost) would be optimum for this reaction. Traditionally, the core-shell bimetallic nanoparticles are synthesized in two steps. In the first step, the core is formed with one metal, and then in the second step the core is coated with second metal as the shell. Thus, the two-step method not only is poor in economic, but also yields mono-metal instead of alloy on the surface, while alloy is needed for high activity in this hydrogenation reaction. Herein, I demonstrated a one-pot methodology for mono-dispersed Cu@Co Core-shell nanoparticle synthesis with alloy of CoCu on the shell as catalyst for hydrogenation reaction. Furthermore, I can not only synthesis the nanoparticle in one-step, but also keep a tight control over the size distribution (relative standard deviation 7.9%).[10]

Thirdly, biomass upgrading to biofuel is not limited to the acid and hydrogenation reaction, and other process can also serve the purpose, such as partial oxidation. The partial oxidation has been proved to be an effective way to produce biofuels from biomass aligned with Fischer-Tropsch synthesis. However, the long-standing problem in partial oxidation is the hot spot formation in the catalyst bed, which can lead to the catalyst deactivation or even explosion. It has been known that the hot spot is formed by the side reaction that is the total oxidation taking place beside the partial oxidation. The key factors guiding the reaction path through total or partial oxidation is unknown, which hinders the catalyst development. To address this issue, I looked into the mechanism of partial oxidation production of syngas ($CO+H_2$). Our research has revealed that the lattice oxygen reducibility of the catalyst is the key factors governing the reaction mechanism. With high lattice oxygen reducibility, total oxidation rather than the partial oxidation is favorite. By tuning the lattice oxygen reducibility of Er_2O_3 with Mn doping, I successfully shift the reaction mechanism from total oxidation to partial oxidation, and eliminated the hot-spot formation.

Read Me More:

1. Wen C, Liu Y, Guo Y, Wang YQ, Lu GZ. Strategy to eliminate catalyst hot-spots in the partial oxidation of methane: enhancing its activity for direct hydrogen production by reducing the reactivity of lattice oxygen. *Chem Commun.* 2010;46(6):880-882.

2. Wen C, Liu Y, Guo Y, Wang Y, Lu G. Synthesis of the rare earth compound nanosheets induced by lamellar liquid crystal. *Solid State Sciences.* 2009;11(11):1985-1991.

3. Wen C, Liu Y, Tao F. Integration of surface science, nanoscience, and catalysis. *Pure Appl Chem.* Jan 2011;83(1):243-252.

4. Liu Y, Wen C, Guo Y, Lu GZ, Wang YQ. Modulated CO Oxidation Activity of M-Doped Ceria (M = Cu, Ti, Zr, and Tb): Role of the Pauling Electronegativity of M. *J Phys Chem C.* 2010;114(21):9889-9897.

5. Liu Y, Wen C, Guo Y, Lu GZ, Wang YQ. Effects of surface area and oxygen vacancies on ceria in CO oxidation: Differences and relationships. *J Mol Catal a-Chem.* Feb 2010;316(1-2):59-64.

6. Liu Y, Wen C, Guo Y, et al. Mechanism of CO Disproportionation on Reduced Ceria. *ChemCatChem.* 2010;2(3):336-341.

7. Media highlights: Pichon A. Catalysis: Lower reactivity means more hydrogen. 2010; http://www.nature.com/nchina/2010/100120/full/nchina.2010.7.html.

8. Media highlights: Saxton C. Enhancing catalytic activity. 2010; http://www.rsc.org/Publishing/ChemTech/Volume/2010/02/enhancing_catalytic.asp.

9. Media highlights: Chinese New Year. 2010; http://www.rsc.org/Publishing/Journals/cc/News/2010/Chinese_New_Year.asp.

10. Wen C, Hattrick-Simpers J, Lauterbach J, Inventors. One-step synthesis of monodisperse transition metal core-shell nanoparticles with solid solution shells, US provisional patent: 61/686,288, 2012.

(3u) Advanced Electro-Catalysts for Energy and Biomass Refinery

Zhiyong Zhang, Chemical Engineering, Michigan Technological University, Houghton, MI

Advanced Electro-Catalysts for Energy and Biomass Refinery

Zhiyong Zhang, Chemical Engineering, Michigan Technological University, MI

Nanostructured transitional metals have attracted enormous attention for catalysis applications, due to their fascinating electronic properties and unique geographic structures. However, it remains a challenging task to construct nanostructured metallic materials serving as efficient electro-catalysts for chemical - electrical energy conversion and biomass refinery, because of the technical difficulty in precisely controlling their size, shape and composition. My Ph. D research work focuses on developing a solution phase synthesis route to preparing advanced electro-catalysts, including 0-dimentional (0-D) Pt,[1, 2] Pd,[2, 3] Au,[4, 5] Ag, Cu, Ni, Pd_xNi_y nanoparticles (NPs),[3] 1-dimentional Pt_xFe_y nanowires (NWs),[6] and 2-dimentional Pd_xFe_y nanoleaves (NLs),[7] and investigate their electro-catalytic functions. By tailoring their size, structure and composition, Pt_xFe_y-NWs and Pd_xFe_y-NLs have demonstrated high catalytic activities and durabilities towards oxygen reduction reaction (ORR) in both acid and alkaline electrolytes;[6, 7] Pd_xNi_y-NPs have exhibited high activity towards ethanol electro-oxidation,[3] and Pt-NPs and Au-NPs have displayed distinct selectivity towards in electro-catalytic refinery of glycerol - a cheap by-product from biodiesel production, into valuable chemicals, including glyceric acid, tartronic acid, mesoxalic acid, and glycolic acid, etc.[1, 5]

References:

[1] Z. Zhang, L. Xin, W. Li, Electrocatalytic oxidation of glycerol on Pt/C in anion-exchange membrane fuel cell: Cogeneration of electricity and valuable chemicals, Applied Catalysis B: Environmental, 119¨C120 (2012) 40-48.

[2] O. Winjobi, Z.Y. Zhang, C.H. Liang, W.Z. Li, Carbon nanotube supported platinum-palladium nanoparticles for formic acid oxidation, Electrochimica Acta, 55 (2010) 4217-4221.

[3] Z. Zhang, L. Xin, K. Sun, W. Li, Pd¨CNi electrocatalysts for efficient ethanol oxidation reaction in alkaline electrolyte, International Journal of Hydrogen Energy, 36 (2011) 12686-12697.

[4] Z. Zhang, L. Xin, W. Li, Supported gold nanoparticles as anode catalyst for anion-exchange membrane-direct glycerol fuel cell (AEM-DGFC), International Journal of Hydrogen Energy, 37 (2012) 9393-9401.

[5] L. Xin, Z. Zhang, Z. Wang, W. Li, Simultaneous Generation of Mesoxalic Acid and Electricity from Glycerol on Au Anode Catalyst in Anion-Exchange Membrane Fuel Cells, ChemCatChem, (2012). doi: cctc.201200017

[6] Z. Zhang, M. Li, Z. Wu, W. Li, Ultra-thin PtFe-nanowires as durable electrocatalysts for fuel cells, Nanotechnology, 22 (2011), 015602.

[7] Z. Zhang, K. L. More, K. Sun, Z. Wu, W. Li, Preparation and Characterization of PdFe Nanoleaves as Electrocatalysts for Oxygen Reduction Reaction, Chemistry of Materials, 23 (2011) 1570¨C1577

(3v) Biochar Characterization As Part of the Biomass Thermochemical Processing Platform

Catherine E. Brewer, Center for Sustainable Environmental Technologies, Iowa State University, Ames, IA

Slow pyrolysis, fast pyrolysis and gasification are thermochemical processes to produce fuel and chemicals from biomass. The char co-products from these processes have much potential as biochars: sustainably produced biomass charcoals used for amending soils and sequestering carbon. As biochar properties vary significantly with feedstock and reaction conditions, biochar characterization is critical for understanding these variations, for obtaining meaningful data from biochar agronomic studies and for determining the most beneficial and economical application of a given char co-product (fuel, sorbent or soil amendment). This presentation describes the biochar characterization work I have done for my PhD research in chemical engineering and Biorenewable Resources & Technology (BRT) as part of Iowa State University's thermochemical processing research group in collaboration with the Departments of Chemistry and Agronomy. The methods used in this research include proximate analysis, CHNS elemental analysis, BET surface area gas sorption analysis, X-ray fluorescence spectroscopy (XRF), FT infrared spectroscopy with photoacoustic detection (FTIR-PAS), solid-state ^{13}C nuclear magnetic resonance spectroscopy (NMR), cation exchange capacity (CEC), and short-term soil incubations. I will also describe my future research interests in engineering biochar to improve soil water retention in arid agricultural soils, producing bio-oil and biochar from high-lignin feedstocks, decreasing catalyst inhibition in the conversion of sugar from biomass sources into chemicals, and developing mobile pyrolysis systems.

(3w) Chemical Product Design Using Chemometric Technique in Property Cluster Space

Subin Hada and Mario Richard Eden, Department of Chemical Engineering, Auburn University, Auburn, AL

Recently, significant research in the chemical industries are focused efficient design/synthesis of new product, this is, the search for the most appropriate new materials or mixture of materials with specifically tailored properties that will exhibit and/or cause the desired behavior. Hence, product design is now being considered as an emerging paradigm in the field of chemical engineering because it requires a different set of tools and skill from other problems encountered in chemical engineering. In product design, we do not know the identity of the final product but we have some idea of how we want it to behave. Examples of chemically formulated products include pharmaceuticals, nano-structured, semi-conductors, household products, proteins, fuel mixtures, and many more.

In this work, a logical and systematic approach to solve chemical product design problems by integrating a larger space of knowledge emerging at the interface of fields that traditionally has not been incorporated into chemical engineering science and practice. Interdisciplinary methods and tools that extend through multivariate statistics, applied mathematics and computer science are encompassed. Methodologies and techniques such as characterization based group contribution method, chemometric technique, reverse problem formulation and property clustering techniques are combined with existing Computer Aided Molecular/Mixture Design (CAMD/CAM^{b}D) algorithms to design chemical products in computationally efficient manner that provide optimum performance in terms of customer requirements.

Two types of product design are targeted: mixture design and molecular design. In mixture design, the property based framework is combined with multivariate statistical technique and applied in a reverse problem formulation on chemical product design problems by systematic and insightful use of past data describing the properties of the raw materials, their blend ratios, and the process conditions during the production of a range of product grades to achieve new and improved products. Projection methods, like principal component analysis (PCA) and partial least squares (PLS) are applied to identify the underlying relationships necessary for simultaneous optimization of all three variables. The method and concept is illustrated using a polymer blending problem.

In molecular design, a multivariate characterization technique like infrared spectroscopic is utilized to describe the molecular architecture of a set of representative samples followed by decomposition techniques such as principal component analysis to find the underlying latent variable models that describe the molecule's properties. Characterization based group contribution method is applied to design new molecules that were not present in the original training set and that satisfies specific product property requirements. The concept and the solution methodology are demonstrated using an example of biodiesel additive formulation.

(3y) Multifunctional Nanoparticles Based Cancer Nanomedicine

Yun Wu, Nanoscale Science and Engineering Center, The Ohio State University, Columbus, OH

Nanoparticles, as building blocks of nanotechnology, provide a particularly useful platform with extremely wide applications. Among various applications, I am particularly interested in multifunctional nanoparticles based cancer nanomedicine from two aspects:

(1) *In vivo*: multifunctional nanoparticles for cancer imaging, diagnosis and therapy. I have combined nanoparticle formulation design, microfluidic synthesis method development with comprehensive bio-evaluation to produce multifunctional nanoparticles with uniform size, structure and composition and high theranostic efficacy. During the poster session, I will highlight both systemic delivery and pulmonary delivery of multifunctional nanoparticles in lung cancer treatment.

(2) *In vitro*: nanoparticles based liquid biopsy device for cancer early diagnosis and surveillance. Although significant progress has been made in the diagnosis, treatment and cure of cancers that originate in human blood (i.e. leukemia), minimal progress has been made for solid tumors. Factors contributing to this disappointment include late presentation of disease, lack of biomarkers for early detection, and molecular heterogeneity. A patient-friendly early detection and surveillance method would substantially reduce the mortality in this serious disease. As a non-invasive method, 'liquid biopsy', which relies on the capture and detection of either circulating tumor cells (CTC) or circulating exosomes/microvesicles in patient blood samples and other body fluids, has shown great potential to achieve this goal. Numerous efforts have been spent in this area, but none are able to simultaneously capture and detect CTC or circulating exosomes/microvesicles on a single device. I have developed a novel and simple tethered lipoplex nanoparticles (tLN) device that offers simultaneous cell sorting and intracellular biomarker detection of both CTC and circulating exosomes/microvesicles from the blood of cancer patients. This facile technique holds great promise in our fight against cancer and other deadly diseases.

Future challenges and my research plans will be also discussed in the poster session.

(3z) Engineering 3D Microenvironments for Neural Tissue Engineering: Directing Survival, Differentiation, and Neurite Growth

Kyle J. Lampe, Materials Science and Engineering, Stanford University, Stanford, CA

Neural regeneration within the central nervous system (CNS) is a critical unmet challenge as brain and CNS disorders continue to be the leading cause of disability nationwide. Three-dimensional (3D) tissue engineering strategies present an innovative therapeutic approach for the delivery of cells, growth factors, and drugs to CNS tissue to restore lost function. Regeneration in the CNS is limited: inhibitory reactive oxygen species result in secondary insult after injury, and signal-transmitting neurons are not capable of proliferation to replace lost tissue. Engineering microenvironments conducive to neuronal cell growth in vitro and regeneration in vivo requires control over known design parameters affecting neural stem cell (NSC) viability, proliferation, and differentiation into functional neurons.

In my thesis research, I used a poly(ethylene glycol) (PEG) poly(lactic acid) (PLA) block copolymer to create 3D hydrogels with tunable stiffness and degradability. Through systematic tailoring of the hydrogel mechanics in both 2D and more physiologically relevant 3D environments, I found that the differentiation of NSCs down the desired neuronal pathway was enhanced within biomaterials of stiffness comparable to that of native tissue. I also discovered that modulating the hydrogel degradation kinetics enabled direct control over the release of lactic acid, a reactive oxygen species scavenger. As a result, these PEG-PLA copolymers directly affected NSC intracellular redox state and promoted cell survival. To demonstrate therapeutic potential, I showed for the first time that degradable PEG-based hydrogels induce a minimal inflammatory response in the brain and can be used to temporally and spatially control the delivery of neurotrophic factors to local brain tissue, a critical step toward directing growth of transplanted tissue within the brain.

As a postdoctoral scholar I have further developed my ability to rationally design new materials by transitioning to the use of protein-engineered biomaterials for dynamic modulation of neuronal fate and *in vivo* delivery of cells. These protein-engineered materials allow careful independent tuning of several functional properties including cell-binding site density, cell-mediated degradation, and mechanical stiffness. With this platform, I have investigated the control of cell-matrix interactions using 3D primary neuronal cultures. These studies have enabled tailoring of the material properties in 3D such that neurons survive and extend neurites over thousands of microns into the surrounding matrix.

I intend to leverage my expertise at the intersection of biomaterials design, stem cell biology, drug delivery transport, and reactive oxygen species kinetics to build a nationally-recognized lab with a multidisciplinary, 3D tissue engineering approach. While keeping an eye on feasibility for clinical treatment strategies, I will focus on applications for neurodegenerative diseases and injuries of the central nervous system. In this role I hope to educate a new generation of diverse young engineers and cultivate their interest in chemical and biomolecular engineering.

(20a) Investigating Belonging for STEM Students

Tamara Floyd-Smith, Chemical Engineering, Tuskegee University, Tuskegee, AL

A study is currently being conducted to investigate belonging for STEM students and its relationship to academic engagement, academic performance and other related factors. The participating institutions are a Research 1 university, a large teaching university, a small private faith-based university, a small private historically black college/university (HBCU) and a small private women's college. The institutions were chosen to represent a range of academic and social environments. Students at the five institutions completed a survey and a subset of students participated in interviews or focus groups. The surveys included Likert scale items on belonging at the class, major and university levels. The survey also included some open ended questions related to belonging. Thus far, the data suggest that outside communities significantly influence student academic engagement and, generally speaking, students' sense of belonging does not increase monotonically by year in school. The results of this study will be presented in poster format.

(20b) An Innovative Multidisciplinary Sustainable Student Design Project to Develop Integrated Biodiesel and Biochar Technology for Sub-Saharan Africa

Jeffrey Seay, Chemical and Materials Engineering, University of Kentucky, Lexington, KY

Developing countries in tropical regions are often rich in biological resources, such as biomass and oil seeds, however, extreme poverty, lack of an educated populace and an absence of a reliable manufacturing infrastructure mean that these resources go largely untapped. Although the vegetable oil needed to produce fuels like biodiesel is available in tropical regions in abundant supply, additional raw materials such as methanol or ethanol and a base catalyst such as sodium hydroxide or potassium hydroxide are also required. This presents a special challenge since these raw materials are typically unavailable to rural people in the tropics. Therefore the motivation for this project draws from the need to develop affordable technology for producing biodiesel as well as all the required raw materials using locally available feed stocks and construction materials.

This contribution will describe the organization and development of a multidisciplinary student design project involving the design, construction and implementation of a novel apparatus to produce biodiesel and biochar from locally available resources in sub-Saharan Africa. This project involved students from Chemical and Mechanical Engineering and included both Senior Capstone Design students as well as underclassmen. The project was implemented through the newly implemented Embedded Study Abroad Program model at the University of Kentucky. This program allows a travel component to be embedded into a traditional core class or elective. This is significantly different from the traditional study abroad model, which typically includes a semester of international travel. For this student design project, a traditional independent study elective was combined with the embedded study abroad model to create a unique design experience for the students.

This project was conceived through a partnership with an NGO called the African Center for Renewable Energy and Sustainable Technology (ACREST), an NGO located in Bangang, Cameroon (www.acrest.org). ACREST operates a small technology center where they work to design and adapt renewable energy and sustainable technologies for the use of rural villagers in Cameroon and throughout West Africa. Recently ACREST has entered into partnerships with U.S. universities, whose students work to develop technologies as part of internships and capstone design projects. The primary objective of the project with the University of Kentucky, Paducah Campus is to develop technologies to allow rural villagers in Cameroon to manufacture biodiesel from locally available vegetable oils such as palm oil or jatropha oil, and biochar from field crop residues and waste biomass.

Seven students and their faculty advisor have recently traveled to Cameroon to implement this unique multidisciplinary project. This contribution will describe the organization of the project and the results achieved by the students in the field. In additional, student perspectives from the field will be included in the presentation.

(20c) Bringing Chemical Engineering to the Masses Through a Fuel Cell Car Demo and Competition

Robert G. Bozic, Chemistry and Life Science, United States Military Academy, West Point, NY

The first West Point Fuel Cell Car Demonstration and Chemical Engineering Car competition was planned, developed, and executed in order to inspire chemical engineering majors, prospective students, and parents about chemical engineering and the West Point American Institute of Chemical Engineers (AIChE) ChemE Car team. The inspiration for this event was based on the desire for an easily transportable hands-on chemical engineering demonstration for West Point Chemical Engineering Faculty and Cadets. The purpose was to enable involvement in a competitive chemical engineering experience built on the foundations of the AIChE ChemE Car competition without becoming overly burdensome due to logistical and safely requirements. To meet the logistical and safety challenge, seven of the Thames and Kosmos X7 fuel cell cars affectionately named after the Army Mules (Spartacus, Buckshot, Ranger, Raider, Trooper, Trotter, and Black Jack) were used. Over 150 cadets, visitors, and guests were introduced to the competition concept and encouraged to participate at increasing levels of complexity. Two levels of competition were developed, "Speed Racer" and "Chemical Engineer". The "Speed Racer" level included chemical safety, basic principles of the mass and energy balance, and emphasis on the chemical engineering nature of the competition. The "Chemical Engineer" level emphasized the basics as well as error analysis, computer aided design using Chem CAD, efficiency calculations, and a more deliberate approach to understanding effects of error on the equations used to estimate the volume of hydrogen gas required to travel the desired distance.

(20d) Game-Based Learning for Fostering the Weak

Marcel A. Liauw, ITMC, RWTH Aachen University, Aachen, Germany

Game-based learning (GBL) in general, and digital GBL in particular, are increasingly being studied. The generation of so-called Digital Natives and the paradigm change in learning and interacting are the background for this. The promises are clear: by self-directed, immersive, problem-based learning in specially designed computer games, contents may not only be learnt easier, but retention may also increase significantly.

The focus of the present paper is how this approach may benefit differently strong candidates. In an extensive literature survey, a handful of references can be found that permit to gauge how strong and weak groups of peers benefit from GBL. Even though those studies may not have been designed as such, and hence lack some rigor, one trend is common and most noteworthy: it seems that weak students benefit more from this new approach than the strong ones. A number of reasons for this observations are discussed and setups are proposed to generate reliable data on this hypothesis.

The work was supported by an ETS award that was financed by the Stifterverband in the Wettbewerb Exzellente Lehre.

(20e) Influence of Simplifying Assumptions On Solubility Studies in Polymers

Marcel A. Liauw, Sarah Jones-Magnor and Lukas Voemel, ITMC, RWTH Aachen University, Aachen, Germany

Reaction engineering calculations are often based on estimations with quite a few underlying simplifying assumptions. Students may know this and still not realize the corresponding impact. If, for example, virtually all simplifications move the estimation result away from the true value *in the same direction*, the systematic error may be all from severe to plain prohibitive.

In an advanced student assignment, the solubility of gases in polymers is studied. Knowing it is crucial for proper understanding of many multiphase polymer systems. When evaluating raw data from proper experiments, the first estimate that is extracted does indeed rely on numerous assumptions as described above. The ideal gas law is assumed to be a satisfactory approximation, as are the isothermicity of the system, stationarity of the final state, or the polymer having a constant volume and no swelling. After this first estimate, the simplifications made are replaced step by step by more sophisticated - and realistic - approaches. For a given model system, the dissolution of carbon dioxide in a polymer, the impact of the simplifications on the calculated solubility data is demonstrated. A critical assessment of literature data proves helpful in gauging the implications on subsequent modeling. Furthermore, the students can deduce what experimental setup is most (or least) suitable to generate high-quality raw data.

The work was supported by the BMBF in the Dream Production joint project, which is gratefully acknowledged.

(20g) Improvements in Computer Methods Courses in Chemical Engineering

Victoria Goodrich, University of Notre Dame, Notre Dame, IN, Joshua A. Enszer, Chemical, Biochemical, & Environmental Engineering, University of Maryland Baltimore County, Baltimore, MD and Rachel B. Getman, Chemical and Biomolecular Engineering, Clemson University

As more core courses in the undergraduate curriculum require significant ability using computer skills, we see a need for improved methods of instruction in computer methods courses required in the chemical engineering curriculum. It is important to provide students with a series of approaches and activities that ensure (1) that students are applying algorithmic thinking and not just learning how to operate in a single computing environment, (2) that students are able to develop their abilities to formulate problems in a computational context, and (3) that students are applying numerical algorithms in meaningful ways and not just following a template. To that end, we discuss here overviews of our computational methods courses and implementation that encourages behavior independent of choice of computer software.

We also share here evaluation of student abilities and perceived importance of computer methods in their future careers to assess the effectiveness of the courses. We share results of a self-assessment survey administered at both the start and end of the Spring 2012 semester.

A THREE STEP APPROACH FOR CHARACTERIZATION OF NON – IDEAL FLOWS IN CHEMICAL REACTORS

Shilpa Mahamulkar, Anurag Kumar, Abhinav Achreja, Nirup Kumar, Preeti Aghalayam
Indian Institute of Technology, Madras, Chennai, India

Introduction

The concept of non-ideal flow in reactors is taught in the Under – Graduate (UG) curriculum at IIT Madras via theory lectures and laboratory tracer experiments. The difference in the Residence Time Distribution (RTD) profile measured in the lab, and the ideal RTD profiles for CSTRs and PFRs, is used to illustrate the effect of 'real' or non-ideal flow in reactors, on RTD. In this work, we propose a different methodology for teaching the concepts of non-idealities in flow reactors, and residence time distributions. The focus here is on providing a more visual basis for the concepts, thus enhancing students' understanding, and comfort levels.

Keywords: Residence Time Distribution, Computational Fluid Dynamics, Bypass, Dead zone, Re-circulation.

Methodology

First, it is proposed that a visual tracer material be used in the laboratory reactor, and snapshots of the movement of the tracer through the reactor be taken at several instants of time, at a host of nominal operating conditions. Next, the routine tracer experiment involving the measurement of outlet tracer concentrations, and evaluation of the RTD profile from the same, is to be performed. The non-idealities such as bypassing, dead volumes, re-circulating zones [1, 2], etc. can be observed in the snapshots, whereas the impact of these on RTD profiles can be directly analyzed via the second experiment.

Next, it is proposed that Computational Fluid Dynamics (CFD) simulations of the same reactor be performed by the students. The steady-state CFD simulations will provide rich information regarding velocity contours all over the reactor while the unsteady simulations give the species contour information at various instants of time [3]. This can be linked to the visual tracer snapshots, for validation, and also, for further understanding. A 'numerical' tracer experiment may then be performed in CFD, using established procedures, to obtain a CFD-predicted RTD profile. This is easily validated against the experiments above.

Analysis

With these four pieces of information, it will be very easy to illustrate concepts to the students. In the past year, we have used this methodology and analyzed a stirred flow reactor of dimensions 21 cm height and 16cm in diameter that is in the Chemical Engineering Lab here. The specific operating conditions, at which bypassing and dead zones (among other non-idealities) are clearly visible, were identified. This was used to illustrate the connection between RTD and non-idealities, very successfully.

In the third year UG class on Chemical Reaction Engineering, the method was demonstrated to the group of 70 students in April 2012. Their performance on tests was found to improve, and their understanding of the concepts deepened after the visual tracer and CFD results were shown to them. In the future, we expect that such CFD simulations should become a routine part of the UG curriculum, due to its increased importance as a design tool. We believe that this simple methodology therefore has the potential to significantly alter student's perspective on reactor modeling, in the future years.

Preliminary results

The dimensions of the cylindrical Continuously Stirred Tank Reactor (CSTR) are: Length (L) = 17cm & Diameter (D) = 16cm. The impeller is at a distance of 5cm from the bottom of the reactor. The dimensions of the impeller blade are: width = 0.2cm, height = 0.5cm & depth = 2.4cm. The radius of the inlet is 2.5cm and is present at a distance of 4cm from the bottom of the tank. The geometry is shown in figure 1. Experimental and CFD studies were carried out on the reactor.

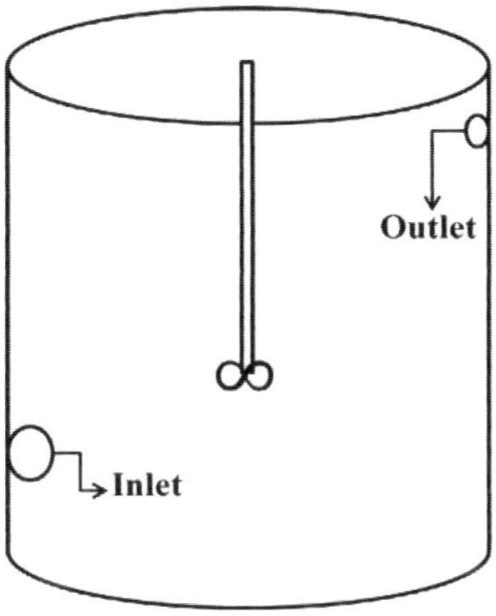

Figure 1: Schematic of the laboratory CSTR

Effect of stationary impeller

To illustrate a case of bypassing, the study in which stationary impeller results in bypassing has been shown here. Figure 2 shows the snapshots taken from the visual tracer experiment at various intervals of time. Figure 3 gives the snapshots of the tracer along a plane containing the inlet and the outlet, at similar instants of time. These snapshots seem to match the visual tracer ones in Figure 2. The virtual RTD carried out using CFD, results in a sharp peak initially signifying bypassing (Figure 4).

Each of these tools helps in analyzing the system better. The visual tracer experiments help us in determining the non- ideality visually without being able to quantify it. The CFD results help in quantifying the non-ideality and determining its exact location in the reactor. CFD is used for optimum designing of chemical reactors. The experimental RTD when compared with the ideal one gives an idea of the deviation from the ideality.

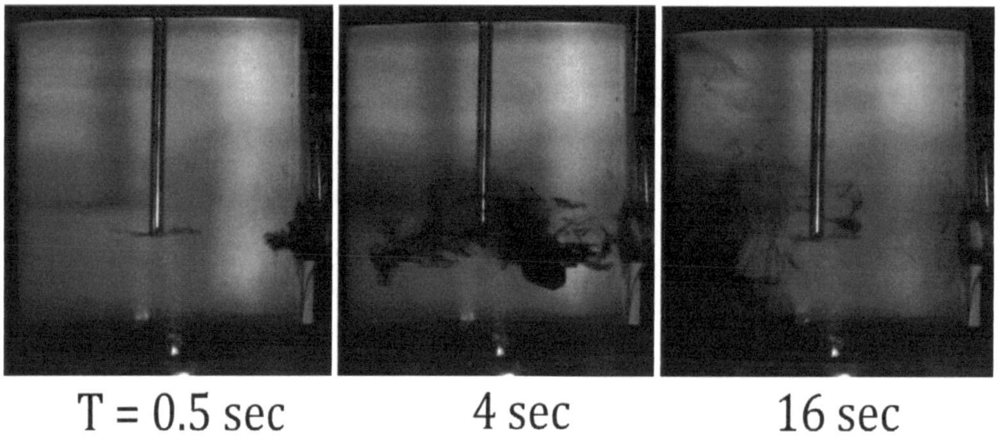

Figure 2: Bypassing seen in visual tracer experiment

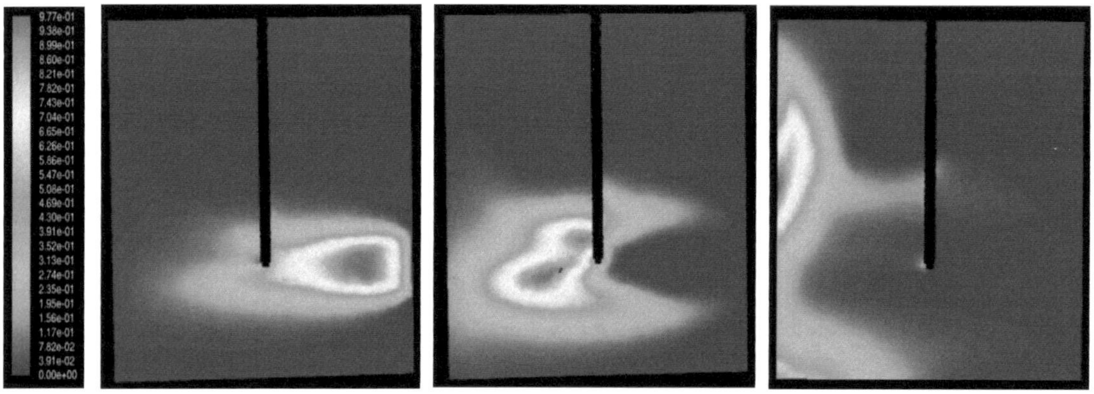

Figure 3: Bypassing seen in CFD simulations

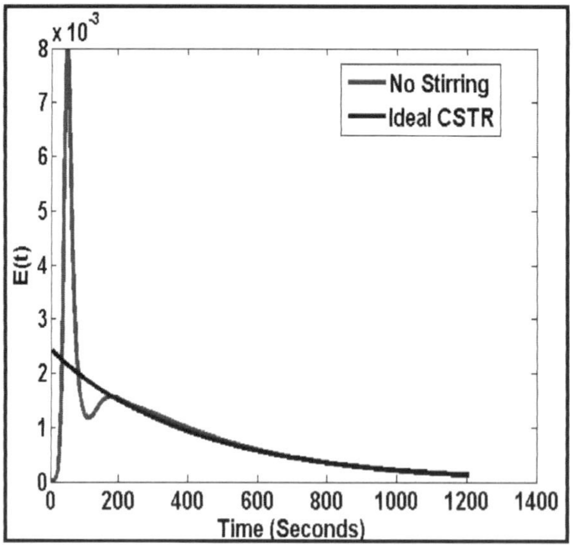

Figure 4: RTD of the reactor showing bypassing

Effect of change in flow rate on the RTD

The flow rates considered for comparison include 300, 500 and 700ccpm. The impeller speed was kept constant at 200rpm. The snapshots were taken at 5s from the time of injection of tracer.

Figure 5: Visual images of the tracer for 300, 500 and 700ccpm respectively

Figure 5 shows that increasing the flow rate leads to faster mixing reducing the mean residence time of the reactor shown in Figure 6.

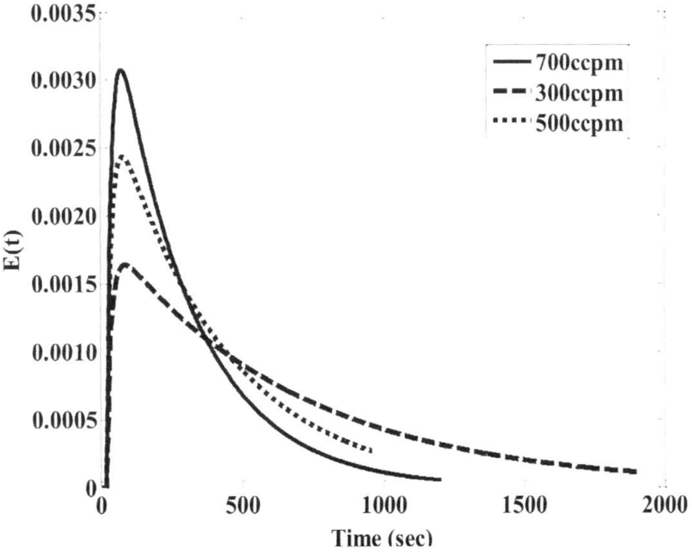

Figure 6: Effect of flow rate in the RTD of the reactor obtained from CFD simulations

As seen from Figure 6, increase in the flow rate does not change the nature of the RTD curve. However, the mean residence time decreases. This RTD curve helps in quantifying the performance of the reactor and comparing with that of an ideal one.

Conclusions

The three steps for characterization of non-idealities in reactors discussed here are: visual tracer experiments, residence time distributions and computational fluid dynamics techniques. The information obtained from each of these steps when put together helps in acquiring a complete picture of the non-ideal nature of the reactor. To explain this methodology, the bypassing effect in a reactor has been explained clearly. The methodology has also been applied to study the effect of change in flow rate on the performance of the reactor.

The method proposed, has been proved to be effective in teaching concepts of non-idealities in reactors.

References

1. Levenspiel O. Chemical Reaction Engineering. New York, NY: Wiley; 1999.
2. Fogler, H.S. Elements of Chemical Reaction Engineering. New Jersey: PHI; 2008.
3. Madeira L. M., Alves M. A., Rodrigues A. E. Teaching non-ideal reactors with CFD tools. Chemical Engineering Education 2004, 38, 154 - 160.

(32a) Characterization of Student Model Development in Physical and Virtual Laboratories

Erick Nefcy, Philip H. Harding and **Milo D. Koretsky**, Chemical, Biological, and Environmental Engineering, Oregon State University, Corvallis, OR

This study characterizes student teams' use of models as they proceed through three laboratory projects in the first quarter of the capstone laboratory sequence in the School of Chemical, Biological, and Environmental Engineering at Oregon State University. Two of the laboratories are physical laboratories, based on the unit processes of heat exchange and ion exchange. Sandwiched between these two laboratories, students undergo a virtual laboratory. The virtual laboratory can be used to simulate complex or expensive tools that are not readily available for use by undergraduate students, but that are more representative of industrial systems. In this study, a virtual chemical vapor deposition laboratory is used. The instructional design of the virtual laboratory project is intended to complement the physical laboratory projects in the curriculum. Students interact with a three-dimensional computer simulation to gather data. In the virtual mode, there is lower cognitive demand required to perform the actual experiments. This aspect affords the students the opportunity to build a rich experimental design based on interpretation and iteration.

Previously, we have reported a graphical method that has been developed, termed model representation and usage maps. These maps characterize student teams' model development as they proceed through a laboratory project. Classifications in the graphical representation of model development include: nature of the model component (quantitative, qualitative, graphical, empirical, statistical), utility of the model component (operationalized, abandoned, not engaged), correctness of the model component, action based on the model component (did it direct the values of input variables for a future run, was a run used to quantify model parameters, was the model qualitatively verified, etc.), and emotional responses to model verification or mismatch. Data sources include laboratory journals, written work products, and, for the virtual laboratory, experimental records available through the instructor interface. In this paper, the model representation and usage maps for 15 teams are developed as they complete both physical and virtual laboratory projects. Analysis of the model representation and usage maps confirms that the virtual laboratory project affords students a richer opportunity for model development, modification, and use of evidence-based reasoning.

(32b) Making a Chemical Process Control Course an Inductive and Deductive Learning Experience

David L. Silverstein, University of Kentucky, Paducah, KY and Gifty Prempeh, Department of Chemical Engineering, North Carolina A&T State University, Greensboro, NC

An inductive approach to teaching chemical engineering courses has been demonstrated to improve student learning in courses such as mass transfer and stoichiometry. One course particularly well-suited to an inductive structure is chemical process control, where experiential learning can also be applied to maximize student learning. This paper discusses the development of inductive laboratory exercises in a three-hour senior-level course at the University of Kentucky Extended Campus Programs in Paducah, Kentucky. Six laboratory experiments were integrated into the course to enable students to make observations for specific cases, draw conclusions, and extend those conclusions to develop relationships for the general case. Three of these experiences were inductive, so that later in lecture periods students developed conclusions reached from observation into general relationships. The other three labs required students to test relationships discussed in class.

The class began with distribution of the syllabus and a visit to the lab to run a pair of experiments in small groups. The purpose was to generate recognition of the key elements for controlling a chemical process so that discussions would be more concrete in the student's minds. Students were required to observe how they manually attempted to control a process, enabling meaningful discussion of proportional control action, sensors, pneumatic signals, actuators, and other key concepts. Over the remainder of the semester, students tested theoretical models, developed empirical models, investigated signal conditioning and calibration, observed the effect of detuning, determined the role in interaction in MIMO systems, and observed a ladder program from a PLC in operation. Student feedback indicated that the exercises substantially improved understanding of process control fundamentals, but the added student time commitment was a significant concern. Students considered the lab exercises more valuable than computer simulations, despite the increased time commitment to run the labs. The potential for adapting this structure within the context of larger classes will also be addressed, along with plans for expanding the role of inductive experiential structure in the course.

(32c) Role of Chemical Engineering in Engineering Education Research

Phillip C. Wankat, School of Chemical Engineering, Purdue University, West Lafayette, IN

Chemical engineers have historically played a very significant role in advancing engineering education; however, since research in engineering education has become much more rigorous in the last dozen years, it is not obvious that chemical engineers will maintain this significant role. There are many specific questions within the five identified engineering education research areas (Epistemologies, Learning Mechanisms, Learning Systems, Diversity & Inclusiveness, and Assessment) that can be best explored by chemical engineering professors. For example, what are the best methods for teaching mass and energy balances. To maintain a leadership role in engineering education research some chemical engineering departments will have to make engineering education a research specialty and offer Ph.D. degrees based on research in engineering education. This will require that a few motivated chemical engineering professors retool their research from a technical specialty to engineering education. In addition, publication venues for professors who are doing engineering education research as a "hobby" need to be maintained.

(32d) Fostering Active Learning and Peer-to-Peer Interactions Among Undergraduates

Michael A Matthews[1], Chris Long[2] and Nancy Thompson[1], (1)Chemical Engineering, University of South Carolina, Columbia, SC, (2)Advanced Controls Integration, GE Energy, Greenville, SC

The benefits of active learning in the traditional classroom setting are well established among engineering educators; however, this learning model can thrive in other settings, namely in a research group. This work presents findings from an educational research project specifically designed to foster active learning among undergraduates and graduate students. Observations and data were collected in the Research Communications Studio (RCS), which uses the research group as a forum for active learning. Distributed cognition is a key concept behind the structure of the RCS. Specific practices that facilitate active participation are described, and a quantitative analysis of group discourse is used to confirm activity levels and to characterize the nature of the group interactions. The methods used to foster active learning can be adapted to a variety of research group settings. This approach has been adapted and institutionalized via several professional development seminar courses, and observations from these courses will also be presented.

(William Corcoran Award paper, 2008.)

(32e) Strategies for Creating and Sustaining A Departmental Culture: Turning Theory Into Action

Lisa G. Bullard, Chemical and Biomolecular Engineering, North Carolina State University, Raleigh, NC, Jason M. Keith, Dave C. Swalm School of Chemical Engineering, Mississippi State University, Mississippi State, MS, David L. Silverstein, University of Kentucky, Paducah, KY and Donald P. Visco Jr., University of Akron, Akron, OH

STRATEGIES FOR CREATING AND SUSTAINING A DEPARTMENTAL CULTURE:

TURNING THEORY INTO ACTION

In this invited paper we build upon work which was originally published in the 2010 ASEE Conference Proceedings. Each engineering department has a unique departmental culture based on its history, faculty makeup, geography, and a myriad of other factors. While some departments have a long history with a well-established culture and traditions, other departments that have been recently formed are faced with the challenge of creating a culture from scratch. The success of a department in creating and sustaining a desirable culture can have a significant impact on recruiting, retention, and general satisfaction of its undergraduate students. This paper reviews the engineering literature and describes different models and activities which departments have successfully used in this area, including freshman courses, professional societies, departmental traditions, professional development courses, and intentional interactions between faculty and students. These types of activities also serve to shape students' expectations about what it means to enter the engineering profession and help new student identify themselves as engineers.

In this paper we discuss some strategies to consider when attempting to create or modify a departmental culture from three perspectives:

- Faculty perspective
- Department Chair's perspective
- College of Engineering perspective

Soft Hydrogel Microparticles of Controlled Size and Stiffness with An Emulsion-Based Method

There is an increasing interest in developing methods to synthesize biomaterials with tunable micro/nanostructure, surface chemistries, and mechanical properties. Here we demonstrate a simple method to synthesize biocompatible hydrogel microparticles (MPs) with tunable size and mechanical properties that are also amenable to surface functionalization. Semi-interpenetrating networks of Polyethylene glycol (PEG)-diacrylate and PEG were prepared by photopolymerization in a water-in-oil emulsion to yield spherical PEG MPs. While the MPs were polydisperse in size, process parameters were systematically varied to prepare MPs with 90th percentile size (D90) ranging from 48 μm to 900 μm. The shear elastic modulus of the MPs was varied independent of MP-size by varying the relative amount of PEG-DA and PEG during synthesis. In addition, by replacing PEG with PEG bis(amine) during synthesis, amine groups were introduced in the MPs. The trinitrobenzene sulfonate assay confirmed that the amines introduced were accessible to functionalization with amine-reactive molecules. Dexamethasone (Dex) loaded-poly (DL-lactide-co-glycolide) microspheres (MS) were included in the emulsion during photopolymerization to prepare Dex-MS containing MPs that released Dex controllably over time. PEG MPs created with the emulsion photopolymerization method were biocompatible when evaluated in vitro using a cell proliferation assay and in vivo using implantation in guinea pigs. The method reported here can be used to prepare biocompatible, functionalizable hydrogel microparticles of tunable size and mechanical properties that may also be conjugated with controlled drug release particles to prepare smart materials with applications ranging from injectable biomaterials to tissue engineering to drug delivery.

CATALYST STUDIES WITH IMPLICATIONS ON THE DESIGN AND FABRICATION OF MICROREACTORS

Eric Snider[*†] *and Dr. Frank Jones* [*]
[*]*University of Tennessee at Chattanooga;* [†]*Georgia Institute of Technology and Emory University*

Abstract

In order to reduce typical industrial processing time and increase product purity in biodiesel production, microreactors have been designed, manufactured, and are currently being researched. Microreactors cut diffusion time from hours to seconds when compared to conventional reaction methods due to scale: the microfluidic channel size is 50 μm by 500 μm. Microchannels are currently coated with a metal base catalyst which can convert impure oil feedstocks (oil + free fatty acids) to fatty acid methyl esters (biodiesel). An experimental microreactor study is being conducted on how feedstock composition, residence time, and reactor temperature affect conversion to biodiesel.

To improve on the current microreactor design, catalyst studies with solid acid and base catalysts are being performed with the purpose of operating at lower temperatures and improving the ability to handle low grade feedstocks (high free fatty acid content). Extensive work was performed at room temperature and the methanol reflux point (65°C) with both catalysts with low-grade feedstock. From this study, a base catalyst was successfully tested with low-grade feeds at temperatures as low as methanol's reflux point. Significant conversions with the acid catalyst were achieved at temperatures as low as room temperature.

Introduction

In today's world, alternative fuels are needed more than ever. Conventional fuels, such as coal, natural gas, and petroleum, are constantly being depleted; however, the world's dependency on these fuels is still growing. Additionally, the price on foreign fuels is ever increasing. For these reasons, the United States and the world are pursuing alternative fuel sources to lessen the dependency on conventional fuels.

One such alternative fuel being investigated is fatty acid methyl esters (FAME), or biodiesel. FAME is primarily produced from reacting vegetable oils (triglycerides) with alcohol (usually methanol) by the transesterification reaction shown in Figure 1. For every one triglyceride molecule, three molecules of FAME and one molecule of glycerin are produced. Free fatty acids (FFA), the primary impurity in raw vegetable oils and waste cooking oils, also can react to produce FAME through an esterification reaction shown in Figure 2.

In order for biodiesel to be produced in a timely manner, these reactions need to be catalyzed. Traditional processes utilize homogeneous or liquid catalysts (such as NaOH or KOH) to lessen reaction times (1). Although these liquid catalysts work well for high purity vegetable oils they also react directly with free fatty acids forming unwanted soap by-product. Also, since the catalyst is in the same phase as the reactants, lengthy and costly purification times result. For these reasons, the traditional catalyst choice is not ideal for producing biodiesel, especially with low-grade oil feedstocks (2).

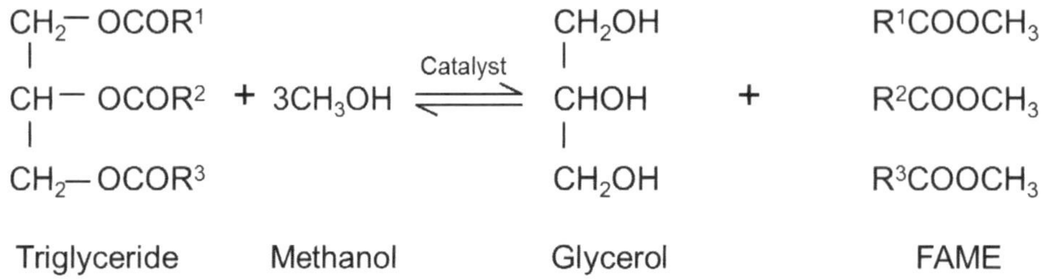

***Figure 1**: Overall transesterification reaction of triglyceride and methanol (3)*

As an alternative, heterogeneous or solid catalysts are being investigated (2). Solid catalysts remain in a different phase from the reactants, changing the dynamics of the reaction. Due to the phase boundaries, reactions involving heterogeneous catalysts can take longer to react and often require higher temperatures when compared to homogeneous catalysts. However, no soap by-product will form in the presence of FFA (2). Also, the catalyst can be recycled by being anchored inside of the reactor, therefore eliminating additional separation costs. For these reasons, a large world-wide research effort is being conducted on heterogeneous catalysis to determine if they are a more economically feasible option when compared to homogeneous catalysts.

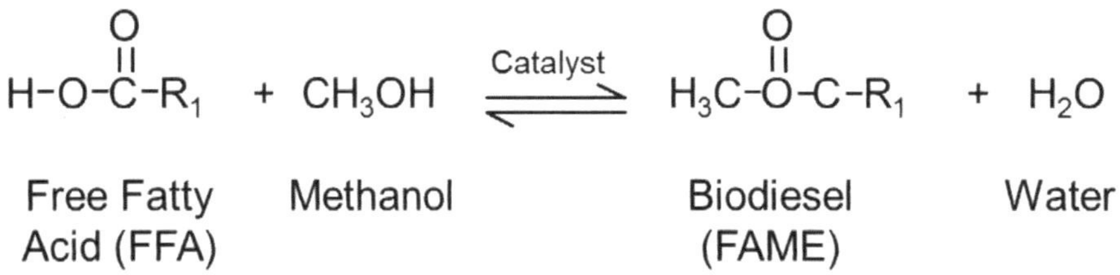

***Figure 2**: Esterification reaction of free fatty acid and methanol (4)*

In this research, two heterogeneous catalysts will be considered: nickel (II) oxide (NiO) and platinum (IV) oxide (PtO_2). NiO is a heterogeneous base catalyst while PtO_2 is a heterogeneous acid catalyst. A catalyst study of these two potential solid catalysts is performed in this research effort to determine their efficacy in producing FAME under different conditions. Experiments are conducted at room temperature and at methanol's reflux point in stirred tank reactors with soybean oil (SBO) and FFA separately. Final results from this study will be used in the design of next-generation microfluidic reactors for biodiesel production.

Microreactors are a novel approach to biodiesel production. Microreactors send reactants through channels of very small cross-sectional area in order to speed up the rate of reaction by bringing all reactants very close to catalyst-coated surfaces. Processing time can be sped up primarily due to diffusion time advantages that microreactors have over conventional reaction methods. Changing from a conventional reactor size scale to the micro level can decrease diffusion times from hours to seconds since $\tau_{Diff} = l^2/D$ (where τ_{Diff} is diffusion time, l is the diffusion length and D is the diffusivity coefficient).

One of the UTC research group's first generation microreactors used in this study is shown in Figure 3. It has a cross sectional area of 50 µm x 500 µm and a nominal length of 5 cm. These channels are etched in silicon and contain triangular packing features (125 µm on a side) to increase catalytic surface area. Each reactor has two entrance ports (one for methanol and one for an oil feed) as well as a single exit port. NiO catalyst is sputtered on the surfaces, and a glass plate anodically bound to close the channel. The second objective of this study was to compare the performance of microreactors to batch reaction methods in producing biodiesel.

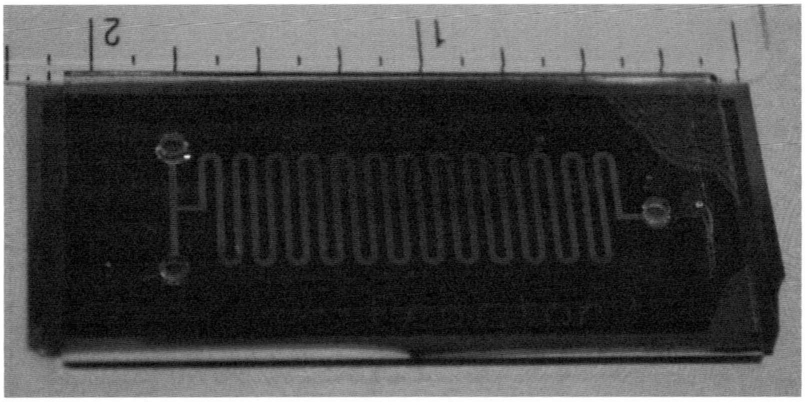

Figure 3: *First generation serpentine microreactor for biodiesel production*

Methods

Stirred Tank Reactors

Experiments were run using 25mL round bottom flasks fitted with water-jacketed reflux condensers. Five experiments can be simultaneously performed using a multiposition stirring hot plate (EchoTherm) and Teflon coated stir bars. Experiments use 10mL of total liquid volume with 72:1 MeOH:SBO and 24:1 MeOH:FFA molar ratios. Samples of approximately 2mL can be extracted at time intervals and conversions measured via gas chromatography (SBO) or acid number analysis (FFA). NiO and PtO_2 catalyst amounts equivalent to 4wt% of the total 10mL volume was used in all experimental studies.

Microreactors

Microreactors require ultralow, steady and controllable flow rates. A Cole-Parmer® Touch-Screen Syringe Pump, which is able to deliver flows as low as five picoliters per minute to the microcontroller, was used in all microreactor work. This project required flows between 250 nanoliters/minute and 2 microliters/minute. For temperature control, heat was provided to the microreactor via a Minco resistance heater. Experimental temperatures ranged up to about 70°C, methanol's boiling point in the slightly pressurized microreactor system.

Acid Number Analysis

Acid number analysis was used to determine FFA conversion. Acid number analysis takes advantage of FFA's acidic nature by using simple acid-base neutralization. In accordance with the American Society for Testing and Materials (ASTM) D974, potassium hydroxide with a normality of 0.1000N is used for neutralizing FFA samples (5). Neutralization of the FFA is noted with a p-Naphtholbenzein indicator. A known amount of analyte (reaction products and unreacted FFA) is added to the titration solvent (100:1:99 toluene: water: isopropyl alcohol volume per volume solution) before the KOH is used for neutralization. From neutralizing the FFA, the acid number will be determined as follows: (Acid Number= (mg KOH)/ (g Analyte)). The acid number of an experimental result is then compared to an unreacted FFA sample to determine conversion (6).

Gas Chromatography

Triglyceride-derived FAME cannot be analyzed via acid number analysis since triglycerides are not acidic; instead gas chromatography (GC) was used. Details for properly setting up a GC for triglyceride analysis can be found in ASTM D6584 (6, 7). For this research effort, a SRI 8610C GC was used equipped with a MXT-Biodiesel TG column (15m in length and .32mm in inner diameter). All reagents and methods for performing GC analysis were in accordance with ASTM D6584 (7).

Results and Discussion

One of the main disadvantages of heterogeneous catalysts is that relatively high temperatures are needed to produce significant conversions to FAME (2). If a catalyst can be found that is effective at lower temperatures, especially room temperature, operating costs can be significantly cut. Trials were run with pure free fatty acid (FFA) using PtO_2 (acid catalyst) and NiO (base catalyst) nanopowder at room temperature. FFA is a major impurity in low-grade vegetable oils and in waste cooking oil. Success converting FFA without producing soap by-products means lower grade feed can be used, which would significantly cut costs in biodiesel production.

Experimental results in stirred tank reactors

Experimental results for the conversion of FFA to FAME in stirred tank batch reactors are shown in Figure 4. Minimal conversions were obtained using NiO nanopowder catalysts at room temperature (about 21°C). Only trace amounts of FAME were obtained after two hours and about 10% conversion after 4 hours. Platinum oxide achieved significantly higher conversions at room temperature than the nickel oxide catalyst was able to reach. After only 15 minutes, almost 40% conversion to biodiesel was reached. Conversions then level off at just over 40% after several hours, indicating equilibrium has been reached at this temperature.

While neither of these catalysts is able to reach complete conversion to FAME at room temperature, platinum oxide shows significant potential. This platinum oxide catalyst shows great promise for the microreactor systems. Microreactors have the potential of processing what would normally take hours in a conventional (stirred tank) reactor in a matter of seconds. If fifteen minutes is required to reach 40% conversion in stirred tanks, a microreactor could reach the same conversion levels almost instantaneously. From this result, platinum oxide is a good catalyst candidate for future microreactors designed to process FFA at room temperature.

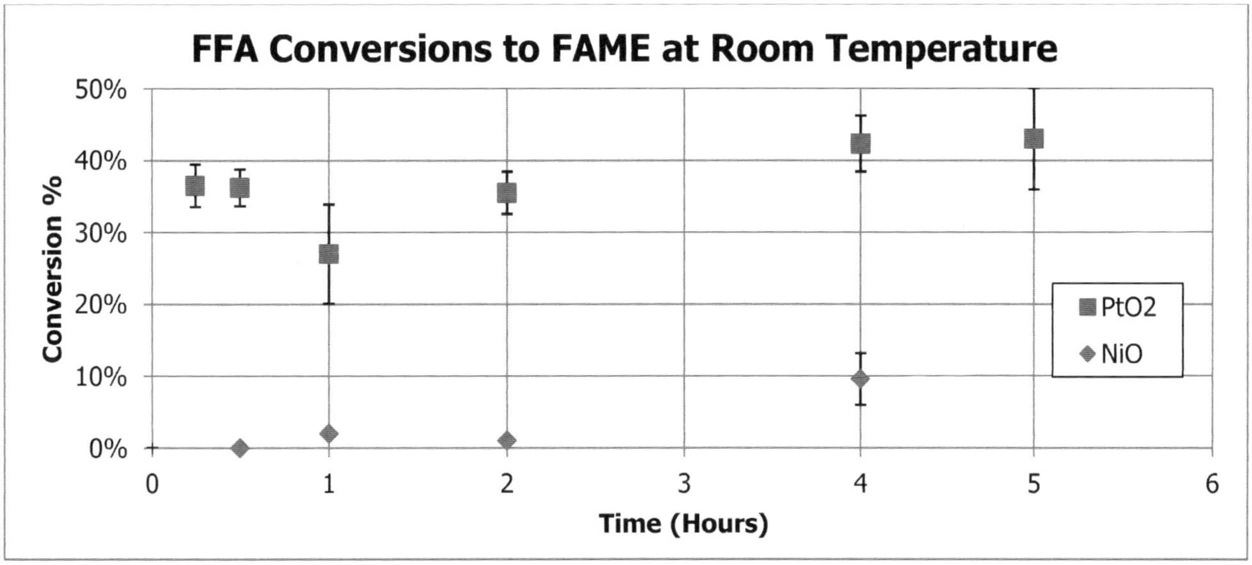

***Figure 4**: Comparison of the use NiO and PtO_2 catalysts in free fatty acid conversion to biodiesel at room temperature in stirred tanks*

Experimental trials were also run at room temperature using soybean oil as a feed using the same catalysts. This is a different (transesterification vs. esterification), more complicated, and slower reaction. It also involves two feed substances that are essentially immiscible. Samples were drawn at 30 minutes, 4 hours, and 24 hours to get an overview of what level of conversions were possible and how quickly they were forming. Using GC analysis, no conversion was detected with either catalyst, even after 24 hour reaction times. The activation energy to begin transesterification seems to be higher than that for the esterification reaction undergone with the FFA feed. Neither catalyst was able to lower the activation energy enough to make room temperature conversion to biodiesel possible with SBO.

At higher temperatures, reaction rates normally increase as activation energy barriers are overcome by the additional heat. Trials were performed at methanol's reflux point of about 65°C. Results for conversions of FFA and SBO to FAME at reflux with NiO and PtO_2 are shown in Figure 5. From this figure it can be seen that NiO was able to quickly reach approximately 15% conversion to FAME with a FFA feed (in 15 minutes). However, conversion seems to equilibrate at that point.

Each catalyst produces significant conversion to FAME when using a SBO feed, but NiO-catalyzed reactions were quicker. Platinum oxide performed better with FFA at room temperature than NiO, but, at reflux, with SBO the result appears to be the opposite. After 4 hours, NiO causes 99% conversion while PtO_2 yields 60% conversion. It is worth noting that SBO catalyzed by NiO at reflux samples was analyzed by NMR spectroscopy previously (3). Some of the differences between the PtO_2 and NiO results could be attributed to the different analytical methods (GC vs. NMR).

Experiments were performed with platinum oxide at reflux using a FFA feed as well, but there was no conversion to FAME. Platinum oxide was potentially oxidizing MeOH to formic acid, and, as a result, PtO_2 was reduced to platinum black (8). Acid number measures the loss of acidity due to formation of FAME, however, formic acid is much more acidic than FFA. Evidently, the higher temperature caused a well-documented instability in the acid catalyst and a loss of selectivity (9).

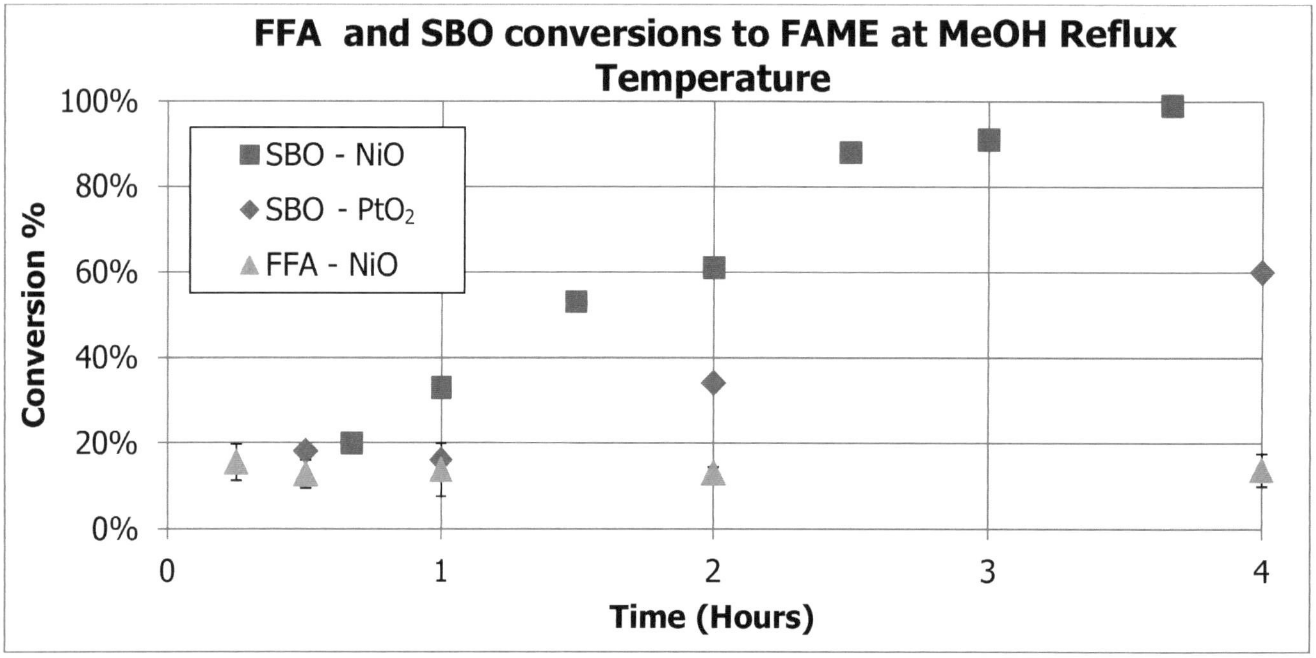

Figure 5: *Comparison of the use NiO and PtO_2 catalysts FFA or SBO conversion to biodiesel at methanol reflux temperature (~ 65ºC) in stirred tanks*

Experimental results in microreactors

While investigating the ability of the microreactors to produce biodiesel, experimental conditions were used that mirrored experiments in stirred tanks as closely as possible. Reaction reflux temperatures were slightly higher in the microreactors, about 68°C to 70°C vs. 65°C, due to the somewhat higher (and variable) pressures in the channels. Also, the molar feed ratio is 24:1 (Me:SBO) vs. 72:1 in stirred tanks due to pumping constraints. Results of the two reaction systems are shown in Figure 6.

SBO and MeOH are immiscible and they are fed into the microchannel at room temperature. The phase boundary, the need to heat up to reaction temperature in the channel, and lack of stirring are constraints on the microreactor performance when compared to the stirred batch system. Also, all reactants must migrate to the catalyst-coated floor of the channels or the surfaces of packing particles. Experiments were run at volumetric flow rates of 1μL/min or approximately a 3.75 minute residence time in the microreactor. In spite of the constraints, the microreactor had a conversion of about 75%. This level of conversion required well over two hours in a stirred batch system.

While results from the microreactor system showed much potential, various design changes are being considered for the next generation. The microreactor performance can be improved by introducing two catalyst coating changes: sputtering on the roof of the channel and nano-structuring the coating. The channel roof has no catalyst coating in the present design. This was done so that fluid flow could be observed through the clear glass roof. In principle, sputtering on the roof will decrease diffusion time to surfaces by a factor of 4 since the maximum diffusion length will be halved. (Recall $t_{Diff} = l^2/D$). In principle, this should reduce the processing time for the case shown in Figure 6 from 3.75 minutes to less than one.

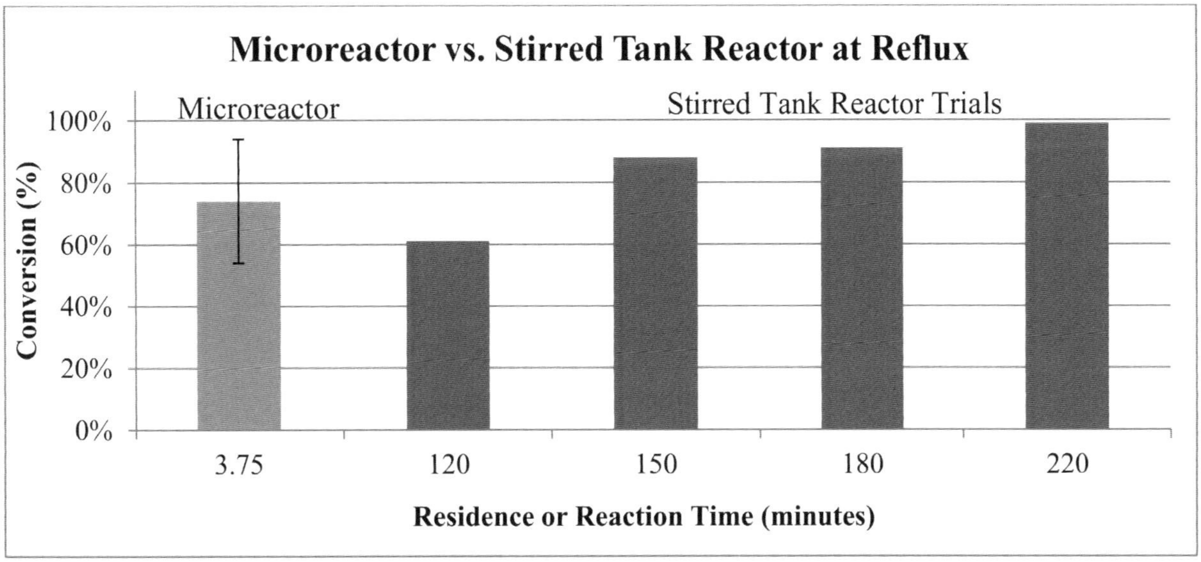

Figure 6: *Comparison of microreactor conversions (green) of SBO to FAME to those in stirred tanks (blue) at methanol's reflux temperature*

The catalyst coating on the surfaces of the present microreactors has an unknown (probably amorphous) lattice structure. At the time of fabrication, the effects of lattice structure on catalytic activity were not considered. Subsequently, various researchers have discovered methods to produce specific lattice structures using magnetron sputtering of NiO on glass (10, 11). It is possible that one structure or some combination of structures will perform best. It has been suggested by other research efforts that lattice structure may have significant impact on conversion (12).

Another means of improving microreactor efficacy is in changing the multiphase nature of the feedstock into a single phase. FFA is a co-solvent for SBO and MeOH. We have found that a 1:1:1 volume ratio of SBO:FFA:MeOH (equivalent to a molar ratio of 1:3:24) is completely miscible indefinitely (8). SBO conversion can potentially be greater if there is no liquid – liquid boundary to overcome. This is especially true since microreactors can have little or no mixing even with flow around packing particles. This is a result of Reynolds numbers significantly less than 0.1 indicating creeping flow.

Conclusions

One goal of this research effort was to conduct a catalyst comparison between nickel oxide and platinum oxide at various reaction conditions tested in batch reactors. The comparison was performed at room temperature and methanol's reflux point using free fatty acid and soybean oil feedstocks to determine which catalyst performs better in biodiesel production. Platinum oxide successfully produces moderate levels of conversion to biodiesel from free fatty acid feedstock at room temperature. Nickel oxide produces no FAME at room temperature. However, at higher temperatures, platinum oxide is permanently modified when using FFA and produced only unwanted side reactions and no FAME. In contrast, nickel oxide catalyst does produce FAME from FFA at reflux, although the levels are low due to equilibrium limitations. At reflux conditions, high conversions are achieved with a soybean oil feed for both platinum oxide and nickel oxide, with nickel oxide being superior.

The ultimate objective of this research effort was to test the performance of the first generation of microreactors in producing biodiesel. The microreactor successfully converted SBO to FAME. In fact, its performance is far superior to stirred tank reactors by significantly reducing processing time. Generally, the microreactor produced equivalent or superior conversions nearly two orders of magnitude faster than stirred tanks. Microreactors can potentially dramatically change biodiesel production costs.

References

(1) Zabeti, Masoud, Wan Mohd Ashri Wan Daud, and Mohamed Kheireddine Aroua. "Activity of Solid Catalysts for Biodiesel Production: A Review." *Fuel Processing Technology* 90 (2009): 770-77.

(2) Lotero, Edgar, James Goodwin, David Bruce, Kaewta Suwannakarn, Yijun Liu, and Dora Lopez. "Catalysis of Biodiesel Synthesis." *Catalysis* (2006): 41-83.

(3) Dacus, Robert W. "A Kinetic Study of the Production of Biodiesel Fuel from Algae Oil." B.S. Thesis. The University of Tennessee at Chattanooga, (2010).

(4) Buecker, Jordan. "A Cost Analysis of Heterogeneous Catalyst Choice on Biodiesel Production." Thesis. University of Tennessee at Chattanooga, (2011).

(5) ASTM Standard D974, 2008, "Standard Test Method for Acid and Base Number by Color-Indicator Titration," ASTM International, West Conshohocken, PA, (2003).

(6) Baig, Aijaz, and Flora T. T. Ng. "A Single-Step Solid Acid-Catalyzed Process for the Production of Biodiesel from High Free Fatty Acid Feedstocks." *Energy & Fuels* (2010): 4712-720.

(7) "ASTM D6584 - 10ae1." Standard Test Method for Determination of Total Monoglyceride, Total Diglyceride, Total Triglyceride, and Free and Total Glycerin in B 100 Biodiesel Methyl Esters by Gas Chromatography. www.astm.org.

(8) Snider, Eric. Catalyst Studies with Implications on the Design and Fabrication of Microreactors. B.S. Thesis. The University of Tennessee at Chattanooga, (2012).

(9) Kiss, A. A., A. C. Dimian, and G. Rothenberg. "Solid acid catalysts for biodiesel production – towards sustainable energy." *Advanced Synthetic Catalysis*, 348 (2006).

(10) Peng, T.C., X.H. Xiao, X.Y. Zhou, W. Wu, F. Ren, and C.Z. Jiang. "Characterization of DC reactive magnetron sputtered NiO films using spectroscopic ellipsometry." *Applied Surface Science*. (2011).

(11) Abbey, Brian, John D. Lipp, Zoe H. Barber, and Trevor Rayment. "Structure-film thickness relationship study of sputtered NiO/Ni bilayers using depth profiling and atomic force microscopy techniques." *Journal of Applied Physics,* 99. (2006).

(12) Strunk, Jennifer, William C. Vining, and Alexis T. Bell. "Synthesis of Different CeO2 Structures on Mesoporous Silica and Characterization of Their Reduction Properties." *The Journal of Physical Chemistry,* (2011).

(38c) Effect of Dendritic Amphiphiles On the Biophysical Properties of Model Biomembranes

Riya Muckom, Chemical and Biological Engineering, Colorado School of Mines, Golden, CO and Amadeu K. Sum, Chemical & Biological Engineering, Colorado School of Mines, Golden, CO

Dendritic amphiphile molecules [$RCONHC(CH_2CH_2COOH)_3$, R = n-C_nH_{n+1}], possess antimicrobial, antifungal, anti-HIV, and anti-STD properties, giving these molecules great potential in the pharmaceutical industry. In order for dendritic amphiphiles to be considered for pharmaceutical use, the molecules should not disrupt the mammalian cell membrane functionality. Using molecular dynamics simulations, the effects of varying concentration and varying tail lengths of dendritic amphiphiles within a lipid bilayer system were observed at an atomistic level. This project analyzes different biophysical properties of simulated bilayer systems, at equilibrium, to determine how dendritic amphiphiles cause changes to the DPPC lipid bilayer characteristics at 300 K and 325 K, corresponding to temperature below and above, respectively, the lipid bilayer phase transition temperature from gel to liquid-crystalline. From the simulations, we observe that at higher concentrations of amphiphiles in gel-state bilayers (300 K), the lipid tails begin to resemble disordered tails as in the liquid-crystalline state. Hydrogen bonding takes place between the amphiphile and DPPC at specific sites, with an average of four DPPC lipids bound to each amphiphile in the bilayer. An area per lipid calculation, done by a Voronoi tessellation, finds that the headgroup of the amphiphile is significantly smaller than that of DPPC, thus higher concentrations of amphiphile decrease the total lateral bilayer area. Diffusion coefficients of amphiphiles in the plane of the bilayer dramatically increase with decreasing amphiphile tail lengths. In addition, as the concentration of amphiphile increased, the diffusion coefficients of both the lipid and amphiphile increase, suggesting a strong influence of the amphiphiles on the bilayer systems. These results provide insight into the molecular interactions of the amphiphile molecules with the lipid bilayer, and provide a means to understand their potential impact in the biophysical properties of cellular membranes.

An Improved Temporary Immersion Bioreactor for Plant Tissue Culture Propagation

Matthew Curtis, The Pennsylvania State University, State College, PA

What is Plant Tissue Culture?

Plant tissue culture allows the study of plants in an aseptic, controlled environment. By using nutrient-rich liquid and solid medium, the inherent variability of soils is avoided, both in terms of nutritional content and bacterial flora. As a result, plant lines can be developed free from bacterial, fungal and viral diseases that are prominent in the plant world. Plant propagation is the practice of generating additional plant biomass in a clonal fashion, thereby bypassing the need to allow for flowering, pollination, seed generation, and germination of additional plants. Examples of plant tissue culture are shown in Figure 1.

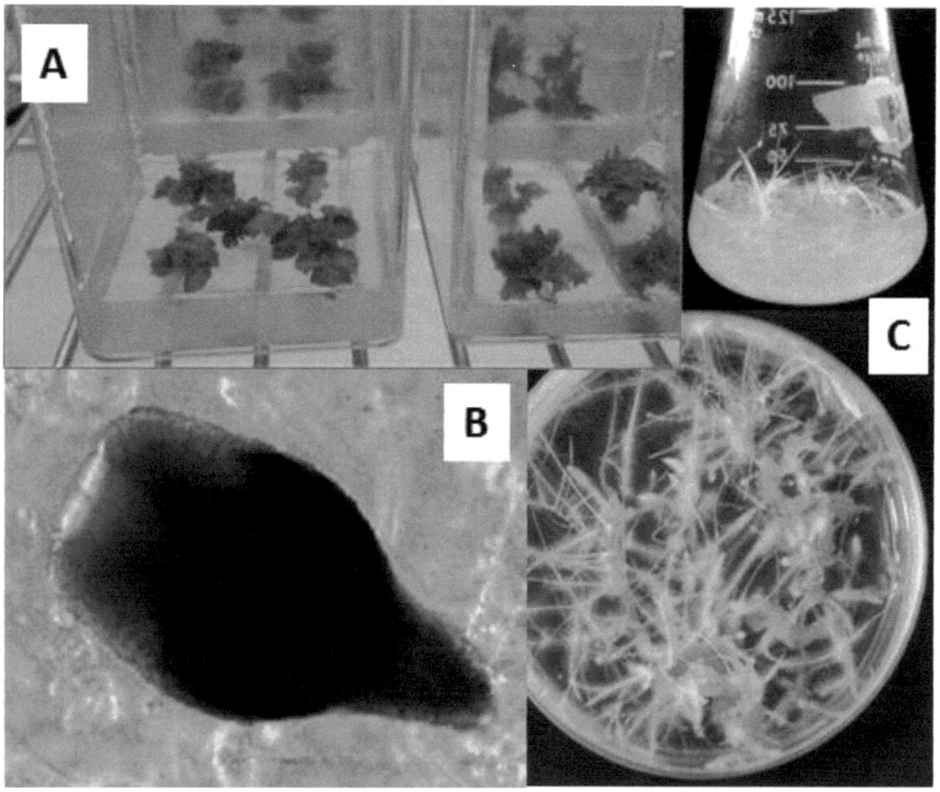

A: Seedless Watermelon Shoot Culture
B: Carrot Somatic Embryo Culture
C: Tobacco Hairy Roots in Solid and Liquid Medium

What can Plant Tissue Culture be used for?

Plant tissue culture is used for a variety of commercial applications to improve plants:

Genetic Engineering

The primary use of plant tissue culture is to improve plant species including genetic engineering of plants. Genetic engineering requires the ability to insert a gene into a single cell, and then regenerate an entire plant from that cell (Li et al. 1998). This requires a variety of controlled hormone and environmental conditions provided by plant tissue culture.

Clonal Selection of Improved Lines

Using plant tissue culture, plants can be subjected to a variety of stressing factors, allowing the direct selection of plants that perform well under conditions such as drought, salt stress, and the presence of heavy metals.

Maintaining Sterile Hybrids and Plant Characteristics Lost through Seeding

Because plants tend to maximize their genetic variability while mating to increase chances of survival, desirable characteristics are often obscured. A superior plant does not necessarily give rise to superior offspring. Additionally, wide crosses often give rise to highly productive plants; however, these plants are unable to reproduce naturally (termed sterile hybrids; eg. seedless watermelon in Figure 1A). Clonal plant propagation allows the maintenance of genetically identical (superior) plants.

Study of Isolated Organs

Plant tissue culture allows the maintenance of isolated tissue types, such as roots, shoots, flowers, leaves, etc. (See Figure 1C). By providing sugar and plant hormones these organs be forced to grow without the rest of the plant (Ramakrishnan and Curtis 2004). This allows the rapid study of differentiated plant tissues without the complications of considering transport between organs.

What aspect of Plant Tissue Culture does my work relate to?

Cultures grow on solid media are limited in their nutrient uptake to the small surface area of the propagule in contact with the media. While growing plant tissue in gyratory shake flasks tends to alleviate this issue, they quickly become limited by the rate at which oxygen can diffuse into the medium. This often results in a slower growth rates, as well as the generation of vitrified tissue due to the constant suspension in water. Furthermore, the constant motion of a shake flask can damage more delicate tissues. To increase the oxygen transfer rate into the tissue as well as prevent vitrification, a technique called temporary immersion is used. Typical temporary immersion systems utilize complex 3-way solenoid circuits to occasionally flood (Figure 2A) then drain (Figure 2B) the tissue-containing reactor with liquid from a separate reservoir. The presence of a thin film of media on the plant tissue contains sufficient nutrients to allow growth, while significantly increasing the rate at which oxygen can transfer from the gas phase into the propagule. We have proposed and developed a novel low-cost system for temporary immersion that utilizes gravity as a driving force, rather than pneumatics. Raising the reservoirs causes flooding of the reactor vessel (Figure 2C), while lowering the reservoir allows draining via the siphon maintained between these vessels. (Figured 2D).

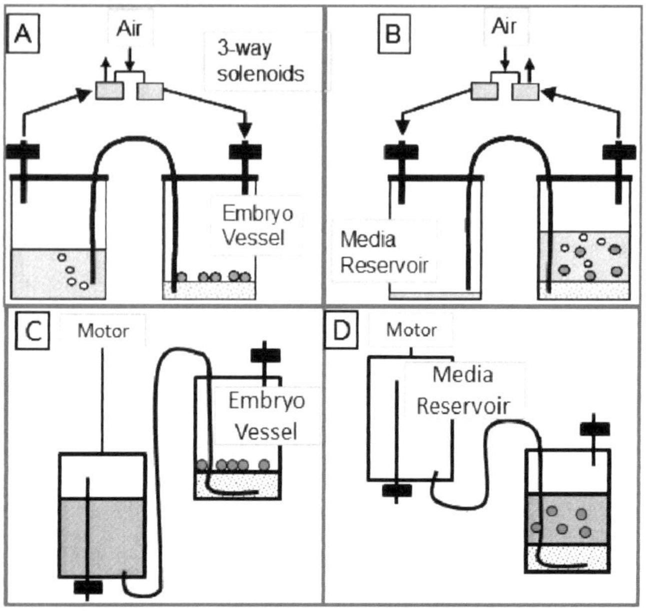

Figure 2: Temporary Immersion Bioreactor Schematics. Frames A and B show a typical gas-flow based temporary immersion bioreactor, where Frames C and D show the improved gravity driven design.

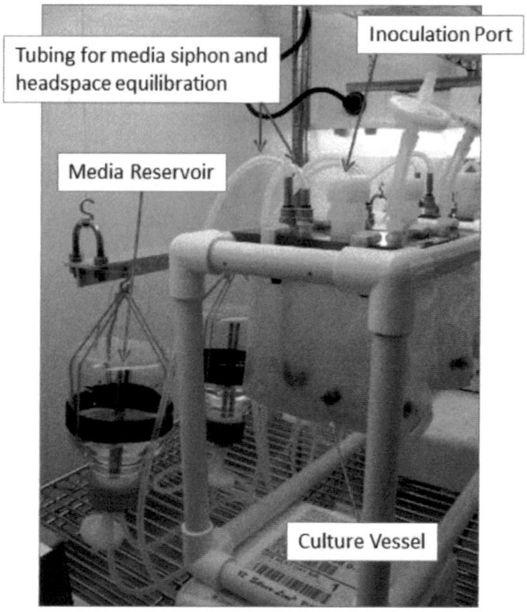

Figure 3: Gravity Driven Temporary Immersion Bioreactor Prototype.

The bioreactor was designed with low cost materials in mind (Figure 3). The culture vessel itself is a simple plastic bag attached to a machined headplate to allow sterile connections between the vessel and reservoir. For consistency of testing purposes, the media reservoirs were raised and lowered with a simple stepper motor.

Troubleshooting of issues lead to the development of several important additions to the design. One of the most critical issues with early designs was a tendency for the plastic bag reactors to become contaminated. This issue was solved by introducing a very low flow rate of gas into the reactor system to provide a constant positive pressure, counteracting the pressure drops caused by the movement of the media reservoirs. Early issues with a liquid distributor caused occasional failures, leading to the development of a suite of LabVIEW monitoring software including automatically uploaded video recordings and text messaged warnings for common issues. The setup was also readily retrofitted with sterile injection ports to allow the aseptic monitoring of growth and nutrient utilization.

Results

Figure 4 shows the comparison between this reactor system and established methods. When growing watermelon shoot tissue (Figure 4A), the reactor performed comparable to solid media controls. While the tissue harvested showed some signs of stress over the control group, this is likely a result of early cycling errors. To begin reactor testing and analysis on a faster time scale, root cultures were tested in the system (Figure 4B). Once again, the reactor performed comparable to the typical culture method, in this case gyratory shake flasks.

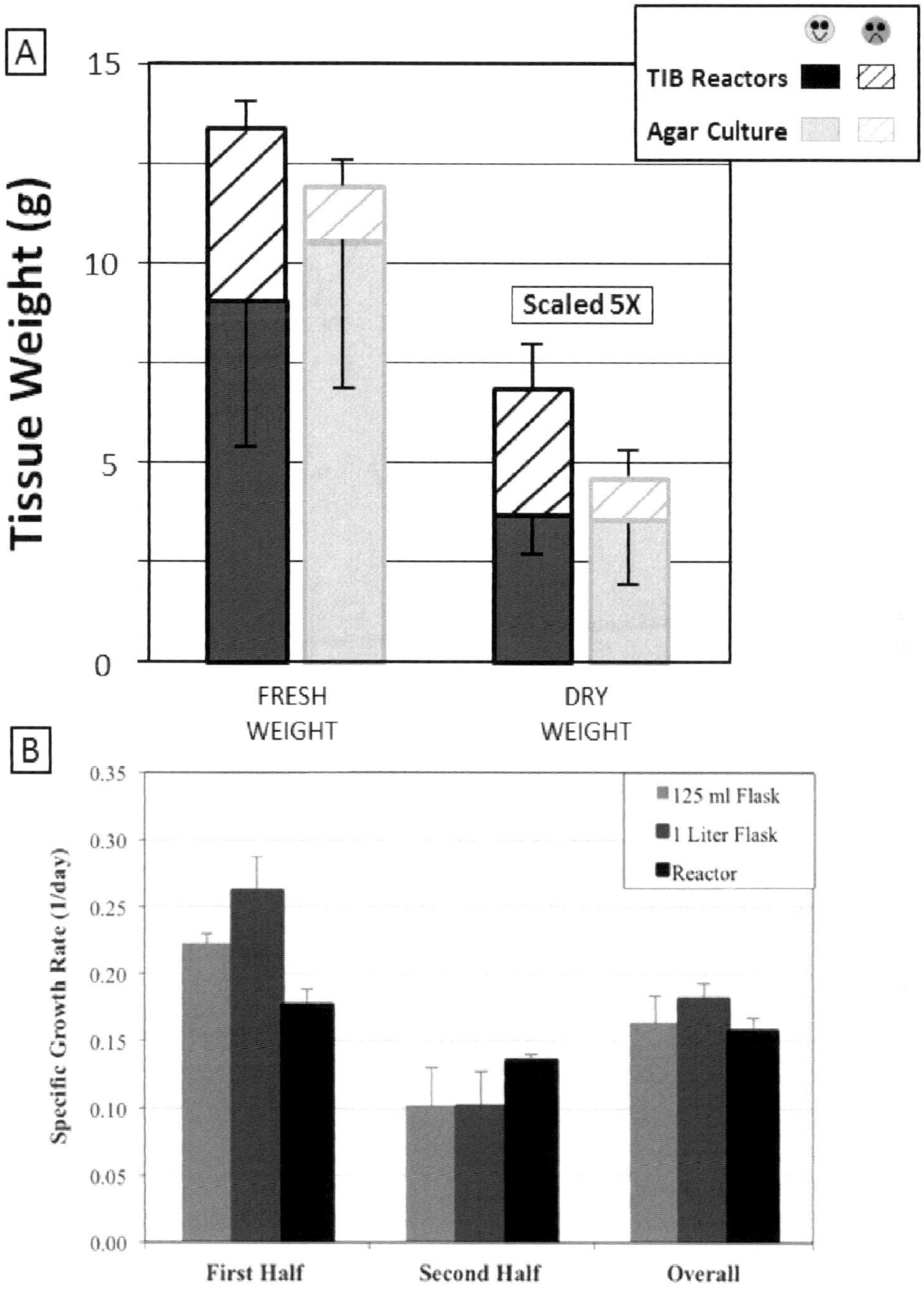

Figure 4: Comparison of Prototype with Typical Tissue Culture Methods. Frame A shows a comparison of the prototype reactor with solid media (agar culture) in the growth of seedless watermelon shoots. Frame B shows a comparison between the prototype and gyratory shake flasks in the growth of hairy root tissues.

Demonstrated Advantages over Typical Temporary Immersion Bioreactors

My talk will describe the many advantages of our bioreactor over conventional plant tissue culture systems. A summary of the superior characteristics are enumerated below:

Reduced Bioreactor Cost:

By using gravity as a driving force, rather than pressurized air, costs associated with switching solenoids and other hydraulics can be avoided. This low pressure design also allows the use of low cost plastic bags in place of solid-walled vessels.

Independent Gas/Liquid Flow:

By using independent gas and liquid flows, it becomes feasible to supplement the gas phase with oxygen to increase heterotrophic growth rates, or carbon dioxide to allow phototrophic growth. A phototrophic bioreactor is highly desirable, as the risk of contamination is significantly reduced without the presence of easily accessible sugars in liquid media.

Improved Control of Culture Media:

The media reservoir design also facilitates rapid switching of media formulations. This could allow the regeneration of plants from an embryonic state, through conversion and hardening into soil-ready plantlets within a single reactor vessel.

Adaptability to Available Technology:

While the setup used for testing purposes has included significant technological interfacing, this technology can be readily adapted to a variety of circumstances, ranging from automated control to manual cycling of liquids.

Linear Scalability:

Without the need to introduce complex gas handling capabilities, the prototype design is much more readily scaled to an industrial greenhouse setting.

Future Work

To provide a demonstration of inexpensive control of gas phase, plant root cultures will be grown on elevated oxygen partial pressures to improve growth rates by overcoming diffusional mass transfer limitations in the root meristems. We anticipate using an order of magnitude less gas during the duration of the run.

Preliminary work has already shown that watermelon shoot cultures can be grown photoautotrophically. This work will be executed with CO2 supplementation to improve tissue growth rates and avoid contamination, thereby setting the stage for large scale implementation of this technology in a greenhouse setting. We are planning to assist in the implementation of this technology with a collaborator in Nigeria for improvement of sweet potato.

A final application of this TIB system is for a novel approach to plant propagation. Recent research has revealed genes which encode proteins that control the process of embryo formation in plants. These transcription factors open up a new possibility of propagating plants by changing normal somatic cells, back into embryos which then develop into whole plants! This improved TIB system will be used to introduce these transcription factors from a bacteria designed in our lab for DNA delivery and transient gene expression (Collens et al., 2004). The specific system under study is Theobroma cacao (chocolate tree) where we hope to introduce tissues such as leaves into the reactor, and then express the transcription factors in that tissue to propagate plants that are resistant to a devastating disease that threatens the livelihood of nearly 14 million subsistence farmers around the world that produce chocolate beans.

References

1. Berthouly M., Etienne H. 2005. In : Hvoslef-Eide Anne Kathrine (ed.), Preil Walter (ed.). Liquid culture systems for in vitro plant propagation. Dordrecht : Springer [Pays-Bas], p. 165-195.

2. Collens, J.I., Lee, D.R., Seeman, A.M. Seeman, Curtis, W.R. The Development of Auxotrophic Agrobacterium tumefaciens for Gene Transfer in Plant Tissue Culture. Biotechnol. Prog. 20(3); 890-896, 2004.

3. Curtis, WR. Growing cells in a reservoir formed of a flexible sterile plastic liner. U.S. Patent # 6,709,862, March 23 2004, associated patent: Curtis WR, Method and apparatus for aseptic growth or processing of biomass", U.S. Patent # 6,245,555, June 12, 2001.

4. Li, Z.; Traore, A.; Maximova, S.; Guiltinan, M.J. Somatic embryogenesis and plant regeneration from floral explants of cacao (theobroma cacao L.) using thidiazuron. In Vitro, Plant 1998, 34, 293-299.

5. Ramakrishnan D, Curtis WR (2004) Trickle-bed root culture bioreactor design and scale-up: Growth, fluid-dynamics, and oxygen mass transfer. Biotechnology & Bioengineering, 88(2):248-260.

6. Shetty, R. Bioreactor design and operation for clonal tree propagation. PhD Thesis, 2005.

(38e) Biodiesel Production without Glycerol Byproduct: Dimethyl Carbonate As Replacement for Methanol

Tamara Frydson Andrade, Lamar University, Beaumont, TX

Biodiesel Production without Glycerol Byproduct: Dimethyl Carbonate as Replacement for Methanol

Tamara Frydson and Tracy J. Benson

Dan F Smith Department of Chemical Engineering, PO Box 10053, Lamar University, Beaumont, TX 77710

Currently, biodiesel is produced as a supplement for petroleum diesel and is comprised of a mixture of fatty acid methyl esters (FAME's). However, glycerol, a low-value byproduct, is also produced. The downstream processing required to separate glycerol from the biodiesel product requires substantial amounts of water, energy, equipment, and process unit operations. This work reports on a biofuel, analogous to biodiesel, that can be produced by reacting lipid oils with dimethyl carbonate (DMC), instead of methanol. The DMC reaction produces 2 FAMEs and 1 fatty acid glycerol carbonate (FAGC) for every triglyceride molecule converted. This reaction scenario generates a more desirable product, without the glycerol byproduct, at lower production costs.

Experiments were conducted using canola oil, DMC, and triazabicyclodecene catalyst. Products were first identified using GC/MS and were found to be fatty acid methyl esters (FAMEs) and fatty acid glycerol carbonates (FAGCs). The FAGCs are miscible with FAMEs, and are, therefore, usable as a fuel. The produced biofuel passed the ASTM D6751 standard for biodiesel. Complete reaction details, including kinetic results, and process modeling using Aspen Plus simulator will be presented.

(38f) Titania Stabilized Pickering Emulsions for Skin Care Products

John Geil, Chemical Engineering, University of Washington, Sammamish, WA

Pickering Emulsions may be formed by adsorbing particles to the interface of the liquid droplet. These particles increase stability of the emulsion by holding droplets further apart than the critical film thickness, at which the droplets coalesce. A study of the use of silane-treated, hydrophobic, titanium dioxide particles to stabilize an isododecane in water emulsion, for the purpose of pigment (titanium dioxide) deposition on skin, was conducted.

Results are shown for the formation of these isododecane in water Pickering Emulsion systems, using shear stress mixing. Mixing method results were optimized to form emulsions without entraining air. Also, results for the silane treatment of bare titanium dioxides particles, to increase hydrophobicity, are presented. As a function of reaction time, the hydrophobicity of titanium dioxide particles can be tuned to force the particles to the interface of the isododecane droplets and water. Furthermore, included is an overview of stability for the Pickering Emulsions. By comparing the coalescence of the particle-stabilized isododecane droplets to naked isododecane droplets, it was observed the particle stabilized emulsion was stable towards coalescence, while the naked isododecane droplets coalesced rapidly.

Using optical microscopy, these titanium dioxide stabilized emulsions are shown to form and enhance stability of the emulsion.

(38g) Effects of Algae Extract On the Growth and Metabolism of Various Microorganisms

Weston K. Kightlinger[1], Kai Chen[1], Daniel W. Crunkleton[1], Geoffrey Price[1] and Tyler Johannes[2], (1)Chemical Engineering, University of Tulsa, Tulsa, OK, (2)Chemical Engineering, The University of Tulsa, Tulsa, OK

The purpose of this project is to investigate the effectiveness of algae extract (ALE) as a nutrient source in cell culture media. Algae have great potential in the production of biodiesel, and algae extract could be harvested as a component of biomass byproduct. This application is likely to make algae extract cheaper to produce than yeast extract – today's most common nutrient source in microbiological media. The algae extract used in this project is made up of the water-soluble portions of autolysed Chlamydomonas reinhardtii cells harvested from a small scale bioreactor. Percent composition of free amino acids, whole proteins, and carbohydrates were assessed and compared to yeast extract using Ninhydrin, Bradford, and Phenol-Sulfuric Acid assays, respectively. Metabolism assays of ethanol production in yeast, lipid production in algae, and protein production in bacteria were also performed using High Pressure Liquid Chromtography (HPLC), Bligh-Dyer lipid assay and Bradford protein assay, respectively. Additionally, bacteria, yeast, and algae cultures were grown using algae and yeast extract and compared for saturation concentrations using optical densities. Algae extract has been proven to be comparable or superior to yeast extract in all tests.

(38h) Assessment of BIO-Ethanol Dehydration Process Alternatives by Process Modeling and Life-Cycle Analysis

Michel Kahwaji Janho[1], Jorge E. Gatica[2], Fernando Daniel Mele[3], María Rosa Hernández[4] and Mauricio Colombo[3], (1)Chemical And Biomedical Engineering, Cleveland State University, Cleveland, OH, (2)Chemical and Biomedical Engineering, Cleveland State University, Cleveland, OH, (3)Ingenieria de Procesos y Gestion Industrial, Universidad Nacional de Tucuman, Tucuman, Argentina, (4)Chemistry and Chemical Engineering, Universidad Nacional de Tucuman, Tucuman, Argentina

Energy is one of the major driving forces in today's economic and civic development, fossil fuels burning account for the vast majority of energy produced worldwide. Burning fossil fuels produces a vast amount of pollutants, this combustion reaction can release a vast amount of carbon dioxide, sulfur dioxide, carbon monoxide, radiation and a vast array of polycyclic aromatic hydrocarbons; these gases are responsible for the pollution that's affecting the planet and life on it, some of them are responsible for global warming while the others cause acidification and the rest can cause cancer. The processes used to acquire these fuels are also a big source of environmental problems; coal mining releases harmful substances which pollute the air, the earth and the water beds surrounding the mines, processing petroleum presents similar problems; and last but not least the fossil fuel reserves are dwindling fast, which affects the economy, security and the development of all countries on earth. It is then imperative that new 'cleaner' and more renewable sources of energy are to be found, studied and implemented. These sources include solar power, wind power and bio-fuels such as bio-butanol and bio-ethanol, studied in this work.

Whatever the feed source, a bio-ethanol production process will include a fermentation section (where the alcohol is formed), a separation section (solid/liquid separation), a distillation section (where the side products are separated out) and a dehydration section (where the azeotrope is 'broken' and anhydrous ethanol is obtained); this work focuses on minimizing the energy requirement of the dehydration step.

Many methods exist to separate the ethanol-water azeotrope (also known as 'breaking' the azeotrope), these methods include:

n Extractive distillation:

n Pressure swing distillation.

n Azeotropic distillation.

n Pervaporation.

Azeotropic distillation and pervaporation are the two separation techniques studied in this work.

This work studies the energy requirements and the effectiveness of the aforementioned separation techniques, computer simulations are heavily used. ASPEN PLUS ® is used as the simulation engine.

Various configurations utilizing the pervaporation module were designed and analyzed.

The first configuration consists of a single module with a feed below the azeotrope.

The second configuration is a 'hybrid' separation train featuring a distillation to achieve the azeotrope and a pervaporation module to obtain anhydrous ethanol.

The top product from the tower is the ethanol-water azeotrope, this is then fed to the pervaporation membrane which separates the azeotrope yielding a purer ethanol product in the retentate stream, while the permeate stream containing water and lost ethanol is sent to a flash drum, the vapor phase containing most of the lost ethanol is recycled to the tower while the liquid phase containing mostly water is discarded as a waste stream.

The third configuration is similar to the second, however it divides the separation between two modules, the first breaks the azeotrope while the second completes the dehydration.

One of the challenges facing this project is that process simulation engines do not offer a module to simulate membrane processes, hence a module that could be linked to ASPEN PLUS® has to be developed in order to study this separation technique. The developed module relies on the FORTRAN® VB link to transfer data between ASPEN ® and Excel

A Life Cycle Analysis (LCA) will also be conducted; this analysis will follow the ISO 1400 series.

The developed module was tested against experimental data for accuracy. The test showed that the calculations performed by the module yield similar results to those find by experimentation throughout most of the studied spectrum, the biggest error found was 13% at an 80mole% feed composition.

The total energy requirement per moles of ethanol recovered in each of the membrane processes were calculated and the results were compared, PFD II was found to require the least amount of energy, these results were then compared to the ones found for azeotropic distillation at a base case producing 96 mole% pure ethanol.

Once all 4 processes are simulated, process optimization techniques will be used to optimize them, such as:

· Heat and energy integration to design a heat exchanger network that could minimize the plant's dependence on outside energy sources to accomplish the heating and cooling required.

· Various types of membranes exist, and the effects of these different types on the energy demand of the pervaporation technique will be investigated.

· The effects of different types of solvents on the azeotropic distillation will also be analyzed.

All optimizations suggested will undoubtedly affect the LCA, this effect will also be taken into account.

(58a) How to Meet New ABET Requirements in Process Safety

Daniel A. Crowl, Department of Chemical Engineering, Michigan Technological University, Houghton, MI

The new process safety requirements were approved by ABET in October, 2011 and are in effect for all chemical engineering departments under ABET review beginning in the Fall of 2012. This paper will first discuss what the new requirements mean for chemical engineering departments. The author will then discuss the various ways that univerisities have used to meet these requirements. This will include a dedicated course, material spread throughout the curriulum or a hybrid approach. The advantages and disadvantages of each approach will be discussed. This paper will also discuss how the unit operations laboratory is very important for introducing aspects of safety culture into the curriculum. The paper will include information on how this is done at Michigan Tech University using a student run safety program. This includes a hazard reporting system, pre and post lab inspections and required safety meetings. This type of safety program closely matches industrial practice and prepares the student to encounter a similar program on their very first job experience.

A Proposed Methodology for Bringing Process Safety and Risk Management into Undergraduate Process Design Courses

Richard Roehner, Ph.D., P.E.
University of Utah, Dept. of Chemical Engineering
r.roehner@utah.edu

Abstract

In order to successfully bring Process Safety and Risk Management concepts into undergraduate courses on Process Design, the instructor must address three specific domains of learning: technical, cultural, and psychomotor. The technical domain is easily addressed and often over emphasized, since it involves presentation of effective quantitative information relating to hazard characteristics, and consequence & likelihood assessments. The cultural domain can be realized through emphasizing the need for the chemical engineer to include the fundamental questions posed in Process Hazard Analysis and Risk Assessment into their design process, along with typical areas such as materials selection, optimization, economics, and design quality checks. Simply stated, the instructor must imprint the students mind with the need for them to consider the existence of hazards, the risk associated with hazards, and possible mitigation or prevention measures when doing basic design work. This will put the "mindset" needed for accomplishing inherently safer design on an equal footing with the more common goals of process optimization and engineering economy. This will also ultimately help prevent the problems inherent to delay in resolving process hazards prior to reaching the common project schedule milestone of Process Hazards Analysis, or worse yet – construction and operation of the process facility. The psychomotor domain is best addressed in undergraduate laboratory courses, but can also be realized through repetitious incorporation of process safety aspects into every problem, case study, and technical concept that is worked in the Process Design Class. Examples and concepts for applying the above methodology will be presented. The ultimate goal here is to make the undergraduate engineer realize it is their job to incorporate process safety and risk management into their designs.

Introduction and Motivation

This is a summary of an approach to incorporation of process safety and risk management in the author's senior level Process Design course being taught at University of Utah. This course, which is often called the "capstone" design course at many schools, can be the first real place where undergraduate chemical engineers get to put together all of their newly learned insights into reaction engineering, transport phenomena, thermodynamics, process control, etc., for the design of equipment, facilities, and products. The author's approach is derived from thirty years of industrial experience which has seen the success and failure of various strategies in bringing the concepts of process safety and risk management into the work place. The hope is that we can benefit from an early exposure of chemical engineers to the fundamental concepts involved in process safety and risk management, while they are being introduced to the basic steps and philosophies of the design process. The goal is to have the students understand that they are responsible for the safety /risks inherent in their designs, and that they have a basic methodology to employ in taking on this responsibility. In this way, process safety and risk management become an integral part of the general design process, not just a process hazards analysis (PHA) done in the course of project execution, or PSM re-validation.

Background

One of the most successful outcomes in industrial process safety/risk management of recent years has been the adoption by companies and employees that safety is every individual's job – not the "Safety Departments". With this realization, chemical engineers can see the need to view all of their activities through the lens of hazard and risk analysis, as they undertake their work (including design). Since the undergraduate chemical engineering student is effectively a blank slate, what better than to introduce fundamentals of process safety and risk management as students are first formally introduced to the design process?

In order to successfully bring Process Safety and Risk Management concepts into undergraduate curricula, it has been suggested the instructor must address three specific domains of learning(1): technical, cultural, and psychomotor. The technical domain is easily addressed and often over emphasized, since it involves presentation of effective quantitative information relating to hazard characteristics, and consequence & likelihood assessments. Excellent materials for this have been put together by American Institute of Chemical Engineers – Safety and Chemical Engineering Education (AICHE-SAChE), which can be easily worked into a design class when discussing process flow sheet, development, plant site selection, reactor design, storage, etc. This paper suggests that the cultural domain can be realized through emphasizing to students the need for the chemical engineer to include answering fundamental questions posed in process hazard analysis and risk assessment in their design process. As a result of placing this emphasis, we build process safety and risk management into solving problems and development of projects by these future engineers. The psychomotor domain is best addressed in undergraduate laboratory courses through completion of Job Safety Analyses prior to and during lab exercises, but can also be realized through *repetitious* incorporation of process safety and risk management considerations into every problem, case study, and technical concept that is worked in the Process Design Class.

Proposed Methodology

Simply stated, the proposed methodology involves having students ask themselves the following simple questions commonly associated with formal steps in hazard and risk assessment as they consider design activities.

Hazards Analysis

- What can go wrong?
- What is the cause?

Risk Analysis

- What are the consequences?
- What is the likelihood?

Risk Resolution

- What will be done about it?

While there are many very effective formal methods for analysis of hazards and risks (i.e. HAZOP, HAZID, FTA, FMEA, LOPA....etc.) they all can viewed as being extensions of these simple questions above. This does not mean the chemical engineers should not be trained in these methods, it is just suggested that the above questions be easily made part of the design process being introduced to students in undergraduate design (whether the class is called process, product, or plant design), thereby making process safety and risk management a part of every design undertaken by the future chemical engineer.

Obviously technical information must be given to undergraduates to allow them to benefit from asking these fundamental questions. They need to be aware of fundamental hazards such as flammability, and toxicity, as well as how to arrive at estimates of consequence and likelihood for scenarios involving specific hazards. But it is the fundamental idea in this paper that students should learn to ask basic questions pertaining to process safety and risk management as they go through their design process. The AICHE-SAChE program offers courseware with specific structure to meet both ABET requirements and desired outcomes for student understanding of process safety and risk management. In particular, the recent program developed by Vaughn, described in Chemical Engineering Education[2], provides an excellent resource for professors to formally have students bring these concepts into their senior design projects. The proposed methodology of this paper can be viewed as setting the proper mindset in the student for going through the formal Design Project Roadmap of Vaughn.

This concept of making process safety and risk management part of the design process fits well with the fundamental use of risk assessment in general, since the best use of risk assessment is to support the decision making process – which is the basic building block of a design process. Students need to learn that simply asking these fundamental questions does not in itself guarantee safety, but it does provide an added input to the decision making process inherent to process design. It is this input from understanding hazards and risks associated with designs that must be used to help balance the economic and technical merits of design alternatives under consideration.

The goal here is that the students in process design courses view process safety and risk management fundamentals just as the other "tools" in their toolbox such as process simulation, economic analysis, and project management, not something that others will come and provide for their designs.

Examples

Finding examples to demonstrate how this suggested methodology can be applied in the classroom and course assignments is easily done by playing upon the decision making process present in many design problems worked by students. Problems such as:

- Which do you recommend – a thermal or catalytic reactor? A classic reaction engineering problem involves comparing the costs between two specific reactor designs. For a high temperature reacting system involving high hydrogen partial pressures a thermal reactor which runs 200 to 300 degrees Fahrenheit higher than a catalytic design can be shown to have additional risk due to the high temperature hydrogen attack (HTHA). Students can learn to include some quantification of this additional risk, along with costs associated with each design to make a recommendation.

- Which is the better separation scheme? Evaluations of separation schemes can include risks associated with hazards unique to process alternatives. Differences in solvents, temperatures, pressures, etc., can be explored not only from a cost estimate basis, but also from associated risks. This becomes the classic example for inherently safer design.

For each decision made in the design process, there are trade-offs. Including an understanding of hazards and risks associated with each design alternative is a fundamental example to students for using process safety and risk management in process design.

Other examples can be found in a design check or reflection step that can be applied to review of the results of student design efforts. These involve crafting a specific design exercise that produces calculated results which expose a hazardous condition with identifiable risks. These are easily found to have safety concerns that produce the classic "a-hah" experience. Examples of these include:

- Finding that there are in fact multiple-steady-state solutions for a CSTR, one of which could lead to high temperature related accident scenarios.
- The realization that a pressure swing distillation alternative creates the need for an expensive pressure relief system or SIL controls to deal with hazards created by direct coupling high and low pressure systems.

The question can then be put to students along the lines of "Did you think of what could go wrong when developing your design, or reviewing your calculations?"

Metrics

Quantification of an educator's success (or specifically the reasons for success) can be an exercise that leads to arguable results. One possible method could involve looking at how students bring this methodology to play, without specific direction, to second semester capstone design projects after practicing the methodology in assigned problems during the first semester of process design. Quantification of the student's assimilation of technical concepts can be handled by traditional assessment methods (i.e. tests, quizzes).

Recommendations

Several recommendations come to mind when looking at how to implement this methodology. As described, example problems and assigned problems with a specific hint or information supplied with the problem description relating to process safety hammer home the importance of process safety and risk management to the design process. A safety minute at the start of every class period in which a student volunteers a specific thought/issue related to personal or process safety can serve to reinforce the need to "switch gears" when re-entering the work/class environment. And finally, it is useful to build an expectation that students will incorporate some aspect of process safety and risk management into homework assignments and capstone design projects turned in for class credit.

Conclusion

The simple concept of incorporating the simple questions of *"What can go wrong? What is the cause? What are the consequences? What is the likelihood? What should be done about it?"* into assignments in undergraduate Process Design classes offers the opportunity for improvements in the performance of future chemical engineering graduates in the workforce.

References

1. Crowl, D., Educating the Next Generation of PSM Professionals, Paper 11 b, AICHE Spring Meeting 2012, Houston, Texas.
2. Vaughn, B.K., An Approach To Help Departments Meet The New ABET Process Safety Requirements. CEE, Vol. 46, No. 2, Spring 2012, 129-134.

(58c) Process Safety At the University of South Carolina

Edward P. Gatzke, Department of Chemical Engineering, The University of South Carolina, Columbia, SC and Vincent Van Brunt, University of South Carolina, Columbia, SC

Since 1990, the University of South Carolina has required a three credit hour chemical process safety course. Topics considered in the course include: toxicology, hazard analysis models, dispersion models, fires / explosions, designing for safety, relive valves, hazard identification, HAZOP analysis, reliability and risk assessment, and case history analysis. Special emphasis is given to the Guidelines for Chemical Process Quantitative Risk Analysis (CPQRA), Risk-Based Process Safety, and SACHE Modules.

Professor Vincent Van Brunt (currently Distinguished Professor Emeritus) originally created and offered this course. Process Safety is typically taken in the Spring semester of a student's senior year. The course is similar to a capstone design course in that it requires students to draw from many various chemical engineering courses to solve and analyze problems. Student evaluation consists of two tests and one final exam for evaluation of student performance, along with numerous (> 20) homework assignments and a course project. Lectures draw from the instructor's many years serving as a consultant for chemical process safety topics. Video and still images of chemical process accidents are used in class to demonstrate the potential impact of engineering safety. Videos are also used for informational purposes for some topics such as active suppression systems and dust explosion safety.

(58d) Capstone Safety and Toxicology Course Methodologies

James Smith Jr., Chemical and Materials Engineering , University of Alabama in Huntsville , Huntsville, AL

Capstone Safety and Toxicology Course Methodologies

by

James E. Smith, Jr., Department of Chemical and Materials Engineering, University of Alabama in Huntsville, Huntsville, AL, Richard F. Garnett, Honda Manufacturing Alabama, Lincoln Alabama, Tara Henriksen, CFEI *Senior Chemical Engineer at CASE Forensics Corp, Seattle, WA* and Robert Jordan, PSRG – Houston TX and HCS Engineering, Decatur, AL·

The Capstone Safety and Toxicology Course will be based upon the Chemical Process Safety Fundamentals with Applications, by Daniel A. Crowl and Joseph F. Louvar (2011) with several interesting adaptations that we will present in this paper and the subsequent presentation. We intend to use presentation materials provided by Dr. Crowl but enhance the materials by adding diversity in four separate areas using several visiting professionals and a virtual visitor to augment the course materials.

Although we will follow the text's table of content, the area of toxicology, Chapter 2 will be the first project area to receive augmentation. For the last 8 years, Dr. James E. Smith, Jr., Professor of Chemical Engineering at the University of Alabama in Huntsville, Huntsville, AL, has been utilizing an approach to expand upon aspects of safety and toxicology that the students seen to enjoy and learn from.

The approach starts by carefully selecting various classes of industrial chemicals including oxidizers, acids, strong bases, toxic gases, carcinogenic, mutagenic, teratogenic and hallucinogenic compounds to name a few. Each student is assigned a unique chemical and initially told to obtain the MSDS sheet and fully expand all acronym typically found in the data. They are also assigned to investigate the fire hazards associated with the chemical, spill mitigation, DOT transportation restrictions, personal protection gear, and the CGA number and its importance to the safe handling of compressed gases utilizing industrial suppliers data from such companies as Matheson or Air Gas if applicable. They are to consult the Hazardous Chemical Desk Reference, by N. Irving Sax and Richard J. Lewis, Sr. for their information on the substance. Finally, they are to consider poisoning and what medical conditions might occur at the various levels typical found in the MSDS sheet. They are to determine how either they or someone they are responsible for might respond to slight to high doses of the compound. The effect on human health from the medical sided should be explored. If the dosage can cause death or disability then it must be investigated and information that hospital poison prevention needs from the EMT considered. Here, we invite either a local EMT Unit or a Registered Nurse from our College of Nursing to discuss issue of poisoning and poison prevention by industrial and household chemicals.

This two-week project is then written into a report at a freshmen level. Since there are numerous chemicals assigned, the interesting part that the students enjoy is the round table discussion, when each student presents a general overview of the chemical they were assigned to the class. In this way, students are exposed to the hazards and toxic effects of numerous chemicals during a single class session. Examples of successful chemicals used in this type of assignment will be part of the presentation.

Our next augmentation has been developed by Mr. Richard F. Garnett PMP, P.E., Honda Manufacturing Alabama, LLC, 1800 Honda Drive, Lincoln Alabama 35096 and he will examine Governmental Rules & Regulation on issues of design. The U.S. regulatory system is designed to protect the health and safety of U.S. citizens and to protect the environment. The chemical design process is impacted by both these mandatory regulations (OSHA, EPA, etc) and voluntary standards (NEC, API, ASME, etc). Leveraging the information provided in Chapter 3 of Chemical Process

Safety text, this lecture will discuss a brief history of safety regulations using the impact of Bhopal on the regulator environment (OSHA 1910.119) as an example. The lecture will provide a brief overview of typical OSHA and EPA requirements and of the National Electric code. ASME B31.3, the ASME Pressure Vessel Code, API 610 and API 520.

Richard holds a Bachelors of Chemical Engineering from Auburn University (1981) and Masters in Management Science from the University of Alabama at Huntsville (1995). He is currently working toward a PhD in Engineering Management at UAH. Mr. Garnett has over thirty years of experience in the Chemical and Automotive industries and has a diverse background in the practice of Process Engineering and Project Management. Before coming to Honda, he worked at D&S Consulting Engineers, Novacor Chemicals Inc. (Decatur, Al) and at AKZO Coatings Inc. (Baxley Ga.). He has been delivering lectures for many years to our senior design course that will migrate to the Capstone Safety and Toxicology Course. His materials will be available during the presentation.

Tara Henriksen, Ph.D, CFEI *Senior Chemical Engineer at CASE Forensics Corp., Seattle WA USA will expand upon issues of fire safety and prevention.* Dr. Hendrickson holds a Ph.D in Chemical Engineering from the University of Utah, and has spent her career studying failure mechanisms and conducting failure analyses in the chemical processing industry. She has investigated numerous failures, which have resulted in fires, explosions, BLEVES, corrosion damage, pollution, personal injuries and fatalities. Dr. Henriksen will contribute to the curricula by presenting 'lessons learned' from case studies she has investigated in the field to demonstrate how process accidents could have been avoided through the implementation of fire safety practices. In particular, students will learn how engineering controls, preventative maintenance, and adherence to appropriate design and testing standards can be used to avoid accidents. Her presentation to the students will enhance materials covered in Chapters 6 and 7. She will present this information by either a prepared DVD or perhaps by SKYPE depending on her schedule. Dr. Henriksen's materials will be available during the presentation.

Finally, Robert Jordan - Senior Process Engineer will address HAZOP related issues expanding upon Chapters 11 and 12. The purpose of this session will be to give chemical engineering students a familiarity with HAZOP studies and what the role of the chemical engineer is as a participant in the studies. The approach will be to first explore a brief history of HAZOP as a technique and the mechanics of the methodology focused on the use of Guidewords/Deviations. Next we will look at the relationship of HAZOP and the Process Safety Management (PSM) Standard. We will review briefly the sections of the PSM Standards with emphasis on those most closely associated with the role of chemical engineers. This will also include a discussion of HAZOP and why it is a superior technique. We will then review and discuss the information a chemical engineer would develop for the study so they know what to bring to the study. This will include information such as: process descriptions, operating limits, and process procedures. Finally the role of the chemical engineer in the study will be examined to include a discussion of the types of questions the engineer should expect. The session will conclude with a demonstration of the HAZOP technique using an example from the CCPS book on Hazard Evaluation Procedures. The background material on the example will be given out prior to the class so the students can review the material and be familiar with the process. There will be no follow-up homework assignments.

Robert holds a BS in Chemistry from the University of Southern Mississippi, a BS in Chemical Engineering from Mississippi State University and an MBA in International Finance from Washington University in St. Louis. He first trained as a HAZOP Leader in 1991 and conducted his first HAZOP Leader Training class in 1996. He has been active in HAZOP Leader training and conducting HAZOP studies both domestically and overseas. He has worked at numerous chemical plants including: National Fertilizer Development Center – Muscle Shoals, AL, Amoco Decatur Plant – Decatur, AL., Amoco Wood River Plant – Wood River, IL, Amoco Chemical Manufacturing Services – Chicago, IL, Amoco WE&C PSM Group – Houston, TX, BP Decatur Works – Decatur, AL and PSRG – Houston TX and HCS Engineering-Decatur, AL. and has a wealth of materials some of which will be shared during the presentation.

Conclusions

We feel that the four topic discussed above, along with the well written text will contribute to the development of a capstone safety course while providing the audience with potentially interesting ideas and concepts they may choose to adapt or suggest modifications to our course as discussion proceed. References

Backer, W. and Mossman, A.L., Matheson Gas Data Book, Matheson Gas Products, East Rutherford, N.J., (1987).

CFR 1910.119 Process Safety Management

Crowl, D.A. and Louvar, J., F., Chemical Process Safety Fundamentals with Applications, Prentice Hall, Upper Saddle River, NJ (2011).

Dreisbach, R.H., Handbook of Poisoning: Diagnosis and Treatment, Lange Medical Publications, Los Altos California (1974).

Sax, N.I. and Lewis, Sr., R.J., Hazardous Chemicals Desk Reference, Van Nostrand Reinhold Company, NY (1987).

(58e) Using a "Level Control" Experiment to Demonstrate Operational Safety Concepts

Peyton C. Richmond, Chemical Engineering Department, Lamar University, Beaumont, TX and Qian Zhang, Dan F. Smith Department of Chemical Engineering, Lamar University, Beaumont, TX

A level control experiment was modified and used to educate undergraduate chemical engineering students about four aspects of operational safety at chemical processing facilities. First, students were taught that safety systems should be based on dedicated instruments separate from the basic control system. For the level experiment, a pressure switch and associated alarms were added to provide an independent measure of the tank level in addition to the pressure transmitter used for level feedback control. Second, students were taught about safety instrumented systems (SIS) which are used to provide additional layers of protection against incidents. To simulate a SIS for the level control experiment, logic was added to protect the outlet pump from being damaged in the event that the level tank ran dry during its operation. The control logic was configured to only activate the relay that supplied the pump with power if the level was high enough to activate the pressure switch. Third, students were made aware of the difference between an indicated process value and the actual process state. They were able to observe during the course of the experiment that the level's percentage full value obtained from the pressure transmitter output was different from the level obtained by visual inspection. Finally, students were exposed to the possibility that the configured range may be exceeded for a particular process instrument. The process point used to display the level's percentage full value was configured to show a "bad value" if the level was out-of-range. Lack of knowledge of how a level instrument responds to over-range and other process conditions was apparently a contributing factor in the BP Texas City Refinery explosion.

Process Safety, Moving From the Classroom to the Research Lab

Instructors with experience with the US Chemical Safety Board, classroom instruction, and application of process safety techniques to research application will describe how a combination of classroom instruction on process safety, along with application of techniques learned in the classroom to academic lab research processes, provides multiple benefits to both students and principal investigators. Benefits of process safety to research operations are many, including but not limited to the generation of P&ID drawings which can be used for both training and management of change as students and processes change, a solid understanding of the error modes and safety controls needed, and the non safety, process quality and reliability benefits which are also derived from a process safety review of laboratory processes. An example of a team approach involving the principal investigator, students, and EH&S staff is included along with discussion of various lessons learned through the application of hazard analysis techniques in laboratory settings. Application of process safety techniques need not wait for deployment after graduation and employment in the industrial sector.

In 2011, ABET took an important step towards making it less likely that a chemical engineering student could graduate from an accredited program without having a good introduction to chemical process safety management concepts and techniques. The chemical engineering curriculum is now expected to educate students in the methods of hazard analysis essential to the safe design and operation of chemical processes. This initiative as part of an ABET accreditation process is commendable.

AIChE, through the Center for Chemical Process Safety (CCPS), the Safety and Chemical Engineering Education program (SAChE) and its many member institutions are providing the educational resources to enable colleges and universities to meet this chemical process safety challenge in the classroom.

The instructional content should include the long established industry process safety practices, as well as a review of OSHA's regulation "Process Safety Management of Highly Hazardous Chemicals (29 CFR 1910.119). The elements of this standard should provide an excellent framework for a comprehensive hazard analysis approach and should be reviewed by all who have hazard identification and control responsibilities, whether or not they are involved with process quantities that are regulated by the OSHA standard. Application of good process safety practices in the non-regulated academic setting can provide faculty and researcher staff an opportunity to marry instruction with practice, training a student will need.

The US Chemical Safety and Hazard Investigation Board (CSB) conducts root cause investigations of serious chemical incidents in the United States. CSB investigation reports and multimedia case studies are excellent educational resources for today's chemical engineering students. Much can be learned and applied in the classroom and laboratory from the CSB products regarding chemical process safety practices and incident investigation techniques, particularly as inputs for consideration in the design, hazard identification and control review, and operational phases of an experiment.

In 2010 the CSB made academic laboratories take notice that smaller scale chemical incidents were also fair game for CSB investigation. Following several high profile incidents in academic laboratories with serious consequences, the CSB indicated it would investigate a campus laboratory explosion. The subsequent recommendations included, among other things, a recommendation to improve hazard assessments in research laboratories.

Based on CSB finding and experience, the following should be included in a laboratory process safety curriculum for chemical engineering students:

- Fundamentals of Chemical Process Safety
  (e.g., Chemical Process Safety: Fundamentals with Applications, 3rd Edition, 2011)
- Occupational Exposure to Hazardous Chemicals in Laboratories Standard
  (OSHA 29 CFR 1910.1450)
- Process Safety Management of Highly Hazardous Chemicals Standard
  (29 CFR 1910.119)
- University's Site-Specific Health, Safety, and Environmental Laboratory Procedures
  (must address physical hazards such as fires, explosions, reactive hazards, etc.)
- Research Laboratory Process Hazard Analysis and Risk Assessment Methods
- Research Laboratory Process Hazard Controls and Mitigation Techniques
- Design and Application of Inherently Safer Technologies
- Chemical Process Incident Investigations and Lessons Learned
  (e.g. CCPS Process Safety Beacon, SAChE materials)

Colleges and universities that implement process safety and hazard review programs as suggested above will certainly reap many educational, research, and student-career benefits. Academic research laboratories that implement hazard review programs can benefit in ways beyond safety.

The performance of hazard assessments in an academic environment should be approached as a teaching opportunity. When conducting a process safety analysis where students are involved, it is important to methodically execute the review, with experience personnel resisting the temptation to "cut to the chase" of standard solutions, but to allow the students to experience the opportunity to ask pertinent questions and see how the team interacts. . This approach develops an understanding and appreciation of concepts which they can apply to future reviews. One should remember that very small scale reviews may involve only the experimenter who has hopefully already learned and developed "process safety thinking" that can be applied to the frequent decisions associated with minor adjustments to their experimental materials and processes. In a properly conducted hazard review, particularly for the novice practitioner, what is learned about the hazard assessment approach is as important as the conclusions reached regarding the particular experiment under review.

In addition, there is no better way for new laboratory personnel to learn the details, the hazards and pitfalls of the experiment, than participation in a hazard review. Where P&IDs are required prior to the start of the experiment, these provide a training tool available for review for future experiments or to manage changes to experimental designs and protocols. Ancillary procedures, such as cylinder change procedures and others which may or may not be included in the hazard review proper, can be identified for further evaluation. In addition to an understanding of key loss control concepts, the participants can also gain a true understanding and appreciation of the need for the control measures selected. In fact, it is well known that buy-in to procedures is enhanced through participation in their development, so compliance with recommended safety procedures determined through the review should be enhanced.

Training and hazard review content should also include an understanding of human factors. Aside from safety benefits, an understanding of error traps and other circumstances where error rates would be expected to increase should allow the student to reduce error rates not limited to those with safety impact, but those which may affect their technical affairs and private life, leading to an overall reduction in adverse outcomes and, ideally, a reduction in stress and increase in productivity leading to a more content human being.

(82a) Keynote: Games to Teach and Games to Test: Developing and Assessing Innovation

David W. Shaffer, Department of Educational Psychology, University of Wisconsin-Madison, Madison, WI

How can we make sure that our kids are learning to be creative thinkers in a world of global competition--and what does that mean for the future of education in the digital age? David Williamson Shaffer offers a fresh and powerful perspective on computer games and learning. *How Computer Games Help Children Learn* shows how video and computer games can help teach kids to build successful futures--but only if we think in new ways about education itself. Shaffer shows how computer and video games can help students learn to think like engineers, urban planners, journalists, lawyers, and other innovative professionals, giving them the tools they need to survive in a changing world. Based on more than a decade of research in technology, game science, and education, *How Computer Games Help Children Learn* revolutionizes the ongoing debate about the pros and cons of digital learning.

[This is the book description of David Shaffer's famous book "How Computer Games Help Children Learn", that has been an inspiration and encouragement to all those in Chemical Engineering who had thought along those lines, or had been ready for it]

(82b) How Real Is Real Enough? Student and Expert Perceptions of an Industrially Situated Virtual Laboratory Project

Debra Gilbuena[1], Ben Sherrett[2] and **Milo D. Koretsky**[1], (1)Chemical, Biological, and Environmental Engineering, Oregon State University, Corvallis, OR, (2)College of Mechanical, Industrial, and Manufacturing Engineering, Oregon State University, Corvallis, OR

Preparation of engineering students for industrial practice is an important outcome of engineering education. We propose one way to help achieve this outcome is to implement a game-based learning system, such as a virtual laboratory.

Virtual laboratories can be examined in regards to the *role and context* within which they intend to place students. From this perspective, two primary classifications emerge. In the first type, the learner is placed in the role of a student in a typical classroom or teaching laboratory. In this role, the learner performs tasks common in traditional classrooms or laboratories except they are performed on a computer. Many virtual laboratories fit into this classification. In this study, we report on a second type of virtual laboratory, where the software and instructional design attempts to remove the learner from the role and context of a traditional classroom or laboratory and ask them to play a different role. Asking participants to assume alternative roles is a common practice in *games*. We focus primarily on a particular type of game in which the learner is placed in the role of a practicing professional. We term this case an *industrially-situated virtual laboratory*.

An *industrially-situated virtual laboratory* can provide a transition between academia and industry, offering students the experience of solving "real world" problems within the familiar setting of the university. While such projects have been described in the engineering education literature, there is less work examining the extent to which these "authentic" and "situated" learning systems need to replicate professional practice. That is, for learning systems intended to prepare students for industrial practice, *how real is real enough?*

There are several ways that providing an authentic experience to students can help them prepare for industrial practice. First, if students believe the task they are completing is likely to be one that they will face as a professional, they will be more motivated to engage in it. Second, contextual similarity allows the students to more readily transfer the knowledge and skills that they have learned in the classroom to problems in the workplace. Third, engaging in a task in which the cognitive and social interactions are reflective of practice helps them develop a more sophisticated epistemological view of engineering.

This paper examines one *industrially-situated virtual laboratory* which has been designed to replicate the experiences of an engineer in professional practice. This learning system is based on a process development task in the semiconductor manufacturing industry where experiments are conducted using a virtual laboratory. Previously we have reported on how students and experts complete the task. In this study, we wish to examine the students' and experts' perceptions of the authenticity of the task.

Two data sources were used. Semi-structured interviews were conducted with three experts who had substantial experience as engineers in industry and with 12 students after they had completed the virtual laboratory project. Second, we use transcripts of think-aloud protocol data from four student teams and three expert teams who completed the project. These transcripts were examined to see references to "industry" and real experiences as the participants were engaged in the project.

In general, participants found the task to be consistent with tasks that participants had engaged in (or expected to) while in industrial practice. Evidence from the interviews identified aspects of the project that are perceived as representative of the "real world." Two of the primary aspects of the project that were perceived to contribute to authenticity included the open-endedness of the problem statement (e.g., no "one" correct solution or solution path exists) and the cost constraint. Other aspects of the project were perceived as less- or non-authentic, e.g., the facilitative coaching sessions. We argue that these less authentic aspects provide more scaffolding than the "real world" and are conducive to learning. Such scaffolding is necessary, and can be implemented without sacrificing the essential perception of authenticity that motivates and situates learning.

We argue that virtual laboratories or games can be used very effectively to help students transition between academia and industry. We propose the degree of scaffolding should be intentionally incorporated into the instructional design, depending on where students are in that transition.

(82c) From Plug-and-Chug to Design Optimization Through Gaming

Margot Vigeant, Chemical Engineering, Bucknell University, Lewisburg, PA

49% of Americans between the ages of 18-24 own a smartphone (Pew Internet & American Life Project, 2011). This, combined with growing rates of tablet adoption, enables students to perform on-the-fly computation that was previously impossible for campuses without required laptops. We now can add very short-term design / optimization 'games' that fit into the same class time previously required for in-class examples. By encouraging students to move rote equation application onto a mobile computing device, students can access higher Bloom levels by comparing results over a wide range of conditions rather than the single condition accessible in a 'by hand' calculation. While this has been possible for many years in computing labs, making use of mobile platforms enables rapidly deployed or spur of the moment 'games' where student groups explore a parameter space for optimal conditions. I will present several examples of these short-term games from Thermodynamics and elective courses.

(82d) Level up! Gamification and Positive Psychology in the Chemical Engineering Classroom

Joshua A. Enszer, Chemical, Biochemical, & Environmental Engineering, University of Maryland Baltimore County, Baltimore, MD

While game-based learning is trending as an upcoming adoptable technology in higher education, there are a number of low- to no-tech ways to incorporate ideas that make games appealing into education: concrete goals, the ability to problem solve and collaborate with others, and the possibility of achieving spectacular successes [1]. These ideas sound familiar, don't they? A number of educators and game designers have discussed some obvious connections, [e.g., 2], but usually discuss game design more than implementation in the classroom. With an entire generation now having been raised in an environment where complex video games have been a part [3], analogies to such games may be a worthwhile effort in engaging students of any discipline. I will describe a small collection of minor connections and actions that can be made in course design and course management software to provide an outlet that may motivate our current generation of students in new ways.

The idea of an "experience points system" is nothing new to video games, and indeed, instructors across the country (though especially in game design) already implement such a system to evaluate student performance [e.g., 4]. Rather than expressing grade computations as general percentages that sum to 100%, converting course activities (in-class work, homework, lab reports, exams) into raw points achieves the same result mathematically. Further, it emphasizes that assignment evaluations are based on evidence of abilities – the more clearly and correctly this evidence is displayed, the more experience points earned; points are not "lost" for doing something "wrong."

With the support of a course management system, "achievements" or "trophies" can be "unlocked" by students who accomplish specific goals in a course. These achievements are virtually limitless in implementation, though it can be easier to program them if they are tied to a numerical computation.

If desired, the extension of course activities can be made to other artifacts common to games and video games: "guild activities" for group work, "bosses" or "final bosses" for exams. This approach is nothing but a rebranding of what is already typical for assessment in courses, but can add fun and motivation to a course.

I will share more detailed examples of implementing these strategies in a sophomore-level course on computer methods in chemical engineering, as well as the result of surveys given to students who experienced this course.

[1] The New Media Consortium. 2012 Horizon Report.

[2] Gee, James Paul. Good Video Games + Good Learning: Collected Essays on Video Games, Learning, and Literacy. New York: Peter Lang Publishing, Inc., 2007.

[3] Schell, Jesse. The Art of Game Design. Boston: Elsevier, 2008.

[4] Sheldon, Lee, and Hoffstein, Jenna. Gaming the Classroom. gamingtheclassroom.wordpress.com

(82e) Enhanced Learning Via Open-Form Laboratory Projects for Process Control

Tomas Co, Chemical Engineering, Michigan Technological University, Houghton, MI

With the increased availability of affordable sensors, data-acquisition devices, transducers, actuators, electronic parts and sophisticated software support, we were able to incorporate an open-form laboratory experience for the students. During the spring semester of 2012, we initiated this experimental approach in the laboratory portion of the undergraduate process control class at Michigan Technological University. Although special projects have always been available as a separate course to the students, the difference in this endeavor was to require all the students in the class, in groups of two, to build their own experiments from the proposal stage, to the specification and purchase of the parts, to the execution of their chosen experiment, and then to final reporting. We took the first three weeks to introduce the most basic elements of control and computing to the students. The rest of the semester was then opened to the students to develop and perform their experiments. Each group was allowed to choose their own experiment and they were given a starting budget of $100. However, if they could articulate and persuade us of the need for extra funds, then we would approve an increase in their budget. This allowed all students, including those who would not have chosen to take on such a project, to discover and apply the concepts of process control that were simultaneously covered in the lecture portion of the course. We observed a very significant increase in personal engagement and learning among the students for the process control as a result of this activity.

In this paper, we will explain the mechanics of our experiment in laboratory education and discuss the important observations we found. We will also include a list of improvements and best practices. Since this was initiated only in the spring semester of 2012, the data will mostly be anecdotal. Nonetheless, I have been teaching this course for several years now, and the difference we saw was very notable. This includes students, who were initially uninterested in the course, devoted several extra hours outside of their allotted laboratory time, plus weekends, to go beyond their initial proposed experiment in order to improve the control performance of their chosen system. In addition, the students were actually conferring and helping each other figure things out since we had not given them any specific procedure or instruction. We did not notice any competition among the groups.

(90a) Energy and Sustainability Modules in Chemical Engineering At Mississippi State University

Jason M. Keith, Bill B. Elmore, W. Todd French, Hossein Toghiani and Rebecca K. Toghiani, Dave C. Swalm School of Chemical Engineering, Mississippi State University, Mississippi State, MS

In this economic climate, departments of chemical engineering are challenged to improve retention to justify appropriate use of state and federal funding. At Mississippi State University, we are attempting to increase student interest by using active learning methods in the classroom in order to improve retention and student interest in the profession of chemical engineering. We are using energy modules, most of which are applied to hydrogen production or hydrogen use in fuel cells, which were created by external support from the CACHE Corporation and the United States Department of Energy. The use of energy technology in the classroom provides us as instructors with a means to apply the chemical engineering fundamentals to a broad problem in sustainable energy. We have anecdotally noted that since most chemical engineering students want a career where they can help others, the theme of sustainability grabs their attention. We report on current uses of the modules across the curriculum.

(90b) A New Course in Concepts, Assessment Tools and Methods in Sustainable Power and Energy Using Novel Course Format and Delivery Methods

Jeffrey Seay, Chemical and Materials Engineering, University of Kentucky, Lexington, KY

Energy and sustainability are important topics in industry, among the public and across college campuses. Therefore it is important to find innovative ways to incorporate energy and sustainability into the curriculum, both in Chemical Engineering as well as in other engineering disciplines. This contribution will describe the development and implementation of a new elective course in Concepts, Assessment Tools and Methods in Sustainable Power and Energy at the University of Kentucky. This course is designed to utilize a mixture of traditional lecture and self-directed web modules to deliver course content to students in a variety of engineering disciplines. Additionally, this contribution will describe the use of an interactive television connection to allow students at the University of Kentucky main campus in Lexington, the University of Kentucky Extended Campus in Paducah, and off campus graduate students enrolled in the course to attend lecture and participate in classroom discussions. This new course is included as part of the Power and Energy Certificate program at the University of Kentucky.

This multidisciplinary course presents an overview of the key topics in sustainability and environmental impact assessment for engineers. Topics included in the course are: assessment of current and potential future energy systems; renewable and conventional energy technologies; supply chain management; sustainability metrics; energy assessment tools; environmental impact assessment; and life cycle assessment. Topics included in the course are presented and their attributes described within a framework that aids in evaluation and analysis of energy technology systems and designs in the context of political, social, economic, and environmental goals. This contribution will also present some of the challenges of developing a multidisciplinary course in sustainability, due to the fact that each engineering discipline approach the topic from a slightly different perspective, and the academic backgrounds are different in each discipline as well. Finally, in addition to the course topics and content, this contribution will include the results of a Pre- and Post- Course assessment of the student's attitudes and perceived knowledge with regard to sustainability, as well as their personal assessment of the novel course structure and delivery format.

The course to be discussed in this contribution was developed as part of the Power and Energy Institute of Kentucky (PEIK) which is partially funded by the US Department of Energy under grant DE-OE0000409

(90c) Energy and Sustainability - Maymester Study Abroad

Edward P. Gatzke, Department of Chemical Engineering, The University of South Carolina, Columbia, SC and John Weidner, Chemical Engineering, University of South Carolina, Columbia, SC

The University of South Carolina has offered a course for the last five years titled "Next Energy." This course covers a wide variety of topics, including: traditional power sources, hydrogen economy, solar power (thermal and photovoltaic), nuclear, fuel cells, wind, geothermal, hydro, and storage systems. The course has been taught to chemical engineers predominately, with some offerings targeting only non-engineering majors from the honors college.

In May 2012, USC ran a version of this course titled "Energy and Sustainability in Europe." Much of the course material is drawn from the "Next Energy" course. However, students spend two weeks traveling to various locations in Europe. Locations include the Fraunhofer Institute for Solar Energy, the "greenest" city in Germany (Freiburg), the Mercedes Benz auto museum, the German Technical Museum, the Technical University of Munich, and the Max Plank Institute for Plasma Physics. This course draws from the entire engineering college, so topics such as building efficiency/design and transportation were added to the course. The program will also predominately use public transportation while in Europe.

A description of course topics and materials will be provided. Insight from the study abroad offering will also be discussed.

Sustainable Materials As Biomedical Materials: A Short Course for Undergraduate Students

Sujata K. Bhatia
Harvard University, Cambridge, MA

The overarching goal of this work is to attract and retain undergraduate students in bioengineering programs. In January 2012, Harvard University implemented an optional on-campus, four-day short course on bioengineering during winter break for undergraduates, with the goal of introducing students to the disciplines of biochemical engineering and biomedical engineering. The course was titled, "Sustainable Materials as Biomedical Materials," and allowed students to examine whether materials created through biochemical engineering could be useful for applications in biomedical engineering. Toward the goal of a sustainable bio-economy, research in biochemical engineering is increasingly devoted to the development of renewably sourced materials, such as bio-polymers and bio-composites derived from corn, soy, kenaf, flax, and cellulose. At the same time, innovators in biomedical engineering are seeking novel materials for implantable medical devices which will be optimally compatible with the human body. A natural intersect exists between these two areas of emerging research: naturally sourced polymers may be ideal for the design of new biomedical implants. This interdisciplinary course operated at the intersection of biochemical science, molecular biology, biomedical engineering, and sustainable processing.

Students received introductory lectures on biomedical engineering, including implantable biomaterials and drug delivery strategies, as well as biocompatibility testing of biomaterials. Students also received introductory lectures on biochemical engineering, including genetic engineering and metabolic engineering. Examples of naturally-derived materials for clinical applications were then presented to students; one such example was the design of polysaccharide-based polymers for wound closure. In the hands-on component of the course, students designed and conducted their own experiments in the undergraduate bioengineering laboratories and design laboratories, to explore the biocompatibility of bio-derived materials with clinically relevant cell lines. For instance, one student examined the proliferation of cardiac fibroblasts on naturally-derived agar hydrogels.

The course attracted a diverse population of students from freshmen through seniors, and was particularly successful in recruiting women and under-represented minorities. The course was designed as an intimate learning experience for a small class. At the conclusion of the course, students completed an anonymous survey which assessed their attitudes and motivations to study bioengineering. The results suggest that the bioengineering short course positively reinforced student perceptions and enthusiasm for studying bioengineering. Further, students' anonymous comments revealed the positive impact of the course. One student wrote, "I think this course was really thought-provoking and inspiring...The course made me more confident in my choice of a bioengineering concentration and I leave excited to pursue more work in my field of choice!" Another wrote, "It has shown me the many things one can work on as a bioengineer, and has made me more excited about the subject matter. It has also strengthened my decision of major." Yet another student commented, "Excellent course – so much to think about. I think the field is great." One student commented that the course influenced his perceptions of future career possibilities: "It was definitely helpful in showing me possible career paths, as well as the breadth of opportunities there are in bioengineering." Importantly, a student stated, "I couldn't have thought of anything more that I would have rather done! Really, biochemical and biomedical engineering excite me a lot!" Students clearly came away from the course with a sense of pride in their chosen profession.

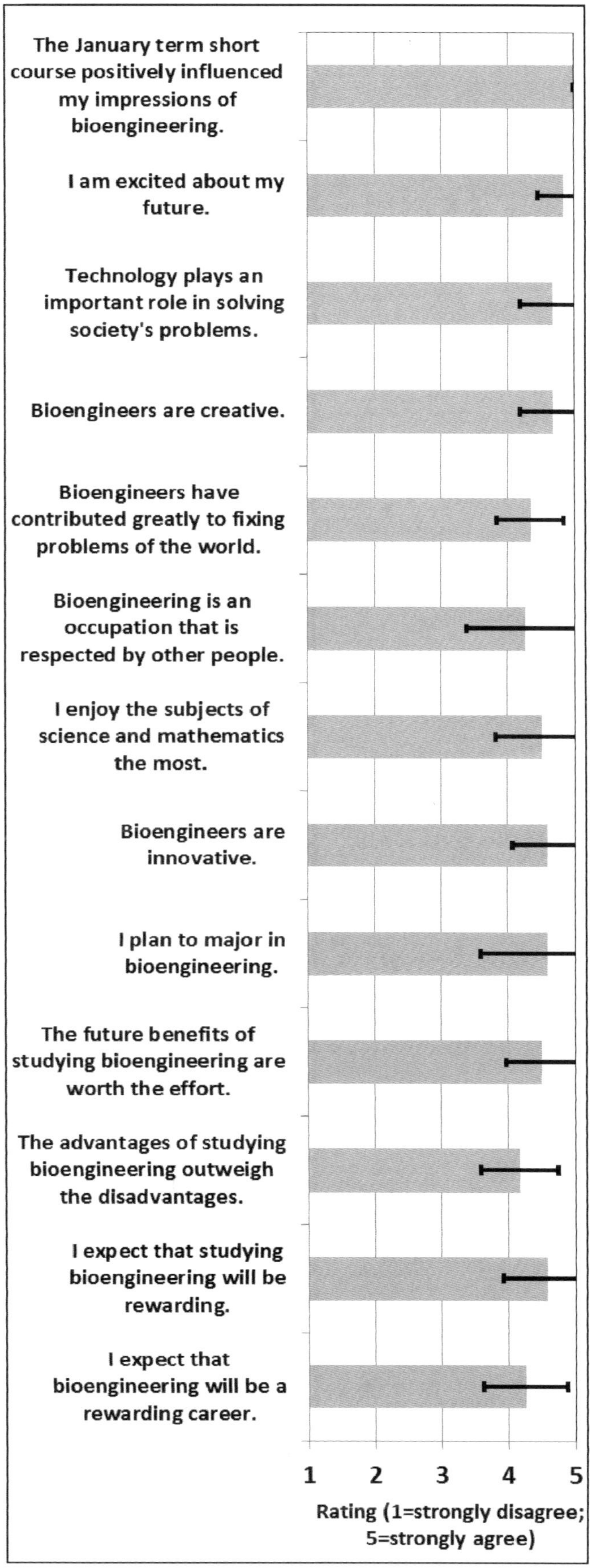
The January term short course positively influenced my impressions of bioengineering.
I am excited about my future.
Technology plays an important role in solving society's problems.
Bioengineers are creative.
Bioengineers have contributed greatly to fixing problems of the world.
Bioengineering is an occupation that is respected by other people.
I enjoy the subjects of science and mathematics the most.
Bioengineers are innovative.
I plan to major in bioengineering.
The future benefits of studying bioengineering are worth the effort.
The advantages of studying bioengineering outweigh the disadvantages.
I expect that studying bioengineering will be rewarding.
I expect that bioengineering will be a rewarding career.
1
2
3
4
5
Rating (1=strongly disagree; 5=strongly agree)

(90e) Incorporation of Process Intensification Into Chemical Reactor Design Through Instructional Modules

Rebecca K. Toghiani, Dave C. Swalm School of Chemical Engineering, Mississippi State University, MS State, MS and Carlen D. Henington, Counseling and Educational Psychology, Mississippi State University, MS State, MS

Process intensification encompasses a variety of concepts and practices that may result in significant savings in capital and operating expenses, as well as enhanced safety, reliability, and risk reduction. Three instructional modules have been developed for the chemical reactor design course that focus on concepts/topics that fall under the process intensification umbrella. The first module focuses on coupling of chemical reactions to realize operational benefits with respect to energy consumption and/or overcoming thermodynamic equilibrium limitations. The second module discusses the combination of reaction and separation through the use of simulated moving bed reactors. The third module focuses on novel reactor configurations, such as the spinning disk reactor and the rotating packed bed, and how their use can improve process efficiency. Design of the modules allows their use either as a lecture in the traditional course with a companion homework assignment, or as a stand-alone activity that students can complete outside of the normal lecture period.

(90f) A New Interdisciplinary Engineering Course – "Nanoscale Transport Phenomena for Manufacturing Nanodevices"

Zhiyong Gu[1], Bridgette Budhlall[2], Hongwei Sun[3], Carol MF. Barry[4], Alfred Donatelli[1] and Jill Lohmeier[5], (1)Chemical Engineering, University of Massachusetts Lowell, Lowell, MA, (2)Plastics Engineering, University of Massachusetts Lowell, Lowell, MA, (3)Mechanical Engineering, University of Massachusetts Lowell, Lowell, MA, (4)Department of Plastics Engineering and Center for High-Rate Nanomanufacturing, University of Massachusetts Lowell, Lowell, MA, (5)Graduate School of Education, University of Massachusetts Lowell, Lowell, MA

Herein we will introduce a new interdisciplinary engineering course - "Nanoscale Transport Phenomena for Manufacturing Nanodevices", which was recently developed in the Francis College of Engineering at the University of Massachusetts Lowell. The course focuses on the principles of nanoscale transport phenomena needed for manufacturing nanodevices and aims to close a large gap between nanoscience and commercial production of nanotechnology products. The course also helps to integrate the interdisciplinary knowledge required for designing and manufacturing nanodevices into undergraduate curricula. To meet these unique needs and challenges, five faculty from three engineering departments (Chemical, Mechanical and Plastics Engineering) have created this interdisciplinary course which was offered for the first time as an elective to seniors in the College of Engineering during the 2011 fall semester. The course was presented through lectures, hands-on laboratory exercises, demonstration experiments, and a final design project. In this presentation, we will discuss the lecture topics and eight hands-on laboratory experiments that were developed into modules to complement lectures in fluid mechanics, heat transfer, mixing, reaction engineering, electroosmosis, electophoresis, and manufacturing methods for micro and nanoscale devices. We will also show the final project designs for the nanodevices that were finished by student teams at the end of the course. Finally, we will show the assessment results from the pre-post student surveys as well as faculty interviews. This new interdisciplinary course will better prepare undergraduates for employment focused on designing and manufacturing nano/microfluidic systems, lab-on-a-chip devices, electronics devices, medical devices, and other emerging technologies. The impact of this senior-level course will significantly enhance the "Nanomaterials Engineering Option" in the Chemical Engineering Department undergraduate curriculum as well as the medical device industry focus in the Plastics Engineering Department, and can be used in the accelerated BS-MS program which is popular in the College of Engineering. The course will be available to the chemical, mechanical, and plastics engineering seniors each year. Our lab modules can also be exported to freshman introductory engineering courses in the College of Engineering. In addition, the microscale fluid mechanics and heat transfer experiments may be incorporated into the undergraduate chemical engineering Unit Operations Laboratory courses.

(106a) NSF Overview

Robert M. Wellek, Division of Chemical, Bioengineering, Environmental and Transport Systems, The National Science Foundation, Arlington, VA

Faculty are invited to attend the NSF Workshop Sessions. Details of the sessions will be available as soon as the program is finalized. Each year NSF conducts these two Workshop Sessions at the Annual AIChE Meeting in order to provide useful information regarding the Foundation and its programs. Session 1 is focused on presentations relating to the NSF programs in the Division of Chemical, Bioengineering, Environmental and Transport Systems (CBET). The presentations will be made by program directors representing these clusters. There will be an opportunity to participate in a panel discussion on current trends in NSF funding. Session II includes a tutorial on proposal writing and round table discussions with individual program directors. We also invite faculty to visit the NSF Information Suite that will be open from from 8:00 a.m. until noon on Monday and from 8:00 a.m. until 5:00 p.m. on Tuesday and Wednesday. The NSF Information Suite provides an excellent opportunity for one-on-one discussions with NSF Program Officers.

(106b) Highlights of CBET Cluster On Biomedical Engineering and Engineering Healthcare

Robert M. Wellek, Division of Chemical, Bioengineering, Environmental and Transport Systems, The National Science Foundation, Arlington, VA

Faculty are invited to attend the NSF Workshop Sessions. Details of the sessions will be available as soon as the program is finalized. Each year NSF conducts these two Workshop Sessions at the Annual AIChE Meeting in order to provide useful information regarding the Foundation and its programs. Session 1 is focused on presentations relating to the NSF programs in the Division of Chemical, Bioengineering, Environmental and Transport Systems (CBET). The presentations will be made by program directors representing these clusters. There will be an opportunity to participate in a panel discussion on current trends in NSF funding. Session II includes a tutorial on proposal writing and round table discussions with individual program directors. We also invite faculty to visit the NSF Information Suite that will be open from from 8:00 a.m. until noon on Monday and from 8:00 a.m. until 5:00 p.m. on Tuesday and Wednesday. The NSF Information Suite provides an excellent opportunity for one-on-one discussions with NSF Program Officers.

(106c) Highlights of CBET Cluster On Chemical, Biochemical & Biotechnology Systems

Luke Achenie, Process and Reaction Engineering, National Science Foundation, Arlington, VA

Faculty are invited to attend the NSF Workshop Sessions. Details of the sessions will be available as soon as the program is finalized. Each year NSF conducts these two Workshop Sessions at the Annual AIChE Meeting in order to provide useful information regarding the Foundation and its programs. Session 1 is focused on presentations relating to the NSF programs in the Division of Chemical, Bioengineering, Environmental and Transport Systems (CBET). The presentations will be made by program directors representing these clusters. There will be an opportunity to participate in a panel discussion on current trends in NSF funding. Session II includes a tutorial on proposal writing and round table discussions with individual program directors. We also invite faculty to visit the NSF Information Suite that will be open from from 8:00 a.m. until noon on Monday and from 8:00 a.m. until 5:00 p.m. on Tuesday and Wednesday. The NSF Information Suite provides an excellent opportunity for one-on-one discussions with NSF Program Officers.

(106d) Highlights of CBET Cluster On Transport and Thermal Fluid Phenomena

Ashok S. Sangani, Division of Chemical, Bioengineering, Environmental and Transport Systems, National Science Foundation, Arlington, VA

Faculty are invited to attend the NSF Workshop Sessions. Details of the sessions will be available as soon as the program is finalized. Each year NSF conducts these two Workshop Sessions at the Annual AIChE Meeting in order to provide useful information regarding the Foundation and its programs. Session 1 is focused on presentations relating to the NSF programs in the Division of Chemical, Bioengineering, Environmental and Transport Systems (CBET). The presentations will be made by program directors representing these clusters. There will be an opportunity to participate in a panel discussion on current trends in NSF funding. Session II includes a tutorial on proposal writing and round table discussions with individual program directors. We also invite faculty to visit the NSF Information Suite that will be open from from 8:00 a.m. until noon on Monday and from 8:00 a.m. until 5:00 p.m. on Tuesday and Wednesday. The NSF Information Suite provides an excellent opportunity for one-on-one discussions with NSF Program Officers.

(106e) Highlights of CBET Cluster On Environmental Engineering & Sustainability

Ram B. Gupta, Division of Chemical, Bioengineering, Environmental and Transport Systems, The National Science Foundation, Arlington, VA

Faculty are invited to attend the NSF Workshop Sessions. Details of the sessions will be available as soon as the program is finalized. Each year NSF conducts these two Workshop Sessions at the Annual AIChE Meeting in order to provide useful information regarding the Foundation and its programs. Session 1 is focused on presentations relating to the NSF programs in the Division of Chemical, Bioengineering, Environmental and Transport Systems (CBET). The presentations will be made by program directors representing these clusters. There will be an opportunity to participate in a panel discussion on current trends in NSF funding. Session II includes a tutorial on proposal writing and round table discussions with individual program directors. We also invite faculty to visit the NSF Information Suite that will be open from from 8:00 a.m. until noon on Monday and from 8:00 a.m. until 5:00 p.m. on Tuesday and Wednesday. The NSF Information Suite provides an excellent opportunity for one-on-one discussions with NSF Program Officers.

(106f) Interactive Question and Answer Session with NSF Program Directors

Robert M. Wellek, Division of Chemical, Bioengineering, Environmental and Transport Systems, The National Science Foundation, Arlington, VA

Faculty are invited to attend the NSF Workshop Sessions. Details of the sessions will be available as soon as the program is finalized. Each year NSF conducts these two Workshop Sessions at the Annual AIChE Meeting in order to provide useful information regarding the Foundation and its programs. Session 1 is focused on presentations relating to the NSF programs in the Division of Chemical, Bioengineering, Environmental and Transport Systems (CBET). The presentations will be made by program directors representing these clusters. There will be an opportunity to participate in a panel discussion on current trends in NSF funding. Session II includes a tutorial on proposal writing and round table discussions with individual program directors. We also invite faculty to visit the NSF Information Suite that will be open from from 8:00 a.m. until noon on Monday and from 8:00 a.m. until 5:00 p.m. on Tuesday and Wednesday. The NSF Information Suite provides an excellent opportunity for one-on-one discussions with NSF Program Officers.

(120a) Hybrid Cancer Therapeutics

Deniz Cetin and **Andrew Pike**, Department of Chemical Engineering, Lehigh University, Bethlehem, PA

Andrew Pike

Deniz Cetin

Project: Hybrid Cancer Therapeutics

Lehigh University

Department of Chemical Engineering

Advisors: Professors Bryan W. Berger & Mark A. Snyder

Abstract:

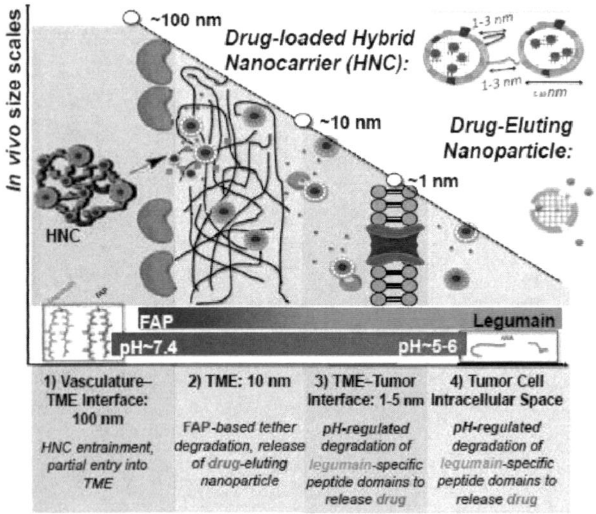

In nano-scale drug delivery, when targeting cancer cells, a recurring challenge is that the more drug-resistant cancer cells survive the initial burst of drug and continue to metastasize into more resilient tumors. There are several size selective interfaces along the delivery pathway that the delivery system will be designed to permeate. For this reason, a delivery system with multiple phases of delivery is desirable. A solution to this is to control the rate of drug delivery through particle size, porosity, and architecture. In addition, pH responsive proteins can be used to target tumor sites and prevent drug release to healthy cell environments. Current approaches include organic particles tethered in order to target specific ligand, which create problems such as chemical instability, inefficiency in use of drug (reduced half-life *in vivo*), eliciting an immune response, and variable porosity. Our project focuses on tethering of mesoporous silica nanoparticles with peptides which change conformation when exposed to environments of different pH values and have certain protease-cleavage sites, with the purpose of creating an initial burst of cancer drug-carrier nanoparticles with sustained release.

Our approach seeks to minimize these problems by using particles of 5 nm in diameter to control the release of drug, while designing the peptide structure to react to changes in the tumor microenvironment. This ensures that the silica nanoparticles are not exposed to the environment and are not releasing drug until the delivery system reaches a tumor environment.

(120b) Improving the Mechanical Properties of Activated Carbon Nanofiber Nonwovens

Breanne Muratori, Department of Chemical, Materials, and Biomolecular Engineering, University Of Connecticut, Storrs, CT, Seetha S. Manickam, Chemical, Materials, and Biomolecular Engineering, University Of Connecticut, Storrs, CT and Jeffrey R. McCutcheon, Department of Chemical, Materials and Biomolecular Engineering, University of Connecticut, Storrs, CT

Activated Carbon Nanofiber Nonwovens (ACNFNs) have a wide range of potential applications because of their unique properties like high specific surface area and good electrical conductivity. Potential applications include use as adsorption media, electrodes in microbial fuel cells, tissue scaffolds, and many others. A critical aspect that hinders ACNFNs from being used more widely in these applications is their weak mechanical strength. The goal of this work is to improve the strength of ACNFNs, without sacrificing surface area and conductivity.

Several different methods of improving strength were explored. The methods of fabrication were examined, looking for areas where the process could be altered, yielding a material with greater strength. The fabrication of ACNFNs is a four-step process: electrospinning, stabilization, carbonization, and steam activation. Alternative polymers for electrospinning, effects of hot-stretching during pyrolysis, and varying carbonization and activation protocol were studied. Samples were tested for mechanical strength, carbon yield, and surface area.

(120c) Nano-Confined CO2 Sorbents for High-Efficiency CO2-Capture

David Palm, Chemical Engineering Department, University of Pittsburgh, Pittsburgh, PA, Karen J. Uffalussy, Chemical and Petroleum Engineering, University of Pittsburgh, Pittsburgh, PA, Robert M. Enick, Department of Chemical and petroleum Engineering, University of Pittsburgh, Pittsburgh, PA; Chemical and Petroleum Engineering Department, University of Pittsburgh, Pittsburgh, PA and Götz Veser, Department of Chemical and Petroleum Engineering, University of Pittsburgh, Pittsburgh, PA

CO_2 capture and sequestration are among the prime challenges of our time, as the need to reduce atmospheric concentrations of this greenhouse gas has become evident. Among the most mature CO_2 capture technologies are liquid amine-based processes; however, even these applications are hampered by high cost and large energy penalty. Grafting or supporting liquid amines on solid supports offers a promising avenue to overcome some of these limitations. In particular, embedding amines into microporous solids constitutes a simple, flexible approach.

In the present project, we are investigating the application of this embedding approach to a novel class of CO_2 sorbents, so-called phase-change sorbent materials. An aminosilicone sorbent ("GAP-0") that has recently shown great promise as a CO_2 sorbent due to its high sorbent capacity and its ability to change from a liquid to a solid upon CO_2 sorption, was chosen as a model compound and embedded into porous silica shells. The resulting material has the advantage of a well-defined particle size, suppressed agglomeration, and fast CO_2 transport. Furthermore, on a fundamental level, the impact of nanoconfinement on phase transitions of materials is poorly understood to-date and the present system poses a well-defined model system for such studies.

GAP-0 was loaded into silica "nanobubbles" with ~30 nm diameter via a straight-forward wet impregnation method. The walls of these nanobubbles are ~10 nm thick and highly porous with average pore diameters of ~8Å. After characterization of the material, the CO_2 adsorption capability of the sorbent was studied via Thermo-Gravimetric Analysis (TGA) during cyclic CO_2 uptake and release over a temperature range between 45-90°C.

Results to-date show that nanoencapsulated GAP-0 does indeed adsorb and desorb CO_2 effectively. However, the GAP-0 compound exhibited unexpected volatility over the temperature range of the experiments, leading to a loss of sorbent material over time. These losses were reduced via optimization of uptake and release conditions (time, temperature), but multi-cycle experiments still showed continual loss of CO_2 sorption capacity. Current work is extending the demonstrated approach onto other aminosilicone-based phase-change sorbent materials.

(120d) Using Porous Carbon Nanotube Membranes for Separation of CH4/CO2 and CH4/H2 Mixtures

Benjamin Bucior[1], Jinchen Liu[2], De-Li Chen[1], De-en Jiang[3] and J. Karl Johnson[1], (1)Dept. of Chemical and Petroleum Engineering, University of Pittsburgh, Pittsburgh, PA, (2)University of Pittsburgh, Pittsburgh, PA, (3)Chemical Sciences Division, Oak Ridge National Laboratory, Oak Ridge, TN

Although separations are typically expensive and energy-intensive, membranes have the potential to separate gases both economically and efficiently. However, conventional polymeric membranes exhibit a tradeoff between selectivity and permeance. Carbon nanotubes (CNTs) exhibit very different transport behavior compared with other nanoporous materials, having transport diffusivities orders of magnitude higher than other matierals. Theoretically, gas mixtures could be separated by pores in a CNT membrane via a size-sieving mechanism. This means that nanoporous CNT membranes have the potential to be both selective and permeable, overcoming the limitations of conventional membranes.

We modeled a theoretical porous carbon nanotube (PCNT) membrane by generating pores on the wall of a CNT. We used molecular dynamics simulations to study fluid transport across the PCNT membrane, comparing models with rigid and flexible bonds. We also calculated the potential energy surfaces of each fluid molecule passing through the PCNT. Our simulations of both rigid and flexible PCNTs showed high selectivity for CH4/H2 and CH4/CO2 mixtures. Separations of N2/CO2 mixtures were also studied.

(120e) Determination of Degree of Polymerization of Cellulose Using MALDI-TOF with a Novel Ionic Liquid Matrix

Michael Mayer, University of Toledo, Toledo, OH, B. Leif Hanson, Instrumentation Center, The University of Toledo, Toledo, OH, Wendell Griffith, University of Toledo, Toledo and Constance Schall, Chemical & Environmental Engineering, University of Toledo, Toledo, OH

Polysaccharides, such as cellulose, have been investigated as a source for production of renewable fuels and other carbon based materials. Characterizing cellulose, especially its degree of polymerization (DP), is an important step in evaluating its deconstruction to monomeric sugars or in its use for materials. Matrix Assisted Laser Desorption Ionization Time of Flight Mass Spectroscopy (MALDI-TOF) has been used to determine the degree of polymerization of many polymersandmore recently carbohydrates. The use of MALDI-TOF for cellulosic analytes has been difficult due to their insolubility in traditional matrix solutions. A new technique has been developed for MALDI-TOF, utilizing an ionic liquid matrix and solvent system. This approach was used to determine DP in an excess of 200 for crystalline cellulose. The technique has potentially wide application in analysis of lignocellulosic substrates.

The Isolation and Incorporation of Chloroplasts Into Silk Matrices

Abstract

Global air pollution is a pandemic phenomenon that threatens both the stability of local ecosystems and the longevity of the human race. Carbon dioxide is the main pollutant of our atmosphere and is produced by both living beings and the burning of fossil fuels; the abundance of carbon dioxide in the atmosphere has resulted in temperatures and sea levels achieving an all time high. A solution proposed is that of creating a carbon dioxide consuming biomaterial. For this study, chloroplasts from spinach leaves were isolated and incorporated into silk matrices to create such a material. Both silk-chloroplast hybrid films and hydrogels were examined using microscopic analysis to determine the optimal environment for chloroplast survival outside of the plant cell. Hydrogels were chosen as the primary medium through which to incorporate chloroplasts into silk matrices as it provided an aqueous biomimetic environment that optimized chloroplast survival and functionality outside of the plant cells.

From there, tests were performed to determine the functionality of both hybrid films and hydrogels outside of the cell. These tests used apparatuses through which cigarette smoke was passed, the carbon dioxide filtering ability of the silk-chloroplast hybrid being a marker of the system functionality. The hybrid films were tested using a less-advanced apparatus than the hydrogels, as it was expected that they would not be as functional. For this test, cigarette smoke was passed through an apparatus and collected onto a pre-prepared silk mass. This silk mass was then processed and used to create films that were then analyzed using microscopy. From the microscopic analysis, it was apparent that the silk films were minimally functional as there was a marked decrease in particular deposits for the filtered group when compared to the control group. However, because of the non-aqueous environment of the silk-chloroplast hybrid film, it is unlikely that the silk-film hybrid would be functional for long-term applications.

The silk-chloroplast hydrogels were then tested using a more complicated apparatus. For this apparatus, the law of mass action was utilized by evacuating water tanks and drawing cigarette smoke into said water tanks, past the hydrogels. The pH of said water was then tested against a control and against pure water to determine the functionality of the silk-chloroplast hybrid hydrogels. It was expected that carbonic acid would form in the water tanks, as it is a product produced by the reaction of carbon dioxide and water. Therefore, the filtered water tanks were expected to have less carbon dioxide reacted with water, thereby having a higher pH. This result was tested using litmus pH strips and the expected result was achieved.

From this research, future investigation of the filtering capabilities of silk-chloroplast hybrid films can be performed to address the growing global concern of atmospheric pollution. Genetic recombination of chloroplasts and the sophistication of hydrogel functionality and test apparatuses are possible avenues of further research for the continuation of this project.

(157a) Proposal Writing Tutorial

Theresa Good, Chemical and Bichemical Engineering, University of Maryland Baltimore County, Baltimore, MD

Each year NSF conducts two Workshop Sessions at the Annual AIChE Meeting in order to provide useful information regarding the Foundation and its programs. Session 1 is focused on presentations relating to the NSF programs in the Division of Chemical, Bioengineering, Environmental and Transport Systems (CBET). Session II includes a tutorial on proposal writing and round table discussions with individual program directors. Both sessions are extremely valuable in helping faculty write competitive proposals for funding. The NSF workshop sessions offer an opportunity for new faculty to learn about NSF, its programs and new initiatives in research and education. There is significant opportunity for faculty to interact with NSF program directors. NSF also sponsors an information suite during the AIChE meeting. The NSF Information Suite is typically open for the first three days of the meeting and provides an excellent opportunity for one-on-one discussions with NSF Program Officers. Faculty are invited to participate in the Workshop Sessions and visit the Information Suite. Feedback from the Workshop Sessions will help NSF improve its programming for subsequent meetings.

(157b) Interactive Breakout Panels

Robert M. Wellek, Division of Chemical, Bioengineering, Environmental and Transport Systems, The National Science Foundation, Arlington, VA

Each year NSF conducts two Workshop Sessions at the Annual AIChE Meeting in order to provide useful information regarding the Foundation and its programs. Session 1 is focused on presentations relating to the NSF programs in the Division of Chemical, Bioengineering, Environmental and Transport Systems (CBET). Session II includes a tutorial on proposal writing and round table discussions with individual program directors. Both sessions are extremely valuable in helping faculty write competitive proposals for funding. The NSF workshop sessions offer an opportunity for new faculty to learn about NSF, its programs and new initiatives in research and education. There is significant opportunity for faculty to interact with NSF program directors. NSF also sponsors an information suite during the AIChE meeting. The NSF Information Suite is typically open for the first three days of the meeting and provides an excellent opportunity for one-on-one discussions with NSF Program Officers. Faculty are invited to participate in the Workshop Sessions and visit the Information Suite. Feedback from the Workshop Sessions will help NSF improve its programming for subsequent meetings.

This talk is part of the invited session "Random Thoughts: Inspired by Rich Felder".

Description: This session honors Dr. Richard M. Felder, Hoechst Celanese Professor Emeritus of Chemical Engineering at North Carolina State University. Rich has been recognized for his sustained contributions to education in the fields of engineering and engineering technology. He is the inaugural winner of the Lifetime Achievement Award in Engineering Education from the American Society for Engineering Education (ASEE). Rich is the co-author of the book *Elementary Principles of Chemical Processes* and has authored and co-authored more than 300 papers on engineering and science education and chemical process engineering. He has presented more than 650 workshops and seminars on effective teaching methods throughout North and South America, Australia, South Africa and numerous European and Asian countries. He has also co-directed the National Effective Teaching Institute under the auspices of the American Society for Engineering Education since 1991.

The session will feature short tributes from Rich's colleagues in the field of engineering education and will also include active learning and creative elements during the session. Rich will present a mini-keynote address at the conclusion of the session.

Speakers include:

James Stice
Ronald Rousseau
Phillip Wankat
Stephanie Farrell
Michael Prince
Armando Rugarcia
Scott Fogler
Lisa Bullard

(202a) AIChE/CACHE Mobile Device APP Competition

Robert P. Hesketh, Chemical Engineering, Rowan University, Glassboro, NJ

This presentation will report on the effort to start the first AIChE/CACHE mobile device App competition The objective of this competition is:

- To give students practice at developing their creative skills.
- To enhance and promote student's interest in new mobile device APPs for use in professional activities.
- To provide chemical engineering students with the opportunity to participate in a multidisciplinary team-oriented software design competition.
- To encourage students to become actively involved in their professional society.
- To show the connection between chemical engineering and society needs.
- To create useful mobile applications for the chemical engineering community.

National Competition

Undergraduate student teams will create an APP for a mobile device. This APP must be deemed to be useful for an AIChE member. Possible application areas include scientific,
engineering, teaching and professional meeting organizers. Teaching apps could include a game designed to teach chemical engineering concepts. The APP that is developed must be able to run on a mobile device such as a smartphone (iPhone, Android, Blackberry etc.) Student teams can consist of any discipline, but at least half of the team must be members of AIChE.
The competition at the AIChE annual meeting will consist of a poster competition where finalists will be selected to give a short presentation to a general audience. There will be two prizes, with monetary awards given to the student team's AIChE chapter:

1. $500 from CACHE for the first place winner of the best mobile device APP
2. $500 from a corporate sponsor for the best mobile device APP that addresses safety.
   Intellectual Property of the APP remains with the students who developed the APP.

Rules

- Student teams must consist of fulltime students
- Student teams can consist of any discipline, but at least half of the team must be members of the AIChE
- Student developed APPs can be constructed using existing software templates and components.
- The final APP must be an original creation of the team.
- Submissions must comply with intellectual property rules (e.g. copyright etc.)
- The student team must have at least one team member presenting at Annual Meeting.
- The App must be able to run on a mobile device that uses one of these operating Systems: Android, BlackBerry, iPhone, Palm, Symbian, or Windows Mobile.
- Timeline:
    - APP and Technical Documentation Submission 1 June 2012 through website upload (details to be announced)
    - Poster Competition – 10/28/12 Pittsburgh, PA

Judging of the overall competition will be based on:

- Creativity and uniqueness of the APP (15%)
- Usefulness of the APP (20%)
- Ease of use of the APP(15%)

- Professional and/or Societal Impact of APP (20%)
- Poster Presentation which includes demonstration of APP on a mobile device at AIChE Annual Meeting (30%)

Judging of Electronic Submission portion of the APP competition

- Submission deadline 1 June 2012; submission instructions will be announced at a later date
- Technical documentation containing
  - Description of the APP and equations used including screen shots
  - Sources of technical information and physical properties
  - Sources of software components that were used in APP
  - Flow chart illustrating how the APP works
  - Sufficient worked-out examples to verifying the utility of the APP
  - Broad impacts of chemical engineering on society
- APP Code

(202b) Integrating the Ipad and Iphone with Mass & Energy Balances

Jason E. Bara, John Patrick McLemore, Ashley M. Parker, Katie E. Jennings and Harvis J. Smith, Chemical & Biological Engineering, University of Alabama, Tuscaloosa, AL

The Chemical Engineering App was conceived to transform the information presented in a typical mass and energy balance course into a fully interactive suite of tools, spreadsheets, equations and data that will enhance students' learning experiences throughout their undergraduate curriculum yet also be highly useful to current chemical engineering professionals. While most of the relatively small number of existing apps geared to chemical engineers merely present basic mathematical formulas into which the user provides one or more inputs (e.g. the ideal gas equation) or perform routine unit conversions, we have taken a much more rigorous approach to implementing problem-solving tools in mobile devices. The major aspect of our strategy is to allow the user to directly access actual data within the calculations, providing a rapid and robust software package for tackling undergraduate course material. Distinct features of our app include thermophysical property data for hundreds of compounds and elements, steam tables, spreadsheet tools, graphing of VLE, Antoine Equation and heat capacity, unit converters, as well as a number of other functions and information encountered in a mass and energy balance course. We have also incorporated engines which can randomly generate an infinite number of mass and energy balance questions for a variety of unit operations as supplementary homework problems.

This presentation will discuss the history and motivation behind the Chemical Engineering App, the design philosophy, student feedback/response, and the impacts, challenges and opportunities in utilizing iPhones and iPads in teaching a mass and energy balance course.

(202c) Using an Operator Training Simulator in the Undergraduate Chemical Engineering Curriculum

Debangsu Bhattacharyya[1], **Richard Turton**[1] and Stephen E. Zitney[2], (1)Department of Chemical Engg., West Virginia University, Morgantown, WV, (2)AVESTAR Center, National Energy Technology Laboratory, Morgantown, WV

An operator training simulator (OTS) is to the chemical engineer what a flight simulator is to the aerospace engineer. The basis of an OTS is a high-fidelity dynamic model of a chemical process that allows an engineer to simulate start-up, shut-down, and normal operation. It can also be used to test the skill and ability of an engineer or operator to respond and control some unforeseen situation(s) through the use of programmed malfunctions.

West Virginia University (WVU) is a member of the National Energy Technology Laboratory's Regional University Alliance (NETL-RUA). Working through the NETL-RUA, the authors have spent the last four years collaborating on the development of a high-fidelity OTS for an Integrated Gasification Combined Cycle (IGCC) power plant with CO_2 capture that is the cornerstone of the AVESTAR™ (**A**dvanced **V**irtual **E**nergy **S**imulation **T**raining **A**nd **R**esearch) Center with sister facilities at NETL and WVU in Morgantown, WV. This OTS is capable of real-time dynamic simulation of IGCC plant operation, including start-up, shut-down, and power demand load following. The dynamic simulator and its human machine interfaces (HMIs) are based on the DYNSIM and InTouch software, respectively, from Invensys Operations Management.

The purpose of this presentation is to discuss the authors' experiences in using this sophisticated dynamic simulation-based OTS as a hands-on teaching tool in the undergraduate chemical engineering curriculum. At present, the OTS has been used in two separate courses: a new process simulation course and a traditional process control course.

In the process simulation course, concepts of steady-state and dynamic simulations were covered prior to exposing the students to the OTS. Moreover, digital logic and the concept of equipment requiring one or more permissive states to be enabled prior to successful operation were also covered. Students were briefed about start-up procedures and the importance of following a predetermined sequence of actions in order to start-up the plant successfully. Student experience with the dynamic simulator consisted of a six-hour training session in which the Claus sulfur capture unit of the IGCC plant was started up. The students were able to operate the simulator through the InTouch-based HMI displays and study and understand the underlying dynamic modeling approach used in the DYNSIM-based simulator. The concepts learned during the training sessions were further reinforced when students developed their own DYNSIM models for a chemical process and wrote a detailed start-up procedure.

In the process control course, students learned how the plant responds dynamically to changes in the manipulated inputs, as well as how the control system impacts plant performance, stability, robustness and disturbance rejection characteristics. The OTS provided the opportunity to study the dynamics of complicated, "real-life" process plants consisting of hundreds of pieces of equipment. Students implemented ideal forcing functions, tracked the time-delay through the entire plant, studied the response of open-loop unstable systems, and learned "good practices" in control system design by taking into account the real-world events where significant deviations from the "ideal" or "expected" response can occur. The theory of closed-loop stability was reinforced by implementing limiting proportional gain for stability limits of real plants. Finally, students were divided into several groups where each group was tasked to control a section of the plant within a set of operating limits in the face of disturbances and simulated process faults. At the end of this test, they suggested ways to improve the control system performance based on the theory they learned in class and the hands-on experience they earned while working on the OTS.

(202d) Improving Process Engineering Tools and Their Application in Chemical Engineering Curriculum

Ajay Lakshmanan, Aspen Technology, Inc., Burlington, MA, Boyd Gochenour, Aspen Technology, Inc., Burlington and George W. Huber, Department of Chemical Engineering, University of Massachusetts-Amherst, Amherst, MA

Accurate modeling can offer valuable insight into understanding the intricacies and trade-offs involved in a process plant. Over the past several decades different decision-support tools have been developed for specific business processes such as process modeling, design, simulation and optimization, control, production planning and scheduling. These tools are highly specialized and were focused on expert users. Advances in computing technology and devices have enabled educators and practitioners to introduce these tools to students and to other mainstream users earlier in the curriculum. To stay in sync with this new user community significant steps have been made in the past decade to improve these tools. This paper discusses how feedback from users in the universities and in the industry impacted the design of chemical engineering simulation and planning tools that are faster to learn and easier to use.

Specific topics that will be detailed include the design of a user workflow based interface for process engineering and planning tools that improves discoverability of "expert" features by the novice user; enhanced collaboration and search capabilities that enable users to locate relevant information within seconds, improved access to process engineering tools any time anywhere through mobile devices; and the joint work with universities to develop training modules for students.

The paper will also detail case studies and best practices at various universities on how professors have applied these tools in their university curriculum to further Chemical Engineering education.

Co-current Parameter Estimation and Model Refinement in Dynamical Systems

ID# 257668

Michael Elly, Chem. Eng. Dept., Ben-Gurion University, Beer-Sheva, Israel

Mordechai Shacham, Chem. Eng. Dept., Ben-Gurion University, Beer-Sheva, Israel

Jose C. Merchuk, Chem. Eng. Dept., Ben-Gurion University, Beer-Sheva, Israel

Introduction

Parameter estimation in dynamical systems has been used since the 1960s, mainly for the estimation of chemical reaction rate parameters. Recently there is a considerable interest in estimation of kinetic parameters in biological models. The computation of such parameters represents specific challenge (as pointed out, for example by Leppävuori et al, 2011) as the models often contain a large number of unknown kinetic parameters that cannot be measured and the number of parameters that can be reliably estimated, based on available experimental data, is often smaller than the total number of parameters. To reduce the number of equations and parameters the model is usually simplified (using, for example the pseudo steady state assumption, Ji and Luo, 2000) and some of the parameters are assigned a constant value a-priory. The presently used parameter estimation techniques (for a recent review see Michalic et al., 2009) may converge to a local minimum. However, even if they do converge to a global minimum the predicted values may differ significantly from the experimental data because of incorrectness of the assumptions that were associated with the model derivation and the assignment of constant, fixed values to some of the parameters.

To alleviate these difficulties we have developed an interactive tool that incorporates the human investigator in the parameter estimation and model refinement loop. This tool is basically a program which uses the, so called, "sequential approach" for parameter estimation. In an outer loop the weighted squared error between the experimental data set and the corresponding model predictions is minimized. In the inner loop an integration routine is used to determine the state variable values at time intervals where experimental data are available. The user can select the parameters that remain fixed and the ones that can be changed by the optimization algorithm.

The proposed interactive parameter estimation program has been implemented in POLYMATH 7.0 (POLYMATH is a product of Polymath Software, http://www.polymath-software.com). In this paper we report on the evaluation of the program using test problems for which the optimal solutions are available in the literature. A systematic procedure for simultaneous model improvement and initial guess variation for attaining a global optimum is also presented.

Basic Concepts

The parameter estimation problem involving an ODE model can be formulated as follows:

$$\min_{\theta} \Phi(\theta) = \sum_{\mu=1}^{n} \sum_{i=1}^{m} \left(x_{\mu,i} - x^{c}_{\mu,i}\right)^{2} / W_{i} \qquad (1)$$

s.t.

$$\frac{d\mathbf{x}}{dt} = \mathbf{F}(\mathbf{x}, \theta, t)$$
$$\mathbf{x}(0) = \mathbf{x}_0 \qquad (2)$$

where **F** is a system of *m* ordinary differential equations, **x** is a vector of *m* dynamic variables, **θ** is a vector of *p* parameters, $\mathbf{x}_\mu$ is a vector of *m* observed values at the μth data point, $\mathbf{x}_\mu^c$ is a vector of *m* calculated variable values at the μth data point and **W** is a vector of *m* weighting factors. In order to scale the variables in the $0 \le x \le 1$ interval the weighting factor:

$$W_i = \max_\mu |x_{\mu,i}| \qquad (3)$$

is used. The ODE parameter estimation problem is solved in POLYMATH 7.0 using a "sequential approach". Using this approach the minimization defined in Eq (1) is carried out in an outer loop while in the inner loop an integration routine is used to determine the state variable values at time intervals where experimental data are available. This approach is also called the feasible path approach because the estimated parameter values always satisfy the model. Presently only the Runge-Kutta-Fehlberg (RKF45, Forsythe et al, 1977) algorithm is used for the integration. This algorithm monitors the estimate of the integration error, and reduces or increases the step size of the integration in order to keep the error below a specified threshold. The accuracy requested is that both the relative and absolute (maximal) errors be less than the truncation error tolerance. The default value of this tolerance is 1.0E-6. The integration is carried out in a piecewise manner from the point t_μ to $t_{\mu+1}$. At this point the μth component of the objective function is calculated and added to the previous components. This process is continued until reaching the last set of data points where $t_{\mu+1} = t_n$.

The Levenberg-Marquardt (LM, see Press et al. 1992) algorithm is used for finding the optimal parameter values. This method is a modification of the Newton method where the second derivatives appearing in the Hessian matrix are neglected. Also, the algorithm switches to the steepest descent method when the Hessian matrix becomes nearly singular.

Graphical representation of the data and the calculated values as well as comparison of the optimal parameter values with the respective standard deviations (σ_k)are used for assessing the adequacy of the model and the parameters.

Example 1 – Catalytic cracking of gas oil

This example demonstrates the case where there is no need to modify the model and no user involvement is required to reach the global optimum. The example involves the modeling of the catalytic cracking of gas oil (*A*) to gasoline (*Q*) and other side products (*S*).

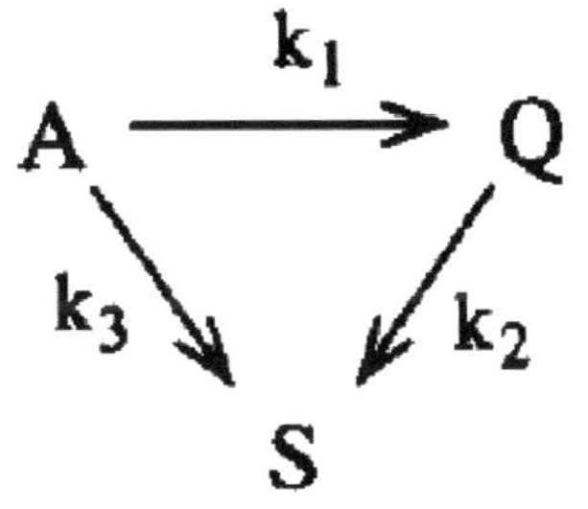

Only the concentrations of A (x_1) and Q (x_2) were measured and the concentration of S is not included in the model. This example was presented by Tjoa and Biegler (1991) and by Floudas et al. (1999). The models and data provided by Floudas et al. (1999) are used here.
The model equations of the catalytic cracking reactions are

$$\frac{dx_1}{dt} = -(\theta_1 + \theta_3)x_1^2$$
$$\frac{dx2}{dt} = \theta_1 x_1^2 - \theta_2 x_2 \qquad (4)$$

where θ_1, θ_2 and θ_3 are the rate constants. The parameter values are bounded $0 \le \boldsymbol{\theta} \le 20$.The experimentally observed values of x_1 and x_2 at times t_μ are shown in Table 1. Part of the Polymath input and the Polymath results are shown in Fig. 1. The Polymath input includes the observed values of the variables in a form of one dimensional arrays (only parts of the arrays are shown in Fig. 1), the model equations (the notation θ_1, θ_2 and θ_3 was changed to p1, p2 and p3) the parameter bounds and initial estimates for the parameter values. In this case the use of an arbitrary initial estimate $\theta_1 = \theta_2 = \theta_3 = 1$ within the parameter bounds was sufficient in order to reach the optimal solution.

The solution is reported in both numerical and graphical forms (Fig. 1). The plot of the calculated curves vs. the observed data points shows excellent fit for this problem. The numerical results for the parameter values and the sum of squares of the error at the optimal solution are compared with the results of Floudas et al. (1999) in Table 2. There are minor differences (in the 3rd or 4th decimal digits) in the solution values for the parameters and the sum of squares. These differences can result from the use of different error tolerances and/or different weighting factors in the objective function.

The standard deviations for θ_1, θ_2 are smaller by two orders of magnitude and for θ_3 by more than one order of magnitude than the parameter values. Thus, all the numerical and graphical indicators show that the global optimum has been found in this case without any need for user's intervention.

Example 2 – Conversion of methanol to hydrocarbons

This example demonstrate the case where the proposed reaction model includes redundant terms (parameters) that need to be removed by analysis of the ratio between the optimal parameter values and their standard deviations. This problem was suggested by Maria (1989) and Floudas et al. (1999). The model and data suggested by Floudas et al. (1999) are used here. The simplified kinetic model for conversion of methanol to various hydrocarbons is:

$$
\begin{aligned}
A &\xrightarrow{k_1} B \\
A+B &\xrightarrow{k_2} C \\
C+B &\xrightarrow{k_3} P \\
A &\xrightarrow{k_4} C \\
A &\xrightarrow{k_5} P \\
A+B &\xrightarrow{k_6} P
\end{aligned}
$$

where *A* represents the oxygenates, *B* is the intermediate $\ddot{C}H_2$, *C* represents the olefins, and *P* represents the paraffins, aromatics and other products. The model of the reaction is formulated under the assumption of simple kinetics and quasi- steady state for *B*.

$$
\begin{aligned}
\frac{dx_1}{dt} &= -(2\theta_1 - \frac{\theta_1 x_2}{(\theta_2+\theta_5)x_1 + x_2} + \theta_3 + \theta_4)x_1 \\
\frac{dx_2}{dt} &= \frac{\theta_1 x_1(\theta_2 x_1 - x_2)}{(\theta_2+\theta_5)x_1 + x_2} + \theta_3 x_1 \\
\frac{dx_3}{dt} &= \frac{\theta_1 x_1(x_2 - \theta_5 x_1)}{(\theta_2+\theta_5)x_1 + x_2} + \theta_4 x_1
\end{aligned}
\qquad (5)
$$

where x_1, x_2 and x_3 are the mole fractions of *A*, *B* and *C* respectively. The parameter values are bounded $0 \le \boldsymbol{\theta} \le 100$. The experimentally observed values of x_1, x_2 and x_3 at times t_μ are shown in Table 3.

The optimal parameter values were first sought for the model containing all five parameters starting at the initial estimate $\theta_k = 1$, for k = 1, 2, ... 5. The results are shown in the first six rows of Table 4. Observe that for θ_4 the standard deviation is considerably larger than the parameter value (σ_4 = 6.831E-05 while θ_4 = 2.587E-07) meaning that the value of θ_4 can be set at zero. Rerunning the problem while setting θ_4 at a fixed value of $\theta_4 = 0$ yields the optimal results shown in Table 4, as the 2nd set of results. In this case the value of θ_5 becomes smaller than its standard deviation. The optimal parameter values for the case when both θ_4 and θ_5 set at the constant value of zero include the θ_3 value that is smaller than the associated standard deviation.

The final optimal model contains only the parameters θ_1 and θ_2 while the additional parameters should be set at zero. Thus the experimental data can justify only the simplified model presented in Eq. 6 for the methanol conversion reaction.
The optimal values for θ_1 and θ_2, obtained by Polymath (Table 4) are larger than the respective standard deviations by more than four orders of magnitude and they are very close to the values reported by Floudas et al. (1999). The model provides good representation of the data with regard to the products *B* and *C* (Figure 2), as for the oxygenates, there is still room for improvement of the model.

$$\frac{dx_1}{dt} = -(2\theta_1 - \frac{\theta_1 x_2}{\theta_2 x_1 + x_2})x_1$$
$$\frac{dx_2}{dt} = \frac{\theta_1 x_1(\theta_2 x_1 - x_2)}{\theta_2 x_1 + x_2} \qquad (6)$$
$$\frac{dx_3}{dt} = \frac{\theta_1 x_1 x_2}{\theta_2 x_1 + x_2}$$

Example 3 – The "Lotka-Volterra" predator-prey model

This example demonstrates the case where there is convergence to a local optimum from an arbitrary initial guess and user intervention is required to reach the global optimum. The "Lotka-Volterra" problem was originally studied by Luus (1998). This problem involves a two parameter predator-prey model similar to those used in ecology. The model determines the population over time of the two different species (the predators, and the prey) given parameters relating to the interaction between them. The state profiles at the solution are cyclic and out of phase with each other. The model and data presented by Floudas et al. (1999) are used here. The models for the predator (x_2) and the prey (x_1) populations are given by

$$\frac{dx_1}{dt} = \theta_1 x_1(1 - x_2)$$
$$\frac{dx2}{dt} = \theta_2 x_2(x_1 - 1) \qquad (7)$$

The parameters $\boldsymbol{\theta}$ are related to the birth, death, and interaction rates of the two species and they are bounded: $0 \le \boldsymbol{\theta} \le 10$. The data provided by Floudas et al. (1999) are shown in Table 5. We first attempted to find the optimum starting at the initial guess $\theta_1 = \theta_2 = 1$. The results are shown in Fig. 3 and in the first three rows of Table 6. Observe that in Fig. 3 both the data and the calculated curves exhibit oscillatory behavior, however the fit between the data and the calculated values is very poor. The relatively large standard deviation values (Table 6) also signal poor fit of the calculated values to the data. The similarity between the trend of the data and the calculated curve may indicate that the model is correct and the poor fit is caused by convergence to a local minimum. Changing the initial guess to $\theta_1 = 5$ and $\theta_2 = 1$ yields much better results. Fig. 4 shows that in this case there is a perfect match between the data and the calculated curves. The numerical results in Table 6 show that with the optimal parameter values found here, the sum of squares of errors and the standard deviations are reduced by at least two orders of magnitude in comparison to the previous case.

Conclusions

A new parameter estimation program for dynamic systems, intended for interactive model analysis and parameter estimation, has been developed. This program was evaluated by solving problems for which optimal solutions are available in the literature. Three examples were presented to demonstrate the use of the numerical and graphical feedbacks provided by the program for analyzing the model and attaining the global optimum in the parameter values. It has been demonstrated that a globally optimal solution for an adequate model is usually characterized by: 1. Similar trend and random residual distribution of the data around the associated calculated curve and 2. Small standard deviation/ abs(parameter value) ratios for all the parameters.

It was shown that if the absolute parameter value is smaller than its standard deviation this usually signals a redundant parameter. Consequently, this parameter and the associated term/s can be removed from the model. In cases where the trend of the data and the calculated curves are similar but there are large discrepancies between the data points and the associated curves this may signal convergence to a local optimum. In such cases variation of the initial guesses of the parameter values within their pre specified bounds is recommended.

Work is currently underway to apply the Polymath ODE parameter estimation program and the proposed interactive model analysis and parameter optimization procedure for identification of the model and parameters for a tobacco mosaic virus (TMV) proliferation problem (Enden et al., 2005). The results of this study will be reported elsewhere.

References

1. Enden, G., Y.H. Zhang, and J.C. Merchuk, A model of the dynamics of insect cell infection at low multiplicity of infection. Journal of Theoretical Biology, 2005. 237(3): p. 257-264.
2. Floudas, C. A., Pardalos, P. M. et al., Handbook of Test Problems in Local and Global Optimization, Kluwer, Dordrecht, The Netherlands, 1999
3. Forsythe, G.E., M.A. Malcolm, and C.B. Moler, Computer Methods for Mathematical Computation, Prentice-Hall, Englewood Cliffs, 1977.
4. Ji, F. and L. Luo, A hyper cycle theory of proliferation of viruses and resistance to the viruses of transgenic plant, Journal of Theoretical Biology, 2000, 204(3), 453-465.
5. Leppävuori, J. T.; Domach, M.M.; Biegler, L.T., Parameter Estimation in Batch Bioreactor Simulation Using Metabolic Models: Sequential Solution with Direct Sensitivities, Ind. Eng. Chem. Res. 2011, 50, 12080-12091
6. Luus, R., Parameter estimation of lotka-volterra problem by direct search optimization, Hung. J. Ind. Chem. 26, 1998
7. Maria, G., An adaptive strategy for solving kinetic model coccomitant estimation-reduction problems., Can. J. Chem. Eng. 67, 825(1989)
8. Michalik, C.; Chachuat, B.; Marquardt, W., Incremental Global Parameter Estimation in Dynamical Systems, Ind. Eng. Chem. Res. 2009, 48, 5489–5497.
9. Press, W.H., P.B. Flannery, S.A. Teukolsky and W.T. Vetterling, Numerical Recipes, 2nd Ed., Cambridge University Press, Cambridge, 1992.

Table 1. Data of observed values for Example 1 (Floudas et al. 1999).

Point No.: μ	t	$x_{\mu 1}$	$x_{\mu 2}$
0	0	1	0
1	0.025	0.7307	0.1954
2	0.05	0.5982	0.2808
3	0.075	0.4678	0.3175
4	0.1	0.4267	0.3047
5	0.125	0.3436	0.2991
6	0.15	0.3126	0.2619
7	0.175	0.2808	0.2391
8	0.2	0.2692	0.221
9	0.225	0.221	0.1898
10	0.25	0.2122	0.1801
11	0.3	0.1903	0.1503
12	0.35	0.1735	0.103
13	0.4	0.1615	0.0964
14	0.45	0.124	0.0581
15	0.5	0.119	0.0471
16	0.55	0.1109	0.0413
17	0.65	0.089	0.0367
18	0.75	0.082	0.0219
19	0.85	0.0745	0.0124
20	0.95	0.0639	0.0089

Table 2. Comparison of the optimal results from Floudas et al. (1999) with the Polymath results for Example 1.

Param. No.	*Value*[1]	σ	*Value*[2]
1	12.24071	0.11935	12.214
2	7.922492	0.0566524	7.9798
3	2.25226	0.1010523	2.2216
sum of squares	0.0026777	-	0.002656

[1]Polymath 7.0 results

[2]Optimal results from Floudas et al. (1999)

Table 3. Data of observed values for Example 2 (Floudas et al. 1999).

Point No.: μ	t	$x_{\mu 1}$	$x_{\mu 2}$	$x_{\mu 3}$
0	0	1	0	0
1	0.05	0.461	0.114	0.018
2	0.065	0.426	0.135	0.035
3	0.08	0.383	0.157	0.045
4	0.123	0.305	0.194	0.047
5	0.233	0.195	0.231	0.084
6	0.273	0.17	0.234	0.095
7	0.354	0.139	0.228	0.111
8	0.397	0.112	0.228	0.134
9	0.418	0.112	0.226	0.168
10	0.502	0.09	0.22	0.148
11	0.553	0.082	0.214	0.157
12	0.681	0.066	0.198	0.206
13	0.75	0.053	0.188	0.206
14	0.916	0.043	0.183	0.214
15	0.937	0.041	0.184	0.213
16	1.122	0.029	0.166	0.23

Table 4. Optimal results for Example 2 with various numbers of parameters

Param. No.	*Value*[1]	σ	*Value*[2]
1	3.998714	0.0002032	
2	0.9478271	8.686E-05	
3	0.760283	0.0001154	
4	**2.59E-07**	**6.83E-05**	
5	0.0685149	4.367E-05	
sum of squares	0.14928		
1	4.329012	0.0155159	
2	0.9410537	0.0064165	
3	0.3091262	0.0309942	
5	**2.55E-06**	**0.002628**	
sum of squares	0.12741		
1	6.122073	0.0004623	
2	1.905657	7.768E-05	
3	**3.39E-07**	**9.99E-05**	
sum of squares	0.16044		
1	5.052398	0.0004623	5.2407
2	1.220247	7.768E-05	1.2176
sum of squares	0.10721		0.10693

[1]Polymath 7.0 results

[2]Optimal results from Floudas et al. (1999)

Table 5. Data of observed values for Example 3 (Floudas et al. 1999).

Point No.: μ	t	$x_{\mu 1}$	$x_{\mu 2}$
0	0	1.2	1.1
1	1	0.799	1.0758
2	2	0.8731	0.8711
3	3	1.2487	0.9393
4	4	1.0362	1.1468
5	5	0.7483	1.0027
6	6	1.0024	0.8577
7	7	1.2816	1.0274
8	8	0.8944	1.1369
9	9	0.7852	0.9325
10	10	1.1527	0.9074

Table 6. Optimal results for Example 3 obtained using different initial guesses

Param. No.	*Initial guess*	*Value[1]*	σ	*Value[2]*
1	1	1.177802	0.1759329	
2	1	0.7200475	0.1198334	
sum of squares		0.79365	-	
1	1	3.192145	0.0038293	3.2424
2	1	0.9385447	0.0012158	0.9209
sum of squares		0.0013234	-	0.001249

[1]Polymath 7.0 results

[2]Optimal results from Floudas et al. (1999)

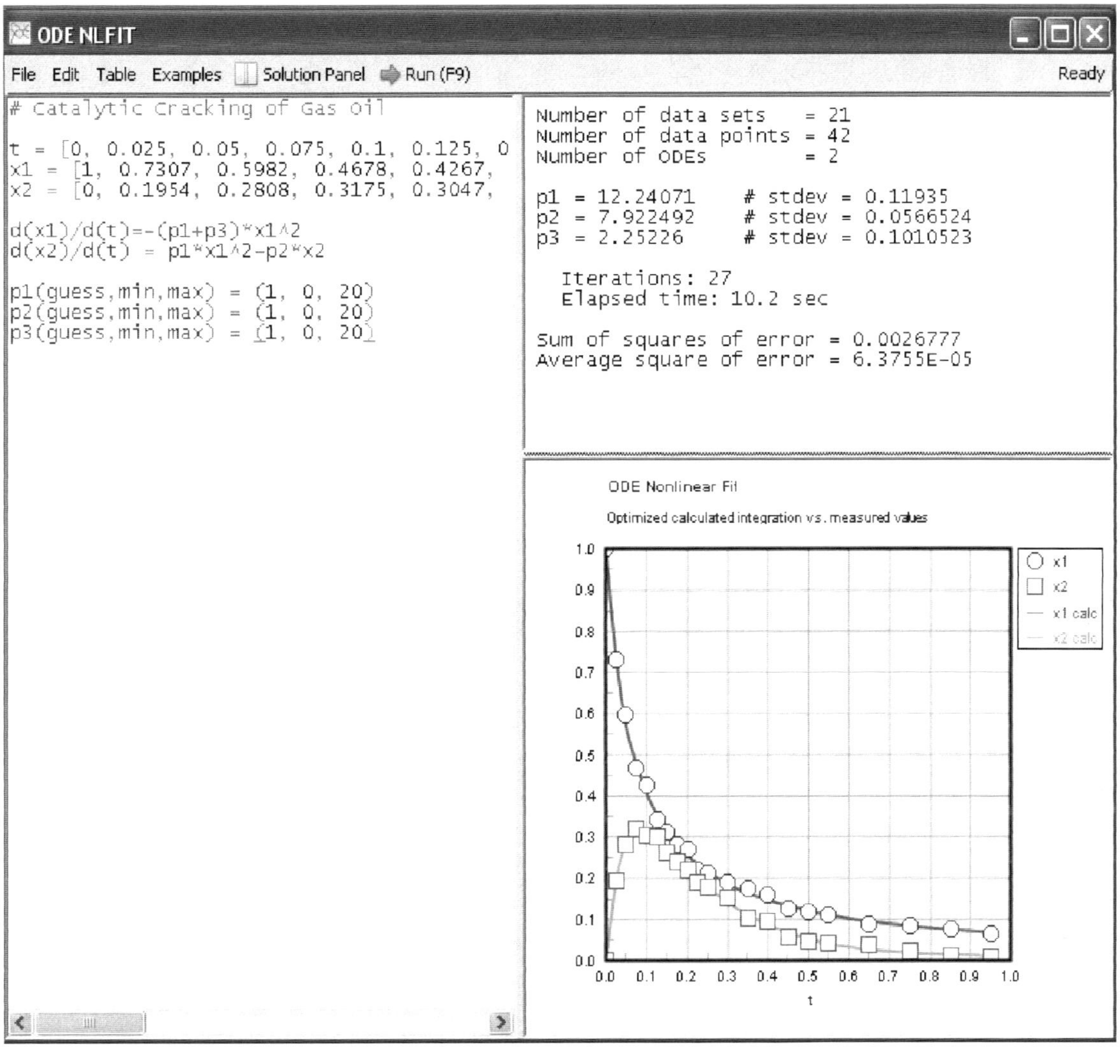

Figure 1. Polymath input and results for Example 1.

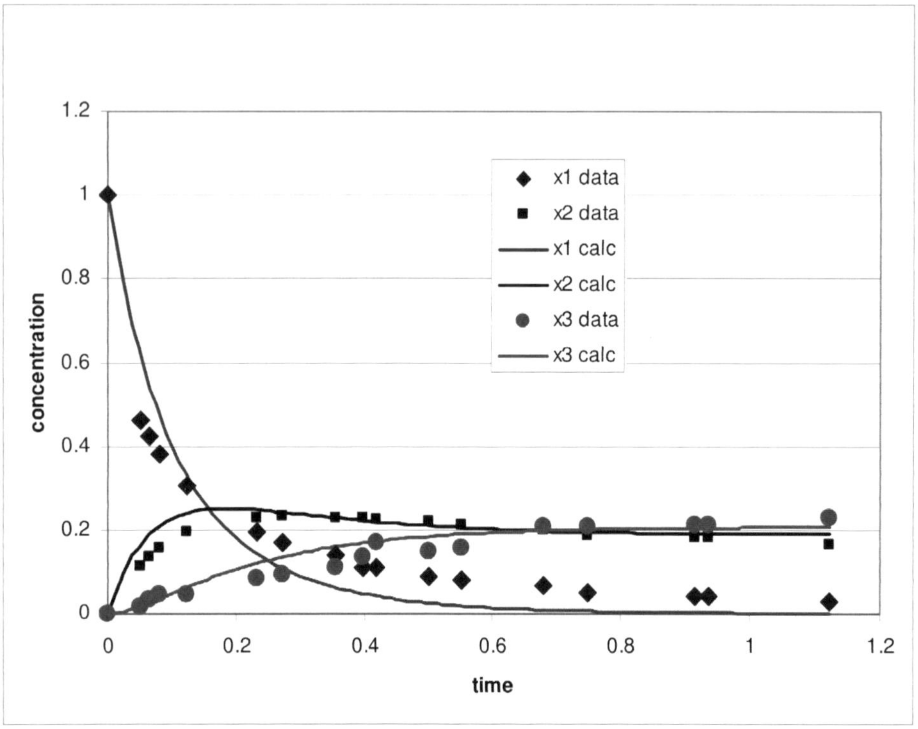

Figure 2. Plot of the calculated curves vs. the observed data points for the two parameter's optimal model for Example 2.

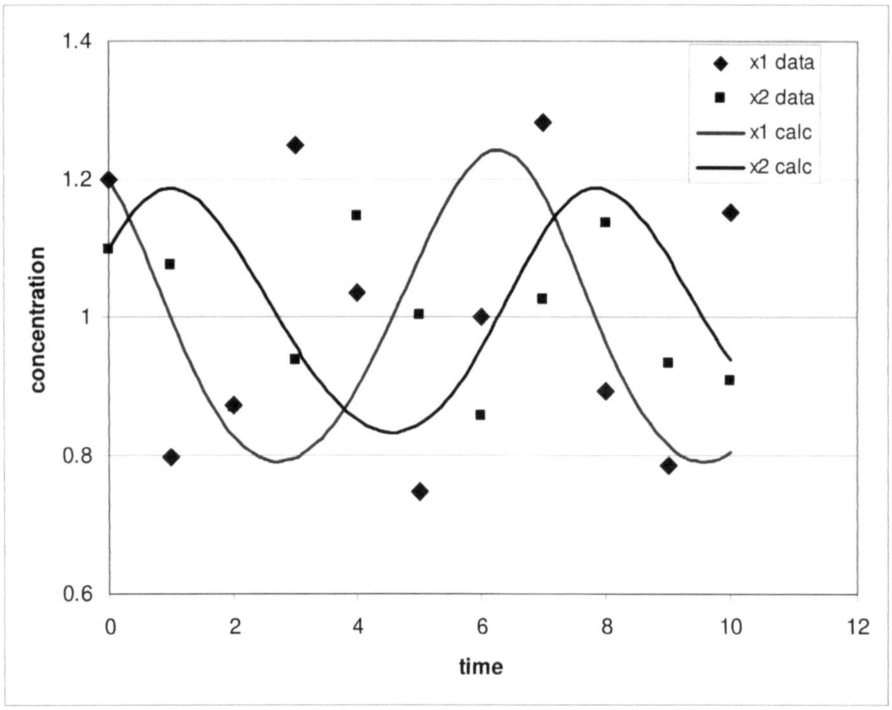

Figure 3. Plot of the calculated curves vs. the observed data points for the optimum found from the initial guess $\theta_1 = \theta_2 = 1$ for Example 3.

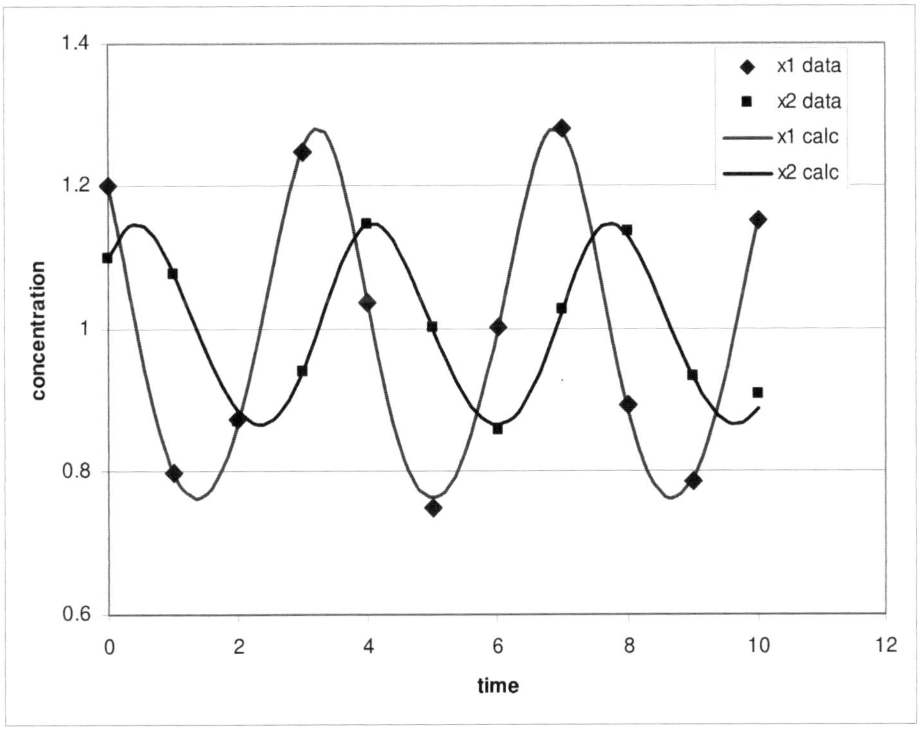

Figure 4. Plot of the calculated curves vs. the observed data points for the optimum found from the initial guess $\theta_1 = 5$, $\theta_2 = 1$ for Example 3.

(202f) A Process Systems Approach to Teaching Distillation

Kody Powell, Chemical Engineering, The University of Texas at Austin, Austin, TX and Thomas F. Edgar, Department of Chemical Engineering, The University of Texas at Austin, Austin, TX

First principles modeling is a valuable instructional tool to help undergraduate chemical engineering students understand how a process works. Distillation is perhaps the most difficult process for students to understand. In this work, a methodology where distillation is approached from a process systems point of view is presented. This approach was implemented in a unit operations distillation lab where students were required to build a first principles model of the process, which they would later use in optimization to maximize the revenue of a hypothetical chemical plant. From a pedagogical point-of-view, a systems engineering approach is beneficial to students as it requires them to gain an intimate understanding of the process through modeling and allows them to develop basic skills in optimization. Furthermore, a systems-engineering approach allows students to see the full scope of the process, while incorporating lab results through model parameter estimation and constraint identification.

The student's response to an open-ended versus a well-defined assignment is also studied. It is observed that, while the open-ended assignment is more difficult and time consuming, the understanding gained by this exercise is far greater. Additionally, this type of assignment compels the student to exercise more creativity and take a higher amount of ownership in the project, resulting in projects that are of higher quality and more diverse. In order for assignments of this nature to be successful, the proper balance must be established between keeping the assignment open-ended while giving students enough background information and guidance for the student to be successful.

In this work, evidence is presented comparing essentially the results from two semesters, one in which the assignment is predominantly instructor-directed, and another in which the assignment is largely student-directed. Although purely qualitative, evidence indicates that the student-directed methodology generally results in higher-quality work and a better learning experience.

(135c) Mini-Session On Gary Powers

Il Moon, Chemical and Biomolecular Engineering, Yonsei University, Seoul, South Korea

This session is in honor of Professor Gary Powers. Professor Powers was active in research centered on design and systems analysis. His work specifically concentrated on process synthesis, risk and reliability, and engineering education techniques.

Process Synthesis
The synthesis of operating procedures for continuous and batch processing facilities is an important task in plant design. Dr. Powers' work in computer programs tested such models and assisted designers in generating high-integrity procedures and flowsheets. Island-driven hierarchical planning was combined with functional models to unify synthesis of the procedures and flowsheet.

Chemical Process Risk and Reliability Assessment
Dr. Powers developed a strategy to assess process hazards and diagnose faults using real-time data using fault trees and digraph models to develop causal and probabilistic relationships between sensed variables annd process hazards.

Powers was also active in the development of a theory for the verification of real-time control systems by combining chemical engineering process models with software engineering techniques. Symbolic verification tests were developed for stages in the control system that range from the initial specification of the user's requirements to final control code and hardware.

(135b) Mini-Session On David Himmelblau

Thomas F. Edgar, Department of Chemical Engineering, The University of Texas at Austin, Austin, TX and Warren D. Seider, Chemical and Biomolecular Engineering, The University of Pennsylvania, Philadelphia, PA

This session will feature reflections on the life and career of David Himmelblau.

Professor Himmelblau taught at the university for 42 years. He authored 11 books and over 200 articles on process analysis, process optimization and fault detection using artificial neural networks.

His book "Basic Principles and Calculations in Chemical Engineering" was recognized as one of the most important books in the field by the American Institute of Chemical Engineers (AIChE).

"Dr. Himmelblau is one of the professors I remember most fondly from my years in the graduate program at UT Austin," said Victor Martinez. "I used his textbook in basic chemical engineering calculations as an undergraduate and it was such a great honor to meet him in person when I arrived for my doctorate. He was such a joy to talk to, always in a good mood and always willing to help."

Himmelblau was president of the Computer Aids for Chemical Engineering Education Corporation (CACHE) as well a director of AIChE during his career. He received numerous awards, including the CACHE Award from the Chemical Engineering Division of the American Society for Engineering Education; the Joe J. King Professional Engineering Achievement Award; and the AIChE Founders Award.

He received his bachelor's in chemical engineering from the Massachusetts Institute of Technology in 1947 after serving as an officer in the U.S. Army during World War II. In 1957 he earned a doctorate in chemical engineering from the University of Washington.

(183a) ABET Accreditation: Updates and Insights

Randy S. Lewis, Chemical Engineering, Brigham Young University, Provo, UT and Douglas K. Ludlow, Chemical and Biological Engineering, Missouri University of Science and Technology, Rolla, MO

This session, hosted by the Education and Accreditation Committee, will provide a brief review regarding the ABET accreditation process. In addition to briefly reviewing ABET General Criteria and the accreditation visit, discussion will also focus on aspects associated with the Chemical Engineering Program Criteria. Invited speakers only for this session, though the other ABET-related session (Best Practices) will accept abstracts.

This session will be chaired by Randy Lewis of Brigham Young University and Doug Ludlow of the Missouri University of Science and Technology. Topics for the session include:

- preparing your program before the ABET visit

- evaluator preparation prior to an ABET visit

- setting up an ABET visit schedule

- post visit interactions

- accreditation issues

- evaluator expectations

- accreditation resources

This will be an interactive session, with short talks by invited speakers (to be determined after the June 2012 ASEE meeting) and time for discussion with the audience.

If you are a new ABET evaluator, or if your program is about to undergo an ABET visit, or if you are interested in engineering accreditation, this session is for you!

(243a) To Teach or Not to Teach, What Is the Answer?

Santiago Faucher, Technologies, Hatch, Mississauga, ON, Canada

The typical university graduate will have spent 17 of their 21 years being taught a prescribed education regimen. The retention of this education is tested through written exams which, in the best of cases, will permit the student to elaborate on the answer and outline the process they took to achieve it. In the worst of cases, the evaluation is reduced to a multiple choice format due to the limited capacity of the education system to do otherwise. In all cases, there is typically only one correct answer or approach thereto.

Fortunately the next 60 years of one's life and career are not so simple and much richer: there is no one right answer, but rather a multitude of possible acceptable solutions to choose from. For Chemical Engineers who have been subjected to problem-based and/or self directed learning, this realization may come early during their college education; for others, the capstone project may present the first opportunity toward this realization. Ultimately it is the professor's approach to learning that dictates when students begin to think more richly towards varied and novel solutions. Should learning be pushed onto the student through professing or be pulled by the student through the establishment of the appropriate problem and reward? Which of these approaches will lead to a deeper retention of the curriculum? As professor of a capstone course, I chose a problem-based learning approach thus taking on the role of mentor and client rather than professor. I responded to student questions with questions. Each question designed to guide but not direct; no answers were provided. My students' response to my unconventional approach fitted well with the model of Kübbler-Ross describing grief management, while their group dynamics nicely fit Tuckman's Group Development Model.

This talk will report on the pros and cons of a problem-based capstone project, the educator and student roles, student and group dynamics, mentoring mechanisms and effectiveness, and finally on an unexpected team performance. As is often the case, both in this course and in life, the question asked in this educational experiment has lead to learning and to a more profound question: should one teach or mentor the chemical engineers of the future?

(243b) Helping Students to Learn by Doing - Capstone Design Experience At Columbia University

Michael Hill and Stanley A. Leshaw, Chemical Engineering, Columbia University, New York, NY

Columbia has transformed its two-semester Capstone Design sequence from a traditional lecture-based format to one which imitates the real-world dynamics of industry. Students meet in small groups or teams for up to one hour each week with an advisor/technical manager/coach who critiques progress, offers suggestions for improvement, and provides direction for the following week. Student teams complete their design projects on a semi-independent basis, receiving individual guidance, but not complete answers, from their advisor. The format also plays to the strengths of adjunct faculty, who have found it easy to adopt. As a result, students "learn by doing" as they address three significant process or product design problems of increasing complexity.

Projects are selected by the instructors to provide the students with sufficient hands-on exposure to several important pedagogical issues in design, such as the interaction of components within a system and the impact of constraints on optimization in design. These projects are also chosen to reflect an awareness of broader societal issues, such as environmental relevance, green energy, and sufficient water.

Student feedback has expressed appreciation for the format, with students frequently citing Design as their favorite course.

(243c) Chemical Process Design and Projects Two Semester Sequence

Alan W. Weimer, Chemical and Biological Engineering, University of Colorado at Boulder, Boulder, CO

The Fall semester 3 Credit course, "Chemical Process Synthesis", along with the Spring semester 2-Credit "Design Projects" course, is the *capstone* course sequence in chemical and biological engineering at the University of Colorado. The 1st semester utilizes a design text for approximately 8 weeks and ends with a "mini-design" team project which serves as the final exam. Topics covered from the text include: (1) an introduction to the design method; (2) health, safety, environmental, and ethics; (3) process creation/simulation, (4) process synthesis heuristics; (5) design methods reviewed for major unit operations; (6) heat and power integration; (7) ideal and non-ideal separation trains; and (8) process economics and profitability analysis (3 weeks of economics). Every class starts with 10 minutes dedicated to current events. The mini-design project is a well-defined problem that can be simulated and students are organized into typically 3-person teams. All teams work on the same design problem. Grading of the mini-design projects is taken very seriously (not hand-waved) in order to "set the bar" for the 2nd semester. For unsatisfactory mini-design projects, the teams must submit a satisfactory "retread" prior to the drop/add deadline of the 2nd semester. First semester grading is 30% homework, 30% two midterm exams, and 40% mini-design project, however, a passing grade is required on the mini-design project independent of all other grades in order to pass the 1st semester class. A peer-review counts as 20% of the mini-design project grade, but does not impact the requirement that the team project itself requires a passing grade - independent of the peer-review. The 2nd semester "Design Projects" course is supported by external typically industrial liaisons. The 3-person team projects are organized in the fall by the instructor and the projects are selected during the first week of class. Students work very closely with their external liaisons. Problem statements are typically very open-ended and may be only one or two paragraphs in length. Teams are required to research the project and to write a project proposal due within 3 weeks. The proposal needs to incluide all required "battery limits". The proposal is graded by the instructor with input by the external liaisons. Teams give two oral presentations during the remaining 8 weeks of class and are required to make an oral presentation at their liaison's facility, if local, during the last week of class, or during final exam week. GoToMeeting is used for non-local liaison presentations. Industry liaisons typically attend the oral presentations and comprise an external panel that asks quesions along with students in the class. Oral grading is done based on certain metrics with input from both the students and the external liaisons. Liaisons that cannot attend are engaged using GoToMeeting. A final written report is submitted and graded with input from the external liaisons. Organizing the external projects is challenging, but the instructor has been able to organize 36 different projects over each of the last two years. Student evaluations are typically excellent regarding the capstone sequence. Many of the industry liaisons are former students who want to contribute to the educational process. Many also recruit at the university and even hire students that they are engaged during the course. The instructor has spent half of his career in industry and half in academia and also directs a very active research program.

(243d) Senior Design At UC Irvine: Process Development and Economics Analysis with PRO/II

Gang (Gary) Xu, R&D, Invensys/SimSci-Esscor, Lake Forest, CA

Industrial standardized simulator PRO/II is applied in UC Irvine senior design class. Students are explore to real time simulation and project design assignments.

With fundamental knowledge of process design and economic analysis learnt from classroom, students get to develop projects by literature searching, PRO/II flow-sheet exercising, data reconciliation, and high-level process optimization from economic and environmental consideration.

Students were also stimulated by choosing their own projects reviewed by instructor. Interesting ones including generation of long-chain hydrocarbon (as fuel easy to store and transport) from natural gas, bio-energy generation from grass, supercritical CO2 decaffeination, sulfuric acid production from SO2, alcohol fermentation from potato (sugar) ... From which practice of fundamental engineering and economic understanding were encouraged.

With all the activities, popular topics such as

1. Azeotropic distillation

2. process optimization by controlling of design variables

3. economic model analysis such as Pareto optimal with engineering invest vs. environmental impact

were also emphasized as key portion of the design projects. Some of these topics, such as Pareto optimal, exceeded undergraduate level. However, basic knowledge of such topics were well understood with reasonable scale of exercise directly connected with projects chosen by the students.

(243e) Conduct Your Capstone Design Class As a Consulting Company

Richard L. Zollars, Voiland School of Chemical Engineering and Bioengineering, Washington State University, Pullman, WA

Capstone design classes are often taught in a similar manner – design a process (ammonia plant, methanol plant, etc.) to meet some production goal. By selecting an appropriate plant the instructor can be insured of covering almost any unit operation to which the students have been exposed at some time during their education. There are a number of shortcomings to such an approach, however. One of these is that the details of such plants, which unit operations are used, in what order, etc., are readily available in the literature. Students will repeat the standard design, go through a laborious series of calculations to size and cost equipment, perform an economic analysis and report their findings. In terms of promoting any in-depth thinking, there is little to offer. In addition, only a small fraction of students will ever be employed in such a situation.

A more common situation for students upon graduation is to be employed as a process or unit engineer. As such, in addition to keeping their unit operating, they have the responsibility of upgrading or debottlenecking their operation. This smaller scale, more focused, design work typically involves only a few unit operations. While few in number the unit operations that are involved are highly variable depending upon the unit in which the engineer is employed.

In an academic setting trying to operate a design class using this narrower focus is difficult without having the problems seem contrived. Thus, eleven years ago, we began operating our capstone design class in a different manner. The instructor for the capstone design class solicited regional industries for projects. With this set of projects available the design class is then conducted like an engineering consulting firm with teams of students vying for various projects. During the design process the students must keep in contact with their "clients" to insure that they are addressing the client's concerns. At the end the student groups must satisfy not only the class instructor, but also the industrial contact person who originally proposed the problem.

The capstone design class at Washington State University is a two semester sequence of classes. The fall semester course covers economic analysis, equipment sizing, safety and team work in addition to starting the design project. The spring semester class is devoted exclusively to the design project. In preparation for the academic year regional industries are contacted for proposed projects. For a variety of reasons (cost, laboratory facilities, safety) the projects are to be paper studies, not requiring work in a laboratory or shop. The projects should require approximately 500 person-hours to complete. In addition to developing such a project the industry is expected to provide part of a person's time to serve as the point of contact for the student design team as well as some financial remuneration to cover expenses for the project.

During the fall semester the students select the group in which they want to work. Ideally, each group is three people but no group will be larger than four people. Once the groups are formed one-page descriptions of the projects that are available for the year are given to each group. Each group then goes through a bidding process for the selection of the project on which they will work. At the completion of this bidding/selection process each group will visit the industrial site to meet with their contact as well as get a fuller understanding of the project. Prior to the end of the fall semester but after the initial site visit each group will conduct a brainstorming session to develop a slate of potential ideas for the solution of the problem they have been assigned.

During the spring semester the students work exclusively on the design project. They develop Gantt charts to help them monitor their progress. These are updated formally three times during the semester. They are expected to maintain regular communications with their industrial contact. A second site visit about half way through the semester is conducted to insure that both the students and industry contact are in agreement with the direction of the project. Final written reports are due after 12 weeks. This is followed by a day of formal oral presentations the following week. Because of the number of projects in any given semester the time for these presentations is short. Thus we have gone to a poster/oral presentation format. The student design groups participate in a poster session first followed by 20 to 25 minute oral presentations. Finally, before the end of the semester, each group is to hand-in a final revised written report which will be sent to the industrial contact.

To date a total of 58 different design projects have been conducted for 15 different industrial sponsors. The industries have ranged from major corporations, to government labs, to state government agencies, to small companies (5 – 10 employees). The range of topics is also quite broad with few projects covering similar processes. There have been some very traditional chemical engineering projects (heat exchanger optimization, hydrogen recovery in a refinery, debutanizer optimization), some others that are still definitely chemical engineering in nature (active metal reduction of nitrates, oxidation of uranium using Fenton's Reagent, purification of soy and palm oils, water treatment/recovery, and copper recovery) while a few others stretch the boundaries of the discipline (recycling of carbon fiber reinforced composites, recovery of neutraceuticals from fruit waste, and converting municipal and/or food waste into biofuels).

As measured by student involvement conducting the design course in this manner has been a success. Student evaluations of the course are uniformly high. More than one alumnus has reported that their experience in the design course was the closest thing they had to actual practice than any other course in the curriculum. The industries that have participated have also expressed their support. Not only have they continued to provide projects, manpower and financial support but also have hired students as a result of the student's work on a project supported by the industry.

Another way to assess the success of conducting the design course in this manner is the quality of the work performed by the students. The quality of the written reports has gone up, in part due to the fact that the reports are going to people who might be hiring the student. The oral/poster presentations have been excellent. More than one person from industry has commented that they meet or exceed the quality that they would expect from their engineers.

The quality of the designs is also outstanding. Approximately 20% of the student design suggestions have been implemented in some from by the company that supplied the project. The quality of the student's work was verified by a situation that occurred recently. One of the companies offered a project for which they had already signed a contract with an engineering consulting firm. They received both the professional firm's report at about the same time as they received the student's report. While the student's report lacked the detail of the professional report (types of welds, fasteners, support structures, etc.) the student design and the professional design contained almost identical unit operations, arranged in the same order, with equivalent material and energy balance calculations.

Integrating Active Research on CO2 Capture in Traditional Process Design

Omkar Namjoshi[1], Paul Nielsen[1], Matthew Walters[1], Bo Lu[1], Siyun Wang[1], Gary Rochelle[1]

[1]The University of Texas at Austin, Austin, Texas, USA

The typical capstone design course in most chemical engineering curricula integrates material from heat and material balances, unit operations, engineering economics, and reaction engineering. At the University of Texas at Austin, we also incorporate aspects from sustainable engineering and from our research program in the design class. Students who take this class are tasked with completing a preliminary design of a post-combustion carbon capture process (capturing CO_2 from a point source) using amine-based solvents or physical solvents.

Themes from sustainable engineering are incorporated at the beginning of the course and reinforced throughout the semester. Students are first given an introduction to climate change, how capturing CO_2 might help mitigate climate change, a brief overview on CO_2 storage sites. These themes are reinforced throughout the class as detailed design progresses and economics are finalized. Students must answer what price CO_2 must be "sold" for the plant to make a profit and also must discuss the environmental impacts of CO_2 capture facility. The link to sustainability gives students valuable insight regarding energy and environmental policy, which is not commonly encountered in a traditional "core" chemical engineering class.

Within the first three weeks of class, students are expected to form teams of three students each and submit a proposal outlining their choice of CO_2 source, capture solvent, and storage site, justifying their choices with quantitative arguments. The teams are expected to use as many resources as possible and are encouraged to get information and guidance from graduate students, research sponsors (which include large oil and gas companies, amine technology suppliers, and power plant operators), and old student reports (which are posted on the course website) to help strengthen their arguments for their proposal (as well as for general help, if it is needed). Students – past and present – have proposed a variety of CO_2 sources, including (but not limited to) coal-fired power plants, integrated combined-cycle power plants, fluidized catalytic crackers, and LNG treating plants. Example of capture solvents include aqueous amine solvents, such as monoethanolamine, concentrated piperazine, and activated MDEA. Students are encouraged to design plants using unconventional sources and solvents; those who do so are evaluated more on effort and a good initial, "first-draft" design. Students who have chosen solvents and sources that have been selected in prior years are evaluated on producing an optimized design that represents an improvement over the prior year's design.

The integration with our research program is achieved by incorporating thermodynamic models built using the AspenPlus simulation software / framework and developed by graduate students into the student projects. Our research group has developed rigorous models for activated MDEA and concentrated PZ and has developed preliminary frameworks for other amine systems; the use of these models by the undergraduate students is a helpful validation tool for the graduate students and helps in improving future revisions of the model.

Assignments are generally assigned each week, with a few exceptions for some of the lengthier and more involved assignments. Each assignment can be best described as "goal directed, mission oriented" – students will have a set number of deliverables for each assignment cycle and are expected to complete them (or explain why they were unable to) and are held accountable to these deliverables. Table 1 gives a list of assignments along with deliverables (note that the deliverables can change from year to year and are group specific) that are assigned to students. A few assignments are individual assignments, and the assessments from each individual assignment are used to differentiate student performance within the group.

The weekly assignment cycle ensures that students do not fall behind on their reports. The final report essentially consists of students putting all of their reports together and addressing any areas for improvement that TA feedback has identified. The grade for the final report is 50% of the overall course grade. This performance bias ensures that students who fall behind are able to catch up (should they choose to do so) and that students who earned good evaluations on prior reports will continue to maintain their performance through the end of the semester.

One unique aspect about course management is that highly motivated graduate students serve as teaching assistants (TA's) for the class. Each teaching assistant supervises three to four student design groups and is responsible for assigning deliverables to the students, mentoring student design groups and ensuring that they have the resources to be successful in completing the assigned deliverables, and holding the students accountable to those deliverables (addressing feedback). A head teaching assistant (generally the most experienced amongst the TA's) is also appointed and, in addition to having his/her own student groups to supervise, mentors and advises the other teaching assistant and also manages course coordination such as AspenPlus tutorial development and ensuring that student evaluation is completed in a way that is fair, consistent, and free from individual TA bias. The supervising professor (currently Dr Gary Rochelle) delivers the lectures during the classroom sessions, is responsible for supervising each TA, and holds a standing weekly "alignment" meeting to collectively set goals and deliverables for upcoming assignments with the TA's as well as address any TA concerns. The teaching assistant experience for this course is an excellent way for motivated graduate students to obtain "real-world" management skills that are useful in either a professional or academic setting. In recognition of this fact, UT's chemical engineering department grants three hours of Supervised Teaching course credit to each student who serves as a process design TA.

The emphasis on sustainability, the integration of research into the student design projects, coursework organization, and emphasis on graduate TA development make UT's design course unique.

Table 1. Plant Design Assignments at UT-Austin

Assignment / completion time	Individual or Group?	Deliverables
Proposal / 2 weeks	Individual	Pick a CO_2 source, capture solvent, and storage site
Proposal / 1 week	Group	Finalize source / solvent / storage site decision, submit one proposal, make improvements to proposal
PFD / 1 week	Group	Develop preliminary flow diagram, outline equipment required, model specifications (e.g. CO_2 compressor pressure)
Simple Heat and Material Balance / 1 week	Group	Develop basic heat and material balance, model CO_2-amine interactions based on an equilibrium model
Rigorous Heat and Material Balance / 2 weeks	Individual	Develop rigorous heat and material balance on one unit operation, model CO_2-amine interactions using rate-based models
Rigorous Heat and Material Balance / 1 week	Group	Combine reports, address feedback
Equipment Specifications / 1 week	Group	Size all equipment (absorption / regeneration columns, heat exchangers) and specify materials (e.g. stainless steel)
Equipment Costs / 1 week	Group	Cost all sized equipment using Aspen Economic Toolkit
Project Economics / 1 week	Group	Find price of CO2 required for plant to be profitable, develop project economics – cash flows, IRR, DCF
Process Safety and Environmental Assessment / 1 week	Group	Identify environmental impacts that the process might have
Draft Final Report / 1 week	Group	Combine all previous assignments into one report, address all feedback. No new information is needed beyond the environmental assessment section.
Final Final Report / 2 weeks	Group	Address feedback on draft final report, also present oral presentation on design to class

(243g) Incorporating Process Simulation Across the Chemical Engineering Curriculum to Improve Student Performance On the Capstone Design Project

David A. Rockstraw and Martha C. Mitchell, Chemical Engineering, New Mexico State University, Las Cruces, NM

The New Mexico State University Chemical Engineering Department recently made changes to the core curriculum to accommodate increased training in the use of chemical process simulation (Aspen Plus®) in parallel with the fundamental topics of the discipline. Over the course of five consecutive semesters, students learn aspects of the simulator, solving the same problems in simulation lab that were previously solved by hand calculation in their courses. The 2012 graduating seniors received twice the exposure hours to the simulator as previous graduating classes, and the 2013 graduating class will have received twice the hours of exposure of the 2012 graduating class. The increased level of understanding of how to properly use the simulator for process design has led to significant improvement in student performance on their capstone design projects in 2012. This presentation will describe the curriculum changes made that have permitted this improvement in student work, and the content and methodologies of the design courses in the sequence.

(253a) Integrated Assessment of Student Success in Achieving a-k Criteria Using Course Management Software

Paul Blowers, Chemical and Environmental Engineering, The University of Arizona, Tucson, AZ and Kimberly L. Ogden, Chemical and envionmental engineering, University of Arizona, Tucson, AZ

Data collection efforts for evaluating programmatic success in training students to be successful at meeting program criteria are traditionally time intensive and difficult. Course management tools offer the opportunity to build program-level data aggregation efforts that allow for computation and numerical evaluation of both groups of students and also individual students. The automation, thought Desire 2 Learn for this effort, allows for more time to be spend on the qualitative descriptions generated by faculty instead of number pulling and crunching.

This talk demonstrates how a pilot project was conducted and provides comments on the pros and cons of how to implement this approach. Pros include easier faculty engagement, regular reminders due to routine data access that data should be collected at the end of each semester, and the ability to compare trends easily from year to year on a statistical basis. A checklist of steps that will streamline future implementation at other programs is discussed, along with a list of suggestions on alternatives that may make sense for different programs. estions on alternatives that may make sense for different programs.

(253b) ABET Preparation and Visit At the University of Toledo

Glenn Lipscomb, Department of Chemical and Environmental Engineering, University of Toledo, Toledo, OH

The Chemical Engineering program at the University of Toledo has held ABET accreditation continuously since its first accreditation in 1964. Our most recent comprehensive review was held September 25-27, 2011.

The evolution of our continuous improvement plan since adoption of ABET's Engineering Criteria 2000 (EC2000) is described. A core component of this plan is *Assessment Day*. On Assessment Day, the faculty meet as a whole to evaluate relevant assessment data and recommend action. The timing of these meetings, data evaluated, and follow-up procedures are discussed.

The preparation of evaluation materials is described as well. Hard copies of evaluation materials were provided on site but the material also was made available pre-visit through our department web site. Supporting material could be accessed by Educational Objective, Student Outcomes, or Course title to facilitate review. The faculty were provided a list of talking points pre-visit to prepare them for their discussion with the evaluator. The components of this one-page summary are described.

Finally, changes we plan to implement based on our most recent review are presented. These include the conduct of alumni and employer surveys as well as how we use social media.

(253c) Assessment of Student Outcomes and Program Objectives: Methods for Effectiveness

Bill B. Elmore, Dave C. Swalm School of Chemical Engineering, Mississippi State University, Mississippi State, MS

Assessment of Student Outcomes and Program Objectives:

Methods for Effectiveness

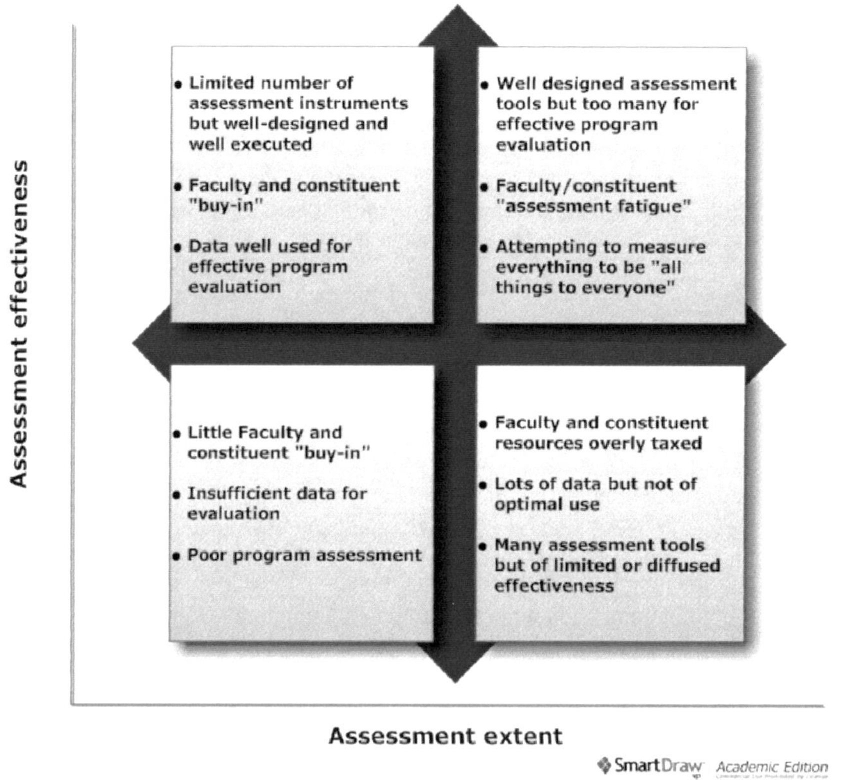

The assessment of Student Outcomes (traditionally aligned with ABET Student Outcomes "a-k") and Program Objectives poses an ever-present demand on resources in any engineering department. Executing successful program assessment and evaluation through department faculty requires careful design of assessment instruments and "points of intersection" between assessment tools and program activities (e.g. courses, constituent involvement, student organizations, etc.). Effective assessment is characterized by efficiency of tools and implementation to garner faculty and constituent "buy-in". As illustrated below, the balance in any assessment program brings a tension between effectiveness and extent of assessment and evaluation. At one end of the spectrum is an extensive use of strong, effective assessment tools. While this may constitute the "ultimate" in the game of program assessment, it is unrealistic for departments relying upon "in-house" delivery by program faculty. The more effective and pragmatic approach is to select or design a few great tools and implement them strategically within departmental resource constraints (e.g. faculty/staff time, available funding, etc.). This talk presents ideas and examples for strategic design and implementation of assessment tools—some successful and some less so. The discussion will include challenges faced with managing ABET program assessment within departmental resources.

(253d) Collecting Evidence for Continuous Improvement: Using Direct Assessment of Student Outcomes

Jennifer Cole, Northwestern University, Evanston, IL

ABET's Criterion 4 (also known as "Continuous Improvement") states that each program must demonstrate a procedure for assessing, evaluating, and implementing improvements to the program. This involves collecting and evaluating data on the extent to which the program educational objectives and student outcomes are being met. One of the most difficult parts of collecting data for continuous improvement is determining what kind of assessment is appropriate and how to make that assessment efficient and useful.

Using a variety of assessment types can give a better picture of student performance. Direct assessments (such as analysis of student work samples or observations of students) and indirect assessments (such as self-evaluations or surveys) are most common. Direct assessments are often harder to obtain.

This paper will focus on how we developed our direct assessment methods for evaluating achievement of student outcomes and how that data was used for continuous improvement at Northwestern University. Details will be shared on the type of direct assessments used to evaluate student performance, how the assessments were implemented in the program, and how the data helped determine where program changes were needed.

(280a) Incorporating Sustainability Into Engineering Research and Teaching

David T. Allen, Department of Chemical Engineering, The University of Texas at Austin, Austin, TX

What does it mean for an engineering design to be sustainable? Design tools that allow engineers to choose inherently safer chemicals, improve energy efficiency, improve mass efficiency, and reduce emissions and exposures are certainly part of an emerging tool set engineers will need for sustainable designs. But, sustainability remains a subjective concept. Given that the concept of sustainability has not been defined in the type of quantitative framework used in engineering design, it is useful to take a pragmatic approach to defining the tools of sustainable engineering. Many engineering educators are incorporating sustainability concepts into the courses that they teach and the research that they perform. Assembling an inventory of what is covered in those courses and research programs represents a first step in defining sustainable engineering. An inventory of university-based engineering research and teaching related to sustainability has recently been completed and the results of this inventory will be presented in three parts. First, the types of engineering courses in which sustainability concepts are being addressed will be described, along with the diverse subject areas covered in these courses. Second, a brief overview of the current landscape of university-based engineering research, related to sustainability, will be presented. Finally, a case study of sustainability-related research, the development of smart, green, electrical grids, will be presented. This research program, underway at the University of Texas, is examining how utilities can balance between carbon emissions, criteria air pollutant emissions, and water use, while still satisfying electricity demand.

(280c) Sustainability Body of Knowledge and Role of Credentials

Darlene Schuster, Institute for Sustainability , An AIChE Technological Community, New York, NY, Deborah Grubbe, DuPont and Erin Chan, Institute for Sustainability, New York , NY

The Institute for Sustainability, an AIChE Technological Community, has developed a body of knowledge to begin to address what a sustainability professional should know and to identify the unique educational content of sustainability. This presentation will review the development of the body of knowledge, present survey results of university courses at the graduate level that address sustainability, and review the new sustainability credentialing program.

The Body of Knowledge (BOK) addresses the following key subject matter areas:

1. Summary Definitions and Principles
2. Professional Competencies of Individuals
3. Corporate and Institutional Sustainability Leadership
4. Environmental Stewardship and Performance
5. Safety Performance Measurement and Metrics
6. Product Stewardship
7. Risk
8. Societal Impacts
9. Economic Growth and Sustainability

10.Value Chain and Supply Chain Management

These categories are crossed matched to various courses being offered at the graduate level at universities. Courses that enable a student to meet learning objectives in the above subject areas are being identified and a capstone design based course will also be reviewed.

(280d) Sustainability in Chemical Engineering Thermodynamics and Separations

Joan F. Brennecke, Department of Chemical and Biomolecular Engineering, University of Notre Dame, Notre Dame, IN

Arguably the best way for chemical engineers to be introduced to sustainability is for these ideas to be incorporated throughout the curriculum. Here we will give examples of how this can be done in conventional chemical engineering thermodynamics and separation courses.

Examples include quantification and reduction of breathing losses from tanks, consideration of liquid-liquid extraction as an alternative to energy intensive distillation for mixtures with relative volatilities close to one, and ways of reducing emissions from power plants by improvements in cycle efficiencies, among many others.

(280e) Pan American Biofuels and Bioenergy Sustainability: A Research Coordination Network

David R. Shonnard, Chemical Engineering, Michigan Technological University, Houghton, MI

The implications of large-scale biofuels and bioenergy production on environmental systems and social conditions are largely unknown. Yet despite this lack of fundamental sustainability knowledge, there currently is a rapid movement toward the development of biofuels and bioenergy production systems that will likely lead to dramatic changes in extant human and natural systems. These changes may be beneficial or detrimental, and may involve a number of issues such as land ownership and management, creation of new employment opportunities, occupational safety, unforeseen transportation and logistical challenges, impacts on water availability and quality, and potentially many others changes to social and environmental systems.

The goal of this RCN project is to create a community of biofuels/bioenergy researchers and industry/government/NGO practitioners concerned with sustainable biofuels and bioenergy development across the Americas. The objectives of this project are to i. establish and sustain an international research coordination network with a focus on biofuels/bioenergy sustainability and with a geographic context of the Pan American region, ii. develop coordinated research programs, and iii. create new knowledge on social, environmental, and economic sustainability implications of large-scale biofuels/bioenergy production from biomass grown in this region. The Pan American Biofuels and Bioenergy Sustainability RCN will consist of researchers at universities and other organizations from North America (United States, Canada and Mexico), Central America, and South America. These participants will have expertise in several sustainability dimensions of production and use of biofuels and bioenergy from a wide range of biomass feedstocks. This networked community will share scientific and educational knowledge through a well designed program that considers the diverse natural, cultural and political settings while creating the framework for developing a comprehensive Pan American biofuels/ bioenergy sustainability assessment to guide strategic research, investment and government policy. This RCN will have two main tangible deliverables; 1. a research roadmap report on sustainability of biofuels and bioenergy in the Pan American region, and 2. a graduate course in sustainability of biofuels / bioenergy to be developed within the network and disseminated to the network through the internet. A series of annual workshops and a major mid-project conference will be the primary means of achieving these deliverables, as shown in Figure 1. The RCN will be connected through a comprehensive cyber-infrastructure based upon the HubZero platform to support real time dissemination of the RCN products, enable real time collaboration across the network and provide for the growth of the proposed RCN.

This presentation will provide an overview of this new network and its activities, and provide a summary of outcomes from the first international workshop to be held in Merida, Mexico May 29-31, 2012.

Thirty Four Years of Teaching Process Mixing to Undergraduates, with Some Learnings

Fluid Mixing is rarely taught to undergraduates even as part of a fluids course. Most graduates acknowledge once in industry the topic of mixing often comes up and seems to be a topic everybody thinks they know something about. For the last thirty four years Dr. Etchells and his colleagues have been teaching a senior elective on the topic based on their industrial experience at several universities. It is a synthesis course bringing together a number of previous topics such as momentum transfer and mass transfer along with aspects of reaction engineering. Topics such as non-Newtonian rheology, multi-phase and turbulence are looked at in a new way. It is a design oriented course with open ended problems and because of the incomplete nature of the field it is constantly evolving. Not everything is understood about fluid mixing. Examples will be given.

(291b) An Academics View On Mixing Education At the University of Alberta

Suzanne Kresta, University of Alberta, Edmonton, AB, Canada

Biography

Suzanne Kresta (BSc, UNB (1986), MSc, Leeds (UK, 1987), PhD, McMaster (1992)) teaches Mass and Energy Balances (CME 265), design (ChE 464) and Mixing (ChE 420 and 620), and serves as co-editor of the Handbook of Industrial Mixing. Her principle research contributions are related to understanding turbulence and mixing in stirred tanks (measurement of turbulence, the use of spatial statistics to define mixing and length scales, nano-particle production, solids suspension, and reactor design) and she has collaborated with colleagues in a wide variety of industries (including oils sands extraction and froth treatment, photographic film, drinking water treatment, mineral processing and metals refining, polymer reactor design, cosmetics, mixing equipment design, fish processing, clean-up of nuclear waste, and fertilizer production). Many of her papers are both widely cited in the research literature and used for industrial design calculations.

Her teaching interests include visual problem solving tools, the use of story telling and case studies, active learning, and explicit use of cognitive levels with learning objectives. She has served as an Iron Ring Warden since 1999, as a Peer Consultant at the University of Alberta since 2004, and has won a number of national and international awards.

Research Interests

The mixing unit operation is far broader than the classical well mixed continuous stirred tank reactor, and is present in most chemical processes. Tightly designed mixing equipment is needed to meet the demanding process specifications of products ranging from Corianä countertops, through pharmaceutical and photographic crystals, cosmetics, fine chemicals, water treatment, and extraction of metals from ore.

Our group is currently exploring the theory of spatial statistics and how it can be applied to mixing problems to provide a rigorous definition of well mixed. The equations underlying the modelling of mixing sensitive reactions are being revisited in this context. Applications projects include the production of nanoscale metal powder, the distribution and suspension of multi-modal particle mixtures, drawdown of floating solids, and dissolution kinetics of partially soluble liquids.

Current Projects

- Continuous production of nanoscale metal precipitates; Shad Siddiqui, Alena Kukukova, Peter Unwin; Umicore, NSERC
- Drawdown of floating solids; Oscar Khazam, Lightnin
- Modelling equations for mixing sensitive reactions; Imran Shah, NSERC
- Segregation dimensions and the characterization of mixing; Suzanne Kresta, Alena Kukukova, Joelle Aubin; NSERC, CNRS
- Multimodal solids suspension and distribution; Inci Ayranci, Lightnin

(291c) Integrating Mixing Education Into Fluid Mechanics and Reaction Engineering Courses At Rowan

Robert P. Hesketh, Chemical Engineering, Rowan University, Glassboro, NJ

At Rowan University mixing is integrated in several courses:

Process Fluid Transport

Chemical Reaction Engineering

Unit Operations Experimental Design and Analysis

As well as taught in a senior elective titled, "**Biopharmacuetical and Industrial Fluid Mixing**"

Topics in Process Fluid Transport:

1 week is spent on Mixing:

Mixing Definition:

Mixing is the reduction of inhomogeneity in order to achieve a desired process result.

Inhomogeneity: concentration, phase, temperature.

Process results: increase mass and/or heat transfer, reaction rate, or product properties.

Typical Tank Dimensions

Types of Tanks with mixers

Fluid flow patterns with different impellers, diameters and liquid viscocity. PBT: Pitched Blade Turbine, FBT: Flat Bladed Turbine.

Impeller Types and Power Numbers

Power and D/T

Scale-up

Blend Time

Homework: Calculate the required blend time for the following tank:

Pitched blade turbine (6 Blade)

Impeller Diameter: 1 m

Operating speed: 84 RPM

Fluid: Water

Tank Diameter = 3.0 m

Tank Volume = 5600 gallons

1. Calculate Reynolds Number
2. Calculate blend time for within 5% of desired value
3. Design a blending tank for a volume of 10,000 gallons assuming =H and =/3 and dimensions scaled with volume
    1. Constant Blend Time
    2. Constant speed
    3. equal power per unit volume
    4. assuming power per unit volume based on Figure 15.

Give the power required, the tank and impeller dimensions and impeller speed for each case.

Topics in Chemical Reaction Engineering

Industrial Mixing in Chemical Reactors, Richard Grenville, DuPont

Biopharmacuetical and Industrial Fluid Mixing

The senior elective is taught by Drs. Arthur W. Etchells and Richard Grenville

In both the biotechnology and the pharmaceutical industries mixing is an important step in development, scale up and manufacture. It is most important in process steps involving multiple phases and complex liquids and chemical reactions. This course will discuss the application of fundamental single and multi-phase fluid dynamics to such processes and the development of industrial equipment designs based on these fundamentals along with the concepts of process scale up and scale down. A major objective of the class is to develop equipment designs based on fundamentals and empirically derived correlations. Emphasis will be on designs for the biotechnology and pharmaceutical industry but having equally applicability to other process industries. Both lecturers have extensive industrial experience with Dupont and other companies.

Arthur W. Etchells is a world recognized authority in the field of mixing for the process industries. For thirty nine years he worked for the DuPont Company and for thirty years as an internal consultant for the many diverse DuPont businesses in the field of fluid flow with emphasis on mixing and slurry transport and retired in 2002. He is a past president of the North American Mixing Forum and winner of their award for contribution to mixing technology. He is now an independent consultant in the field.

Richard K. Grenville is the lead consultant in mixing for the Dupont Company. He has twenty five years experience in the field. He has extensive publications in the field, has given many seminars to companies outside of Dupont and is a member of the North American Mixing Forum governing board.

(291d) Ready to Use Module for Introducing Mixing in a Fluids Mechanics Course

Richard K. Grenville, DuPont Engineering Technology, Wilmington, DE

As Chemical Engineers we are concerned with the transformation of raw materials into valuable products. These transformations may require the mixing and reacting of the raw materials and the separation of the desired products from by-products. Also, as these transformations take place, the materials may need to be heated and/or cooled. All the disciplines required to understand the operation and design of mass and heat transfer equipment are covered in the under-graduate Chemical Engineering curriculum.

Mixing has been identified as a critical operation for the Chemical Industry (see for example, Smith, 1990, *Trans IChemE.* 68, pp. 3 - 6) yet it is not, generally, taught to Chemical Engineering students. There are a number of reasons for this. Among those given are:

It is the University's job to "teach the basics" and the engineer will learn "practical skills" once he/she has started work.

In order to add a course on mixing, some other subject would have to be removed from the curriculum.

But the same comments are not made about heat exchangers or distillation columns.

If we accept that there is no "market" for a full mixing course at most Universities, how can we ensure that the students get some exposure to this technology before they graduate? We believe that a working understanding of mixing can be taught by adding lectures as mixing modules to the classic Chemical Engineering courses. For example:

An agitator impeller is a pump and it can be characterized in the same way as the pump in a piping system that students will have studied in Fluid Mechanics.

Design of agitators for heat and mass transfer can be covered in the traditional courses.

Mixing issues relating to fast, competitive reactions can be covered in the traditional Reaction Engineering course. When does the assumption that a reactor is "perfectly-mixed" fail?

The content of several proposed modules will be presented

(332a) Coursecasting to Conquer Content in Thermodynamics

J Richard Elliott, Chemical and Biomolecular Engineering, The University of Akron, Akron, OH

The content in the sophomore thermodynamics course can become overwhelming when it is the only thermodynamics course in the curriculum. Recent additions relate to molecular perspective for preempting misconceptions, biological insights and applications, and active learning. Each addition faces tough constraints regarding traditional content on energy, entropy, and equilibrium and their linkage throughout the curriculum. Fortunately, coursecasting can enhance the efficiency of classroom contact while retaining the benefits of verbal guidance. Its best use is in illustrating sample calculations. Class time is used to introduce concepts and illustrate brief examples. More in-depth examples are best treated through coursecast recordings that include verbal queries and prompts to pause. Statistics on the viewing habits of students indicate which examples have the highest impact and how students respond to project due dates and test preparation. A companion workbook guides students through reading exercises, ConcepTest preparation, and drills for test preparation. Student comments indicate that the incorporation of coursecasting is the most appreciated study aid in years.

(332b) Using Podcasts to Teach Nanotechnology Across Three Engineering Departments

Dr. Srinivas Palanki, Department of Chemical and Biomolecular Engineering, University of South Alabama, Mobile, AL

In the past 15 years the field of nanotechnology has matured to the level where devices based on nanotechnology are emerging and making an impact to today's society. This field is truly interdisciplinary and requires a synthesis of knowledge from science and engineering principles to develop nanotechnology devices. In this talk, we present our experience in utilizing recent innovations in engineering education, coupled with video creation technology, to develop a set of modules that can be integrated into several sophomore, junior and senior level core courses in three different engineering departments (Chemical & Biomolecular, Mechanical, and Electrical & Computer Engineering). The modules were developed by faculty members who have research experience in nanotechnology and were integrated with material covered in core engineering courses. Each of these modules involves problem solving skills at the levels of knowledge, comprehension, application, analysis, synthesis and evaluation. This allows instructors to use parts of the same modules in multiple courses. In this talk, the following specific issues will be addressed:

• Details of 12 podcast modules that were developed in an asynchronous, on-line learning format that teach science and engineering aspects of nanotechnology at multiple levels

• Implementation of these modules in several core and elective courses in Chemical& Biomolecular, Mechanical, and Electrical & Computer Engineering Departments.

• Assessment of outcomes.

Using student-produced videos to enhance learning engagement in a Chemical Engineering Thermodynamics Course

Douglas K. Ludlow
Department of Chemical Engineering
Missouri University of Science and Technology

ABSTRACT

The traditional term research project report and presentation in the first chemical engineering thermodynamics course has been replaced with an assignment in which students produce short "YouTube" type videos that describe and demonstrate some underlying principles from thermodynamics. The students participate in this active learning project which allows them to further understand basic thermodynamic concepts and the same time as developing rudimentary video production skills and practicing good communication skills. The short video demonstrations developed are viewed by the class. In addition, video produced in previous semesters are available to students currently enrolled in the class in an asynchronous manner to help to reinforce key concepts on an as needed basis.

This project addresses four focus areas, namely 1) enhance student professional development, 2) increase faculty-student interaction, 3) promote active learning and 4) improve attainment of learning outcomes. Oral and written communication skills have always been a critical component of a well rounded engineering or professional education. With the ease and availability of video recorders and platforms such a YouTube, visual communication using small videos is becoming more prevalent. More practicing professionals will be expected to develop short informational videos to share concepts, document operations and train coworkers. The video project aids in this professional development. As the students prepare their video presentations they interact more with the faculty member in a small group setting to provide feedback and direction in the development of a short video on some thermodynamic concept. The researching of the thermodynamic concept, the development of the "script" for their video presentation and the production of the video all lend themselves to active learning versus merely taking notes in lecture or reading the textbook. Also, reviewing the thermodynamic concept videos developed by their classmates also enhances the understanding and retention of the underlying concepts.

INTRODUCTION

Often times in engineering science courses the instructors incorporate a semester long project to help the students explore some concepts in more depth than would be covered during the course of the regular lectures. The pedagogical "hope" of the instructor is that this exercise in delving more deeply into some topic area might spark some further interest in the student for the topic area. Another of the educational objectives of these term projects is to have the work carried out by small groups of students so that they can experience small group dynamics and hence develop competencies to work in groups in an industrial setting. In addition, often these semester long projects are culminated with some oral presentation to the instructor and the other students in the class. The educational objective of this activity is the professional development of the students and to expose the other classmates to additional information not covered in class. This manuscript describes a variation on this semester long research project, report, and presentation.

The idea for video project was first attempted during the fall 2010 semester in an elective course, "Chemical Process Hazards Management." That course is modeled after the HAZardous Waste OPerations and Emergency Response (HAZWOPER) regulations set up by Federal Government (29 CFR 1910.120) and is used to familiarize students with various implications of the regulations and to get hands on experience in mitigating hazardous releases and the corresponding hazardous response practices. As part of the course, four or five member groups of students would develop a scenario of a hazardous material (Hazmat) release and what would be the necessary response to mitigate the Hazmat incident. In previous years the students would develop their scenario, work out what would be the necessary response, write it up in a report and then give an oral presentation on the scenario. In the fall 2010, I gave the class an option to prepare both a written report and give an oral presentation or to create a short "YouTube" type video as if they were a local TV channel reporting the incident and the clean-up operations. I did not give them any directions on how to go about making the videos, but just a short list of things that needed to be covered in either the report or video. All groups chose to do the "YouTube" videos and there was a marked increase in student interest and enthusiasm about the traditional semester project.

After this success and enthusiasm on part of the students in an elective course, where they were in the course because of interest in the topic on their part, I wanted to see if it would be as successful in a required core course in the chemical engineering curriculum – thermodynamics. Thermodynamics is an important fundamental science with many applications in both engineering and the sciences. Typically at least one thermodynamics course and many thermodynamic concepts are required in most engineering and science undergraduate curriculums. Even though thermodynamics is a well developed and mature field, students often have difficulty obtaining a clear understanding of the underlying concepts and principles. Repetition and presentation of the underlying concepts using different means and examples aid students in learning, understanding and being able to apply the key thermodynamic concepts that they will find useful in their future careers.

This assignment was first attempted in the summer 2011 semester without much success. The primary difficulty was that the shortened format of the summer semester did not allow the students enough time to formulate and produce a video. Also with fewer students in the class there seemed to be a lack of critical mass of students who had some experience with creating videos. The assignment was revised and given again during the fall 2011, spring 2012 and fall 2012 semesters. During fall 2011 and spring 2012 semesters the assignment was enthusiastically completed by both classes. In the current fall 2012 semester the students are engaged and currently working on their videos. A modified assignment, as will be described below, was given during the summer 2012 semester and again in the fall 2012 semester. During the summer 2012 semester videos were not produced again do to the relative small class size and shortened time.

The semester long project was completed by small a three-to-four member group of students to develop a short "YouTube" video presentation describing and demonstrating some thermodynamic principle. The researching of the thermodynamic concept, the development of a "script" for their video presentation and the production of the video all lend themselves to active learning versus merely taking notes in lecture or reading the textbook. Also, reviewing the thermodynamic concept videos developed by their classmates enhances the understanding and retention of the underlying concepts.

ASSIGNMENT

The students were given the assignment as shown in Appendix 1. For the assignment they are asked to form a three-member group to create a video presentation about some aspect of thermodynamics. There were minimal instructions given about how to create the video. The students had the entire semester to develop a concept, create a script and make the video. During the semester there were milestones to aid the students in moving along such as an early date for them to form groups, review the previous videos and to submit a topic. Many of the groups came to me as they developed their concept and I made suggestions. Most of the groups worked autonomously. The last few class periods of the semester were reserved for presentation of the videos. Each of the videos was rated by all of the class members using the rubric shown in Appendix 2, using personal response (clicker) technology. The videos were also graded separately by me and the graduate teaching assistant using the same rubric.

During the summer 2012 offering of the course, I added an additional assignment. For this assignment, I have the students review the previous thermodynamics videos produced and rate them according to technical content with a score from 1 to 5 and also to rate the video on organization and quality. This assignment is given to force the students to review the various videos earlier in the semester to help give them ideas of types of projects and to see which video techniques they prefer. I tabulated these ratings so that each video on the web site has a composite rating in terms of technical content and the quality of the video. This will further help guide students in future semesters when they are looking at videos for asynchronous review of concepts. This assignment is located in Appendix 3.

RESULTS

Forty two students were enrolled in the fall semester thermodynamics course and they produced fourteen student video projects. The videos were viewed over three class periods at the end of the semester and the videos were rated by Dr. Ludlow, the graduate teaching assistant and by five different students for each video. The technical content and quality of the videos varied. Some students produced a video with little or no talking, just some PowerPoint slides and pictures with some music. Others prepared videos with PowerPoint slides, pictures and/or video clips with a voice over. One group created a video that was completely animated. Finally several groups produced a video with all live actors. Some of these videos where parodies of popular TV shows.

Sixty students were enrolled in the spring semester and they produced nineteen videos. The students had access to, but were not required to view the videos produced in the previous semester and those who did took tips to improve their videos. The nineteen videos produced generally were a better quality than the first batch and there was a similar mix of formats, however none produced a completely animated video. Also, due to my encouragement, fewer of the videos were ten minutes long. Because of the size of the class four class periods at the end of the semester were taken up viewing the videos. All of the videos were graded using the rubric (Appendix 2) by the instructor, a GTA and separately by all the students in the class using personal response technology.

During the fall 2012 semester there are fifty four students enrolled in the class and seventeen groups have been formed. During this semester the students were required to view all thirty three of the previously produced videos and to rate them using the rubric in Appendix 3. The groups have chosen their topics for their videos. Four class periods have been set aside at the end of the semester for viewing the videos produced and all will be graded by all the class using personal response devices using the rubric in Appendix 2.

The topics for the videos have varied and each semester there are always overlaps of some topics, although I try to limit it to no more than two groups working on the same topic. However, even if two groups create a video on the same topic each group approaches the topic differently. The video topics are shown in Table 1.

Table 1 – Student Produced Video Topics

Fall 2011	Spring 2012	Fall 2011
Carnot Engines	Aerosols	Boyles Law
Coal Fired Power Plant	Carnot Cycle	Counter Current Heat EX
Cubic Equations of State	First Law	Diesel Cycle
Dippy Bird	Hand Warmers	Equations of State
Heat Effects	Heat Capacity	First Law
Ideal Gas Processes	Latent Heat of Vaporization	Ground Source Heat Pumps
Ideal Gas Processes (2nd)	Light Bulb	Mpemba Effect
Nozzles	Nozzles	Nozzles
Mpemba Effect	Otto Cycle	Potato Gun
Second Law	Potato Gun	Potato Gun (2nd Group)
Second Law (2nd group)	Pressure Cooker	Refrigeration
Sensible and Latent Heat	Refrigeration	Thermodynamics of Fireworks
Boston Molasses disaster	Rolla Knows the First Law	Thermodynamics of Metabolism
Thermoof making Pizza	Second Law	Thermo of a Process Disaster
	Steam Engines	Second Law
	Thermodynamics of Cooking	
	Thermodynamics of Metabolism	
	Turbines	
	Using Residuals	

During the fall 2011 and spring 2012 semesters I conducted an anonymous survey on Blackboard to evaluate the students' opinion for the project. The students were asked a series of question asked to rate on a 0 to 4 scale with 0 be strongly disagree and 4 being strongly agree. The results of the two surveys are summarized in Table 2.

Table 2 – Student responses to video production assignment

Survey Question	Result
Questions assessing student interest	
"Dr. Ludlow should continue to assign the video production project"	3.3±0.7
"I enjoyed watching the videos prepared by my classmates"	3.1±0.8
"I thought watching the videos in class was a waste of valuable class time"	0.95±0.8
Questions concerning asynchronous viewing of videos	
"Having the videos available on the web for review during the semester aided me in my learning"	2.8±0.9
"I felt like I learned something by viewing and reviewing the videos prepared by my classmates	2.7±1.0
Questions dealing with student learning	
"I feel that I would have gained more knowledge about thermodynamics from completing a research project rather than doing a video project"	1.0±1.0
"I feel that the video project helped me to better learn some concepts of thermodynamics	2.0±0.9

The survey indicated that most of the students found the video production project to worthwhile and not a waste of time. They also indicated that having the videos available asynchronously was advantageous. They were less positive about the amount of learning from completing the project, but on average felt it was more useful than doing a research report and presentation.

In response to the question "How could this video project assignment be improved?" I received some of the following feedback

- A few more guidelines – suggested resources
- Set minimum time and also lower maximum time (9 – 10 minute videos seemed too long)
- Wanted available on Blackboard (this was first semester response even though no videos had been produced yet)
- Allow all class to grade each video (after first semester allowed all students to grade using personal response devices)
- Numerous responses – noting that it was a fun and great assignment

As a result of the first listed response, after the first semester I have added some suggested resources. Also I now include sites where they can access music and pictures so that no copyright laws are broken. As a result of the second group of responses I have encouraged subsequent groups to make more short and concise videos. Students during the fall semester commented that they wanted to see some videos on the web/Blackboard even though none had been produced yet. During the spring semester survey there were no responses like this. In response the comments by students about allowing all students to grade versus only a handful of students, I changed the procedure during the spring semester so that clickers were used so that all students could rate every video.

In the anonymous surveys I also asked for feedback concerning what should NOT be part of the video project. Mostly I received comments that they enjoyed the project and suggested no changes. Some changes suggested were to allow only one group to cover a topic. Subsequently I have tried to encourage a broad spectrum of topics, but with a couple of dozen videos being made it is hard not to have some duplication of topic. In both surveys there have been a couple a students who want to see the students in the videos and not just a voice over of some slide.

CONCLUSIONS

An alternative to the traditional semester research project with a written and oral report has been replaced with a project where students produce short videos on some thermodynamic concept. The project has been received enthusiastically by the students and they feel that they learned as much as they would have done by writing a report. The videos produced are made available on the internet to subsequent classes and the students can use the videos to review some topics covered in the class and learn about things that are not covered in the regular lecture courses.

ACKNOWLEDGEMENT

I would like to acknowledge the Missouri University of Science and Technology Center for Educational Research & Teaching Innovation (CERTI) for their support in developing this project and for the CERTI grant which allowed me to spend some time developing the concept. I would also like to acknowledge the Missouri S&T Educational Technology (EdTech) office for the assistance in creating the web space for the student produced videos.

Appendix 1

ChBE 141 –Chemical Engineering Thermodynamics I
Thermodynamic Concept Video Project

For the term project you will work as a three –member team to create a presentation that describes some thermodynamic concept or calculation. This presentation will be placed in a "You-Tube" type video format. Please do not post on You-Tube!

The catalogue description of ChBE 141 – Chemical Engineering Thermodynamics is the "Development and application of the laws and fundamental relationships of thermodynamics to industrial chemical processes. Emphasis is placed on estimation of thermophysical property values for applications to chemical process engineering."

The topics to be covered in the course as specified in the ABET evaluation process are:

- Develop a mathematical description of the state of material in a system and the thermophysical properties of the process streams in a system.
- First law of thermodynamics
- Volumetric properties of pure fluids
- Heat effects
- Second law of thermodynamics
- Thermodynamic properties of fluids
- Application to flow processes
- Production of power from heat

The purpose of the short video project is for you to research, summarize, teach and give a real world example of some concept associated with: the mathematical description of the state of material; the mathematical estimation of thermophysical properties; applications of the first law of thermodynamics, volumetric properties of pure fluids, the thermodynamic properties of fluids; and heat effects of processes. The video needs to present factual information and descriptive techniques.

Not only will you benefit from participating in the process to produce a short video covering some thermodynamics principle, in the future other students will have access to the short YouTube like videos via Blackboard so that the current and future students can benefit and learn in the asynchronous manner on an "as needed" basis.

To help you plan your video you need to complete the Video Review assignment individually (See separate assignment sheet). The web site is: http://classes.mst.edu/chemeng141 or you can access this site from Blackboard. You can view the previously prepared videos to get some sense what can be done. Hopefully yours will be even more creative.

Project specifics:

1) By "You-Tube" format it is meant that the video will be less than 10 minutes in length (the time limit of You-Tube videos). Preferably the video will be much shorter than 10 minutes, 3 to 5 tend to not become boring.
2) The first part of the video needs to have title that includes "Missouri University of Science & Technology, the topic and the names of those who made it.
3) The video can consist of pictures or videos of processing equipment, PowerPoint slides with voice over, chalk board presentation, talking head, demonstration, etc. or some combination of all.

4) If you use material off of the web, you need to acknowledge that and give references in the video (typically in the trailer after the presentation). We have had some instances in the past where students have been a little too liberal in borrowing music/video and have violated copy right laws and we had to remove the videos from the S&T "YouTube" website. For instance the "Nozzles" and "Boston Molasses" were removed from the "YouTube" site because they violated copyright requirements. There are two sites you can go to get pictures and sounds that clearly will not violate any copy right laws. The sites are: http://search.creativecommons.org/ . When the creative commons opens you use "flicker", the other site is http://www.freeplaymusic.com/ .
5) The central idea is to present some thermodynamic concept or calculations that are related to chemical process equipment and then teach about it and demonstrate it in a format so that future students can access the video to learn more about the concept.
6) The grading will be based on scientific and mathematical correctness, clear communication and creativity – it would be nice if they are entertaining!!

Each of you will grade the other group's video using a grading rubric. The grading rubric is attached. After each video presentation we will use the clickers for you to key in your evaluation of the video just viewed.

Your video needs to either sent as an email attachment (if not too large) or you will need to bring it on a flash drive so that Dr. Ludlow can load all the videos onto the desk top. The video is due on ###### in class.

In order to not have to watch too many videos on the same topic, you will need to email me your proposed topic as soon as your group decides what they are going to do or at least by **######.** If we get too many on the same topic I may suggest some changes. Of course this is on a first come has the highest priority basis – so let me know (via email) what your proposed topic is.

Dr. Ludlow will decide on the final order that the videos will be shown. This will primarily be to minimize overlapping concepts on the same day and also on the relative lengths of the videos. We will typically watch 5 or 6 videos during the class periods of **######.**

Some suggestions on creating the video.

There really is no standardized "tip sheet" on how to go about editing your video since you each will have access to different cameras, software etc. So here are a few suggestions from previous classes.

You want to use a digital camera, cell phone etc. These typically create video format files in .mov .mod .flv .avi formats. Depending on the cameras you use you might get a different format. Depending on the editor you use you may need to convert from one format to the other.

You can also get video and sounds to include in your video from YouTube or other video sites. One way to legally get sounds from a website is to plug a 3.5 mm cable into the speaker output and place the other end into the microphone to record the sounds. There are probably many "shifty" ways to do so – please don't tell me and I would prefer that you do not do them.

Apple, Windows 7 and others have video production programs. You should ask around to see if anyone has access to such programs. Adobe Premiere CS5 was highly recommended.

Appendix 2

ChBE 141 – Chemical Engineering Thermodynamics I
Video Presentation Grading Rubric

Category	1	2	3	4	5
Introduction and Concept Description					
Q1 Introduction & Concept Description	Little or no introduction of concept or calculation, minimal motivation of need or application		Poorly defined concept or calculation, minimal background and motivation of need or application		Complete introduction of the concept and motivation and need for the application
Q2 Usefulness of information	Little usefulness for the topic covered		This video contained some interesting concepts		Video contained useful information presented in an interesting way
Science and Mathematics					
Q3 Thermodynamic Principles	Little or no description of the thermodynamic principles involved		Description of the thermodynamic principles involved		Complete description of the thermodynamic principles involved
Q4 Technical Approach	Superficial or little technical understanding of material presented		A general understanding of technical aspects of the material presented		Clear understanding of the technical aspects of the material presented
Presentation and Organization					
Q5 Presentation Organization	Poor organization, repetition or omission, rambling or overly terse, no connection among different parts		Acceptable organization, each section generally appropriate length, adequate linkage between parts		Well organized presentation, structure is logical and evident to listener. Linkages evident.
Q6 Presentation Style & entertainment	Video poorly prepared; excessive pauses, filler words, and colloquial speech; poor visual aids		Acceptable presentation style with few pauses; appropriately technical speech; good visual aids; effective and professional speaking style		Excellent presentation style; highly professional speaking and well prepared visual aids. Entertaining

Appendix 3

ChBE 141 –Chemical Engineering Thermodynamics I
Thermodynamic Concept Videos - Review Project

Background:

In previous semesters, 3 member teams of students in ChBE 141 have worked as a team to create a "YouTube like" video presentation that describes some thermodynamic concept or calculation.

The purpose of the short video project is for students to research, summarize, teach and give a real world example of some concept associated with: the mathematical description of the state of material; the mathematical estimation of thermophysical properties; applications of the first law and/or second law of thermodynamics, volumetric properties of pure fluids, the thermodynamic properties of fluids; and heat effects of processes. The video needs to present factual information and descriptive techniques.

You have access to the videos produced previously by students via the internet or Blackboard. The idea is that you have access to the videos to learn about the topics in the asynchronous manner on an "as needed" basis.

Video Review Project:

The videos produced so far are located at the web site "http://classes.mst.edu/chemeng141". You can access this from a web browser or you can easily access this in the "content" section of the ChBE 141 on Blackboard. For this assignment you will be required to review all of the thermodynamic videos (currently located under Movies F11 and Movies S12). Previously, I and a GTA have rated the videos – so I have a sense of their relative quality. Via this assignment I want to gather the ratings from additional students since we all may view things differently.

For each video you will give it two scores from 1 to 5. The first score you will rate the technical content of each video.

	1	2	3	4	5
Technical Content	Little or no description of the thermodynamic principles involved. Superficial or little technical understanding of material presented		Description of the thermodynamic principles involved. A general understanding of technical aspects of the material presented		Complete description of the thermodynamic principles involved. Clear understanding of the technical aspects of the material presented

For the second score you will rate organization and quality of the production.

	1	2	3	4	5
Organization and Quality	Poor organization, repetition or omission, rambling or overly terse, no connection among different parts. Video poorly prepared; excessive pauses, filler words, and colloquial speech; poor visual aids.		Acceptable organization, each section generally appropriate length, adequate linkage between parts. Acceptable presentation style with few pauses; appropriately technical speech; good visual aids; effective and professional speaking style		Well organized presentation, structure is logical and evident to listener. Linkages evident. Excellent presentation style; highly professional speaking and well prepared visual aids. Entertaining

Assignment Deliverables

You will need to turn in a spreadsheet which lists every video, the technical content rating score and the quality rating score. This assignment is worth 5% of the semester grade. Your grade in the course that will be assigned to you will be based on the number of ratings you complete and whether it appears that you have made an honest judgment (i.e. not given them all the same rating score, or your evaluation is significantly different from other's ratings).

(332d) Text Messaging As a Tool for Engaging Chemical Engineering Students

S. Patrick Walton[1], Daina Briedis[1], Stephen Lindeman[1], Amanda P. Malefyt[1] and Jon Sticklen[2], (1)Chemical Engineering and Materials Science, Michigan State University, East Lansing, MI, (2)Center for Engineering Education Research, Michigan State University, East Lansing, MI

This generation of student is more interconnected than any generation to date. From a variety of social networking opportunities to the pervasive use of mobile devices, students are fully comfortable interacting with people that, in some cases, they have never even met face-to-face. Current modes of communication among instructors, however, still typically default to, in some order or preference, face-to-face meetings, email, and phone calls. As such, there may be a disconnect in the ways students would prefer to interact with their instructors and the ways available to them. It would seem, then, that to maximize student engagement, retention, and support, instructors should consider interacting with their students using means that the students prefer.

We investigated whether allowing students to use text messaging for course communication would enhance student instructor communication and concomitantly student engagement. We found that students did not prefer using text messaging for communicating with their instructor but that the availability of text messaging may have encouraged engagement via other methods such as email and face-to-face contact.

(332e) Interactive Online Instruction Using Course Management Software and Approaches That Engage Students Actively and Asynchronously

Paul Blowers, Chemical and Environmental Engineering, The University of Arizona, Tucson, AZ and Gregory E. Ogden, Chemical & Environmental Engineering, University of Arizona, Tucson, AZ

Innovative instruction has moved from traditional lecture format to formats that engage students in novel ways, often building on their learning styles and the availability of technologies to support learning. A series of big question arise when one attempts to port a successful in-person active learning experience to an interactive online environment. How does one build student accountability when there is no face to fact interaction? How does one involve students in their real-time asychronously from you to build successful skills? How can one build rapport with interactions that may not be optimal for success of students? All of these issues become more critical as education moves more globally and also to realms where engineering may not have considered going in the past.

Through an Intel and University of Arizona grant, two faculty used their strengths to build a sophomore course that has interactive elements within an online environment. Course management software (CMS) allows faculty to monitor student progress while also building connections with students with feedback about progress and strenths and weaknesses of answers submitted, some that are automatically evaluated through the CMS, and other feedback coming through synchronous office hours, and other feedback through asynchronous email feedback. The interactive interface requires students to watch short videos and the respond to prompts through the CMS and then respond to the feedback.

This effort describes the techologies used, their limitations, and lays out a series of stages that will allow faculty who are considering building interactive and engaging online learning experiences to follow up and accomplish their goals. As this is a pilot project, clear discussion of problems encountered in this incarnation will be highlighted and discussions held regarding how to move past limitations of technologies, time commitments, and knowledge of how to proceed that will enable faculty to engage in these ways as education changes over time and online offerings for credit become more prevalent.

(332f) Developing a Graphical User Interface for Your Favorite Computational Science Project Using Eclipse Ide for Java Developers Package with Its Windowbuilder Plug-in Is Rewarding

Paul F. Harten, Sustainable Technology Division, U.S. Environmental Protection Agency, Cincinnati, OH

Developing a Graphical User Interface (GUI) for a computational science research project can be a very difficult task. Writing software that opens up new windows to interact with users, prompts them to input specific parameters needed, utilizes those parameters for a particular computational run, and displays the results of a run is extremely different from writing software for the computational run itself.

Instead, this difficult task can be made much easier by using the freely downloaded and popular Eclipse Integrated Development Environment (IDE) for Java Developers package with its WindowBuilder plug-in. Extensive help pages and tutorials are available in the downloaded Eclipse package and WindowBuilder plug-in to make the software writing process acceptable. Once completed, your favorite computational science project with its new GUI can be made available to the general public at no charge. With this additional user interaction, the total project can become one of the most rewarding efforts a computational scientist produces.

(332g) Simulation Modules for Improving Learning in Process Dynamics and Control Courses

Mary M. Staehle, Chemical Engineering, Rowan University, Glassboro, NJ

Process Dynamics and Control (PD&C) is typically the first course in the chemical engineering curriculum that addresses dynamics in chemical processes explicitly. Developing an understanding of process dynamics is of paramount importance for industry-bound chemical engineering students, yet students often feel as though the subject is oddly disconnected from their curriculum and challenging to comprehend. Furthermore, most PD&C instructors have found it prohibitively challenging to implement new pedagogical strategies into PD&C courses. The suggested course content is expansive, experimental set-ups are expensive, and dynamic experiments are time-consuming. Several simulation modules have been proposed as alternative means of experiential learning in PD&C courses, but as a whole, while these modules may improve the computational abilities of the students, they do not promote learning of the course material.

We have developed a series of MATLAB/SIMULINK-based modules that use simulations and computer-driven exercises to introduce students to computational tools for problem solving and build intuition regarding the identification and control of dynamic processes. With these modules, students are required to identify control system components from a written description and then to design, build, test, and analyze control systems using a variety of control strategies. Preliminary data suggests that the addition of these modules to a traditional PD&C course enhances learning and increases student confidence in the subject material. In this presentation, we will describe and demonstrate the modules, discuss the specific learning objectives, and present feedback from students.

(332h) Using Components in the Bag, Student Assemble and Run Simple Experiments in Undergraduate Lecture Course On Process Dynamics and Control

Pál Tóth and **Mikhail Skliar**, Chemical Engineering, University of Utah, Salt Lake City, UT

Using Components in the Bag, Student Assemble and Run Simple Experiments in Undergraduate Lecture Course on Process Dynamics and Control

Pál Tóth and Mikhail Skliar

Chemical Engineering, University of Utah

Last year, we completely transformed the instruction in an undergraduate class on *Process Dynamics and Control* from the traditional approach, consisting of lectures and homework assignments – which occasionally include computer simulations to test the designed controllers, etc. – to a teaching paradigm that emphasizes project-based learning. The subject of process dynamics and control is difficult to teach because of the breadth of covered topics, reliance on concepts expressed in a fairly formal manner, and the need to teach students how to relate abstract concepts to the outcome of practical designs. The undergraduate students often view this senior level course as an abstract exercise in mathematics, and usually have difficulty applying what they have learned to the actual process control. It was clear for a long time that to bridge the theory and practice of process control, this course would benefit from the addition of experimental projects. However, all previous attempts at introducing experiments into this lecture course were only marginally successful. For example, a few years ago we introduced projects that used professionally designed educational experiments only to find out that time sharing of the same equipment located in our instructional laboratory (Senior Lab) is difficult and the value to learning based on a "pre-packaged" experiment was not that high. This "experiment with experiments" was scrapped the year after it was tried.

Our new approach is a concept of "experiments in a bag." In this approach, groups of three students receive a Ziploc bag of components, including a low cost microcontroller, resistor, thermistor, transistors, etc., and are asked to put together their own experiments, which are small enough to carry in a backpack and run at any continent time and location after connecting the experiment to a laptop. It is assumed that students have no experience in prototyping electronic systems but are capable of following basic bread boarding instructions, such as those given at

http://implicit.che.utah.edu/Teaching/4203/GuideAssignment1.htm. Students are asked to assemble a simple circuit shown in Fig. 1 which describes a thermal control experiment in which the temperature of resistor R2 is measured using thermistor R4. As part of the assignments, students are asked to: (1) using the provided driver, establish the communication between the microcontroller, which serves to acquire data and control an average current through R2 using pulse-width modulation, and their Matlab code; (2) calibrate the thermistor to provide temperature measurements of R2; (3) acquire test data and identify the response models of the experiment, and (4) design and test simple temperature controllers for the process.

Our initial experience with the introduction of experiments into this traditional lecture course is very positive and sets an example for future development of lecture courses with incorporated experimental components. The ability for the student to have an experiment that they implement themselves, carry in their backpacks, and run at any convenient time and place are the key contributing factors to the success. The students' own assessment of this new approach was remarkably positive, despite a substantially higher workload needed to complete experimental assignments.

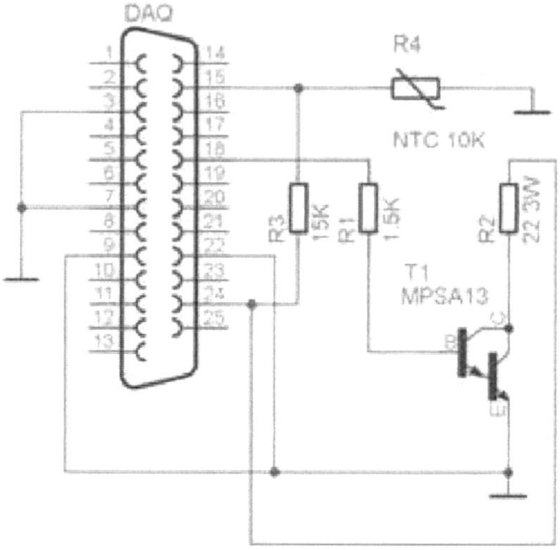

Figure 1: Schematic of the student-assembled thermal control experiment

(382a) Packed Bed Absorption -- Experiments From Afar

Charles Lemonds, Chemical Engineering, WPI, Worcester, MA and Jim Henry, Chemical Engineering, University of Tennessee at Chattanooga, Chattanooga, TN

A senior WPI student completed an 8 week Major Qualifying Project in renovating and facilitating the remote operation of a packed-bed absorption column at UTC. This presentation describes the procedures used and the resulting equipment. It also describes the remote usage of the equipment and shows the range of results achievable.

The remote user can operate and watch the equipmet as the experiment is running via the internet. The system is set up for absorbing CO2 from an air stream using counter-flow water. The remote user can operate the gas and liquid flow rates and observe the measured flow rates, pressure drops and concentrations of the gas stream. The live cameras and the pressure drop data allow students to learn about flooding conditions and behavior.

This equipment is available 24/7 for approved users. Automatic lighting allows for watching the behavior even at nighttime. Demonstration of the capabilities will be performed live during the presentaiton.

The Institute of Advanced Studies Canada-Mexico: Accomplishments and Challenges

Hugo de Lasa (1), Benito Serrano (2)

(1) Chemical Reactor Engineeirng Centre (CREC), Department of Chemical and Biochemical Engineering, Faculty of Engineering, University of Western Ontario, London, Ontario N6A 5B9

(2) Universidad Autónoma de Zacatecas, Zacatecas, México

Abstract

The Institute of Advanced Studies Canada-Mexico (IASCM-IEAMC) is headquartered at the Chemical Reactor Engineering Centre (CREC) at the University of Western Ontario (UWO). CREC was founded in 1982 and the IASCM-IEAMC has benefited from its successful partnerships with universities and industries across Canada, as well as from its International collaborations with various countries in Latin America, Europe, Africa, and Asia. These joint projects, liaisons and partnerships have propelled the development of the IASCM-IEAMC initiative. The focus of the IASCM-IEAMC has, however been mainly the creation of collaborative research alliances in Mexico.

The central objective of the IASCM-IEAMC has been the training, and exchange of undergraduate and graduate students as well as PDFS, professors and researchers between Canada and Mexico. The understanding is that the exchange of ideas, students, faculty and investigators will be mutually enriching for research, teaching and industrial progress in both Canada and Mexico. A main topic of research covered under the IASCM-IEAMC is and has been the development and implementation of clean alternative technologies for the protection of the environment , a subject which in many ways is common to Canada and Mexico as well as one of global concern.

1. Introduction

For anyone who has had the opportunity to visit CREC-The Chemical Reactor Engineering Center, in the newly built LEED Gold Claudette MacKay-Lassonde Pavilion at the University of Western Ontario, you will perhaps notice that it is not only a place burgeoning with excitement for green research but also a setting where multiculturalism prospers.

In its 29 years of existence, CREC has been instrumental in the forging of academic and professional ties with Latin America, Europe, Africa, and Asia and has attracted many international exchange, graduate students and PDFs to study in its world class laboratories with top researchers. Therefore, it is no surprise that CREC, headquarters the Institute of Advanced Studies Canada Mexico (IASCM-IEAMC) an initiative which began in 2006.

This compelling research environment mutually enhanced by both an interest in green research and a welcoming atmosphere for international cultures has also been expanded through CREC's research partnerships with Universities across Canada, such as University of British Columbia, Laval University and Ecole Polytechnique de Montreal as well as others around the world. CREC's focus, however, has been the creation of alliances and research connections in Mexico.

These research liaisons and partnerships, conferences and seminars, and exchange of students, between Mexican and Canadian Universities have led to the creation and continuity of the IASCM-IEAMC. Furthermore, this institute has continued to flourish since its inception.

2. Mission of the IASCM-IEAMC

The IASCM-IEAMC is an enterprise of unique vision for the interaction between a network of universities in Mexico and in Canada. It is a framework for the undertaking of cutting edge research in the areas of environmentally sustainable chemical processes.

Fig 1-Logo of the IASCM-IEAMC

Mexico and Canada share an important number of issues in environmental protection such as the need to develop and/or use alternative energy sources and fuels. This series of commonalities extends to the realms of health, socio-cultural, legal and immigration issues. The IASCM-IEAMC is a forum and think tank to deal with all these issues relevant to Canada and Mexico.

The IASCM-IEAMC initiative has the following goals and objectives:

a) To develop collaborative research in areas of strategic importance for both Canada and Mexico. This would consist of developing, testing and implementing technologies for the protection of the environment; the use of clean and renewable energy resources; as well as research on related socio-cultural, health, legal and migration issues.
b) To contribute to the training of Mexican and Canadian Master and PhD graduate researchers and university professors, in their respective fields.

c) To promote the creation of spin off companies in the areas of green process engineering by involving experts from industry and academia from both Canada and Mexico.

2) Antecedents

The seeds of the Institute of Advanced Studies Canada-Mexico (IASCM) were firmly planted back in the early 80's. The initiative of the Institute can be credited in good part to both Dr. Hugo de Lasa and Dr. Benito Serrano, presently co-presidents of the Institute. Since 1982, Dr. Hugo de Lasa has been working on many research projects focusing on the development of environmentally sustainable processes and establishing a solid research group at the University of Western Ontario (UWO), in London, Ontario, Canada. Students and researchers from Canada as well as from various areas around the world such as Latin America, Europe, Africa and North America have been part of this unique research group undertaking projects in chemical reaction engineering.

de Lasa's interaction with Mexico started in 1987 with the acceptance of Dr. Alberto Soria Lopez as a Ph. D. student at UWO. He graduated in 1991 and went back to Mexico to become a Full Professor at the Universidad Autonoma Metropolitana. In 1994, Dr. Benito Serrano-Rosales joined de Lasa's research group as a graduate student to pursue a Ph. D. degree in Chemical Engineering. Dr.Serrano-Rosales' studies focused on photocatalytic processes applied to waste water treatment, a subject of utmost importance for both Mexico and Canada.

Upon graduation in 1998, Dr. Serrano-Rosales re-joined the Universidad Autonoma de Zacatecas (UAZ) as a Professor and researcher. He kept a close relationship with Dr. de Lasa throughout various collaborative research projects. Dr.Serrano-Rosales has become a champion of the initiative of the IASCM and has provided, together with the Universidad Autonoma de Zacatecas (UAZ), the most valuable support and exemplary partnership for the creation of the Institute.

Figure 2. The Universidad Autonoma de Zacatecas. The Mexican leading partner of the IASCM project

Around the same time in 1998, Dr. Miguel Salaices-Arredondo of Mexico, started his Ph. D. studies in Chemical Engineering at the Chemical Reactor Engineering Centre (CREC), UWO, under de Lasa's supervision. This promoted a close, fruitful collaboration between Dr. Serrano-Rosales, Dr. Salaices-Arredondo and Dr. de Lasa with very positive outcomes such as outstanding developments in photocatalysis, the publication of several papers in prestigious journals and the publication of the book "Photo-catalytic Reaction Engineering", in 2005. Dr. Miguel Salaices-Arredondo was the first graduate student to graduate from the IASCM-IEAMC.

As part of the academic and research exchange, Dr. Serrano-Rosales participated in various projects as Visiting Professor at UWO and also acquired novel Photo-CREC units developed and manufactured at CREC-UWO to carry out research activities at UAZ. This led to the productive development of research projects performed simultaneously at UWO and UAZ.

Figure 3. The University of Western Ontario. The Canadian leading partner of the IASCM project

This productive relationship fostered by the willingness to undertake collaborative projects, prompted Dr. Serrano-Rosales and Dr. de Lasa to propose a formal agreement of collaboration between their home universities. As a result, an Agreement of Academic Collaboration between UWO and UAZ was signed on September 13, 2003 by the Presidents of both universities. During that event, there were also a series of lectures related to common relevant issues between Mexico and Canada, as well as networking among the attendees. The outcome of this event was the commitment to create an inter-university Institute devoted to studying common problems in Mexico and in Canada. Soon after that, a name was coined for the initiative: The Institute of Advanced Studies Canada-Mexico (IASCM). The formal proposal for the IASCM was signed in Zacatecas, México, on May 12, 2005, by the Governor of the Zacatecas State, the UAZ President, the UWO President representative, the Canadian Ambassador for Mexico and, the President of the Academia Mexicana de Ciencias. This event was called the *"Canada Day",* and marked the formal initiation of the activities of the IASCM.

3. Partner Institutions of the IASCM-IEAMC

During the February to May 2010 period the IASCM-IEAMC expanded its membership. Agreements for research collaboration were signed between the following parties:

- University of British Columbia, Faculty of Applied Science, Vancouver, BC
- Ecole Polytechnique de Montreal, Montreal, Quebec
- Universite Laval, Quebec City, Quebec
- University de Sherbrooke, Sherbrooke, Quebec
- University of Saskatchewan, Saskatoon, Saskatchewan
- The University of Western Ontario, London, ON

and

- the Institute of Advanced Studies Canada-Mexico(IASCM-IEAMC)

Furthermore, the current participating institutions include six Mexican Universities. These are:

- Universidad Autonoma de Zacatecas
- Universidad San Luis Potosi
- Instituto Tecnologico de Celaya
- Universidad Autonoma Metropolitana
- Universidad de Guanajuato
- Universidad Michoacana-San Nicolas Hidalgo.

These agreements promote long-term research and collaboration between the IASCM-IEAMC and the participating Canadian and Mexican universities. The understanding is that the exchange of ideas, students, faculty and researchers in the context of the IASCM-IEAMC will be mutually enriching for research and teaching as well as for industrial progress in both countries. Although, the focus of this exchange will be compatible with the orientation of each University, the primary area of research will be green energy and green sustainable chemical processes.

These agreements were made for a 5 year period. Each university has appointed a Program Director. For the initial period of the agreement, the Research Directors are as follows:

- Prof. Nicolas Abatzoglou, Universite de Sherbrooke
- Prof. Jamal Chaouki, Ecole Polytechnique de Montreal
- Prof. Ajay Dalai, University of Saskatchewan
- Prof. John Grace, University of British Columbia
- Prof. Serge Kaliaguine, Universite Laval
- Prof. Hugo de Lasa, University of Western Ontario
- Prof. Benito Serrano Rosales, Universidad Autonoma de Zacatecas
- Prof. Benito Serrano Rosales and Prof. Hugo de Lasa, IASCM-IEAMC

On April 30, 2010, the two institutions championing the IASCM- IEAMC, the Universidad Autonoma de Zacatecas and the University of Western Ontario signed a renewal of the agreement extending the academic collaboration for another five years.

3. Graduate Student Training Program Objectives

At the centre of the IASCM- IEAMC is the training of graduate students. Objectives of the IASCM- IEAMC graduate student training program are as follows:

a).To mentor and train students in a multi-institutional program with the participation of professors from chemical and materials engineering, chemistry, business, biology, and health sciences from the participating Canadian and Mexican universities. This will offer exciting opportunities to attract and train a cohort of outstanding Mexican and Canadian students who will have the skills to become leaders in meeting the green technology challenges of the coming decade. Students will be supervised via a model, termed *triangular collaboration* which is comprised of the following: i) a Mexican or Canadian graduate student, ii) a Canadian Professor-co-Advisor and iii) a Mexican-Professor co-advisor. Currently, five students are working under this triangular arrangement. Several of these PhD students have already participated as co-authors in refereed publications in prestigious journals and international conferences, which show the high value of the proposed training. Some of these students have also received prestigious research and teaching awards. One PhD student already graduated from this program.

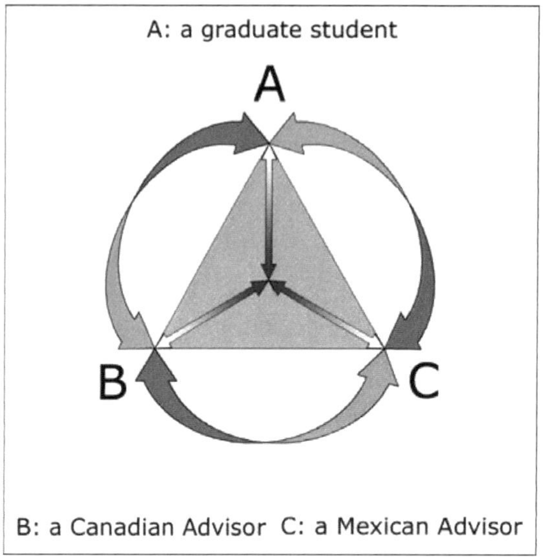

Figure 3: Schematic representation of the triangular collaborative approach used by the IASCM

The arrangement of *triangular collaboration* helps in this process, Mexican universities as well as Canadian ones. Mexican professors are involved as co-advisors in teams of international reputation, giving them credibility with Conacyt and the SIN (National System of Researchers form México).

b) Graduate Students with residence in Mexico will spend periods of time at Canadian partner institutions working on complementary projects and/or gaining access to expertise and instrumentation not available at the home institution. There will be a similar program for graduate students with residence at Canadian universities. They will also spend time at Mexican partner institutions developing research. This will allow them to carry out projects in collaboration with researchers of diverse technical and cultural backgrounds. This mobility and exposure will help them develop the necessary communication skills to efficiently deal with multi-disciplinary teams all working together for a common objective. This will also help them see and tackle given problems from different angles opening their minds to out-of –the-box solutions.

c) Graduate students will participate in research seminars and graduate courses related to Clean Energy and Green Technology that will be broadcast at the partner institutions.

d) Graduate students will benefit from the invitation of engineers, scientists, and business people, from industry and academia to spend time at each of the partner institutions. This will give them a chance to meet and network with reputable researchers and business leaders. It will allow them to be exposed to jobs in academia and industry as well as to develop the necessary contacts and knowledge to start up their own spin off companies that may result from their research work.

e) The partner institutions in Canada and Mexico will organize 3 day symposia on "Clean Energies and Green Technologies" with lectures and papers by Principal Investigators (PI) and top guest lecturers from Canada and abroad.

f) From all of the points mentioned above, graduate students will have had the opportunity to develop the skills necessary to effectively present their ideas in front of specialized audiences. Furthermore, this will help them develop the skills to communicate with people in industry and academia as well as be prepared to lead R& D teams in companies by designing, simulating and implementing new environmentally sustainable processes and clean energy.

g) In addition, the IASCM-IEAMC will encourage the development of professional opportunities and the creation of jobs both in Mexico and Canada for Mexican and Canadian students who have completed their Masters and PhD degrees, to return to their home countries, to become a factor of change in their communities.

4. Graduate Students

During the 2002-2010 period, the research supervision which took place under the umbrella of the IASCM-IEAMC, included 8 PhDs and 1 MESc graduate students with Conacyt-Mexico Scholarships. These students were: A. Ortiz Gomez, Hernandez Barajas, R. Hernandez Enriquez, E. Salaices, J. Moreira, S. Escobedo, J.M. Garcia, G. Navarro, P.Valades. Four new graduate students with Conacyt-Mexico Scholarships are expected to start their PhD programs in 2010-2012. Since January 2009, one PhD student, Gladys Jimenez-Garcia with main residence at the Universidad Michoacana, San Nicolas Hidalgo is enrolled in a PhD program with Dr. Hugo de Lasa and Dr. Rafael Maya Yescas (Mexican professor) both acting as co-advisors under the graduate research training model of triangular collaboration.

5. New CREC Labs available for collaboration under the IASCM-IEAMC

UWO has had a successful collaborative partnership with UBC and Laval which was recently significantly strengthened through a new $18.2 million Canadian Foundation for Innovation-Leading Edge Fund (CFI-LEF) infrastructure grant entitled "*Green Technologies and Environmentally Friendly Processes*" led by Prof. de Lasa. This infrastructure project enabled the construction of a new 10,000sq.ft. research facility in the UWO Green Building named the Claudette MacKay-Lassonde Pavilion. It is divided into four integrated "research nodes" and also includes a shared testing and analytical laboratory. The facility contains over $10 million of new state-of-the-art equipment. This allows research spanning from the synthesis, characterization and testing of nanomaterials and nanocatalysts for fuel cells, solar cells, and biomass gasification, to the building and modeling of small-scale and large scale prototypes. This equipment which is to be used in partnership with the extensive laboratory facilities at UBC's Clean Energy Research Centre, Laval's Catalyst-Reaction Engineering Lab and

Ecole Polytechnique's Fluidization Centre provides a world-class infrastructure for the training of students through the IASCM-IEAMC.

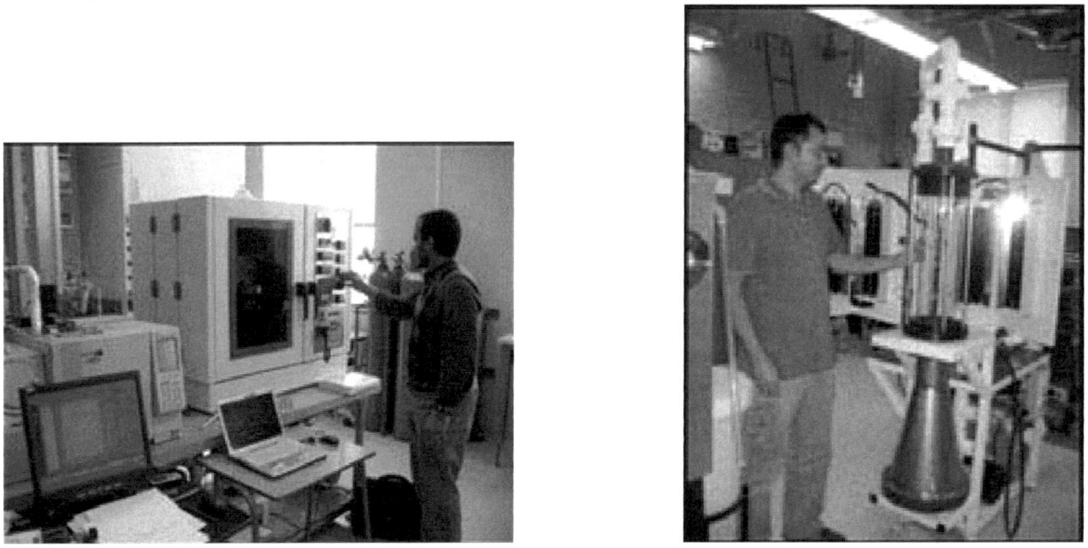

Figure 4. Images show technical staff and students operating new state-of-the-art equipment in the CREC labs of the newly built Claudette-MacKay-Lassonde Engineering Pavilion.

6. Academic Exchanges

The research exchanges at the CREC laboratories, UWO under the umbrella of the IASCM-IEAMC included of several Mexican students and three visiting professors since 2004. David Carrillo Reveles, an MESc student from the Universidad Autonoma de Zacatecas, stayed at CREC for a 3 month period, during the summer of 2010, developing research under joint supervision of Professor Benito Serrano and Professor Hugo de Lasa. His area of research covered catalytic gasification of biomass in the CREC Riser Simulator. He is expected to graduate from UAZ in 2011.

(a) (b)

Figure 5(a). David Alejandro Carrillo (centre right) with fellow Mexican Conacyt graduate students Enrique Salaices(far right) , Gabriela Navarro(left of David) and Juan Manuel Garcia (far left). Figure 5 (b) J.M.Garcia, S.Escobedo, E. Salaices, J. Moreira,,Conayct PhD students currently developing PhD research at CREC.

The research training of Mexican students in the summers of 2004 through to 2009 included the following: J. Moreira and S. Escobedo in 2004; M. A. Aguayo, N. Gamoa and Mariana Michel in 2005; G. Gerling and E. Padilla in 2007; Rogelio Jimenez, Andres Tavizon Lopez and Oscar Abel Luevano Rivas in 2008 and Miguel Angel Badillo in 2008 and 2009. These Mexican students from the Universities of Zacatecas and San Luis Potosí have participated by performing experimental work in various fields of chemical engineering and chemistry. Their research in part, has consisted of studies in the purification of water, using a novel photocatalytic reactor named "Photo-CREC-Water"

As an outstanding outcome of these stays, Miguel Angel Badillo was designated co-author in one refereed manuscript published in the "Fuels" Journal. This clearly shows the excellent opportunities offered for the initiation and training of summer students under the IASCM-IEAMC. Following these stays at Western, most of these students proceeded with graduate programs in prestigious Mexican and Canadian universities.

As part of the activities of the institute, two professors from the Universidad Autonoma de San Luis Potosí, Dr. Elisa Leyva and Dr. Edgar Moctezuma spent a sabbatical year at Chemical Reactor Engineering Center (UWO) to collaborate on research projects related to photocatalysis and organic chemistry. I

Figure 6. Assembled in front of the Spencer Engineering Building, from left to right: Gabriela Navarro, Undergraduate Student-Biochemistry UASLP; Lluvia Itzel Graduate Student UASLP; Professor Edgar Moctezuma UASLP; Denisse de Loera Graduate Student UASLP; Professor in Chemistry Elisa Leyva UASLP; Socorro Leyva, Chemistry Professor UASLP; Professor Hugo de Lasa UWO.

Figure 7. During one of the weekly review meetings, from left to right: Dr.Alan Weedon UWO , Denisse de Loera Graduate Student UASLP, Dr. Michael Brandt UWO, Lluvia Itzel Graduate Student UASLP, Professor in Chemistry Elisa Leyva UASLP

Figure 8. From left to right: Erika Padilla undergraduate student UASLP, Aaron Ortiz PhD Student UWO, Jose Luis Muñoz Research Engineer CREC-UWO, Luis Guillermo Gerling undergraduate student UASLP.

7. Special Events and Conferences

Since its inception, the partner institutions of the IASCM-IEAMC have organized and hosted a number of conferences, seminars, and sessions in order that the research developed and that the information about the institute be disseminated. These events have led to additional universities joining the IASCM., bringing maximum benefit to both coutries. Furthermore, many of these events have been followed by and have triggered excellent publications in reputable journals, patents and books.

a) The creation of the IASCM-IEAMC was strongly motivated by the ECI conference "CRE X: Innovations in Chemical Reactor Engineering" held in Zacatecas between August 28-September 2, 2005.

Figure 9. Manuel Felguerez Museum of Abstract Art, Zacatecas, Mexico

b) *IASCM-IEAMC Seminars* Universidad Autonoma San Luis Potosi (UASLP), Mexico and Instituto Tecnologico de Celaya, Celaya Gto, Mexico on November 6 and 7, 2006 respectively. These seminars were organized to bring to the attention of the Mexican Graduate Student community, the importance of pursuing doctoral studies in Canadian Universities. During the first phase of the IASCM-IEAMC establishment, it was felt that the promotion of human resources training at the postgraduate level was an issue of critical importance.

b) *1rst IASCM-IEAMC International Seminar on Water Treatment and Water Purification Technologies.* Instituto Tecnologico de Celaya, Celaya Gto on March 29 and 30, 2007. This seminar counted with the participation of more than 215 students, academics, industry and government representatives from Canada and Mexico. There were 16 invited lecturers with a first-class and balanced representation of governmental and technical experts from both countries. This international seminar also included a round table discussion with industry representatives and a poster session with more than 50 graduate student contributions. . A CD of abstracts was published following the event.

c) *2nd IASCM-IEAMC International Seminar on Green Technologies for the Production and Sustainable Use of Energy* Instituto Tecnologico de Celaya, Celaya Gto, on March 25-27, 2009. This symposium counted on the participation of more than 250 students, academics and industry and government participants. The plenary lectures were presented by experts from both Canada and Mexico. The technical program also included a round table discussion and a poster session with more than 60 graduate student contributions. A CD of abstracts was published by the Instituto Tecnologico de Celaya.

d)*2nd International Conference on Climate Change and Sustainable Development*
The "Green Reaction Engineering: Green Technologies, Sustainable Processes and Renewable Energy" event was held on April 29,2010 in Zacatecas, Mexico as part of the 2nd International Conference on Climate Change and Sustainable Development . It counted with the participation of the six Canadian and the six Mexican Partner Institutions of the IASCM-IEAMC and with additional financial support of the IDRC-Canada. The technical sessions were comprised of 14 presentations by members and directors of the Partner Institutions. The conference was successful in bringing to the forefront issues and developments in sustainability relevant to both Mexico and Canada.

Figure 10. Canadian and Mexican Participants at the 2nd International Conference on Climate Change and Sustainable Development

e) Since the birth of the Institute, presenting to audiences in Canada and Mexico was an important components in the activities developed. Canadian Lecturers were requested to participate in the annual Mexican *AMIDIQ* (Academia Mexicana de Investigacion y Docencia en Ingenieria Quimica-Mexican Academy of Research and Teaching in Chemical Engineering) conference. Dr. Paul Charpentier and Dr. Hugo de Lasa, UWO chemical engineer professors were invited as plenary lecturers during the 2006 and 2007 AMIDIQ conferences, held in Manzanillo and Iztapa respectively. These meetings counted with the participation of more than 600 graduate students and professors from more than 30 Mexican universities.
f) On Sunday, October 24, 2010, a Canada-Mexico Partnership Development Session took place between 16:00-17:30pm at the 60th Canadian Chemical Engineering Conference in Saskatoon, Saskatchewan, Canada which ran between October 24-27, 2010. The presentation was entitled "*The Institute of Advanced Studies Canada-Mexico: A Big Challenge, Great Benefits.*" All conference delegates were welcome to attend this information session presented by Hugo de Lasa from the University of Western Ontario,

Benito Serrano from Universidad Autonoma de Zacatecas and Rafael Maya-Yescas from the Universidad Michoacana de San Nicolas de Hidalgo, Mexico.

Figure 11. Representative of the IASCM-IEAMC assembled during the CSChE Poster Session. Edgar Ramirez-Jimenez, Instituto Politecnico Nacional, Mexico, Gladyz Jimenez Garcia and Rafael Maya-Yescas, Universidad Michoacana, San Nicolas Hidalgo, Mexico, Dr. Benito Serrano Rosales, Universidad Autonoma de Zacatecas, Mexico, Dr. Hugo de Lasa, University of Western Ontario

Following the presentation in Saskatoon Hugo de Lasa and Benito Serrano were invited to write a formal article for the Canadian Chemical Engineering Journal-section *"New Trends"* about the IASCM-IEAMC.

8. Books and publications

The events held by the IASCM-IAEMC since its inception, have created momentum for the writing and publication of several publications and a book. They are as follows:

- Hugo de Lasa, Benito Serrano, Miguel Salaices, "Photocatalytic Reaction Engineering" , Springer 2005
- Hugo de Lasa, Benito Serrano(Editors), "Photocatalytic Technologies", Volume 36, Elsevier
- Industrial and Chemistry Engineering Research, Special Issue to Honour Hugo de Lasa

9. Activities with Partner Institutions

In 2006, the IDRC (International Development Research Centre) of Canada approved a seed grant to launch the institute's activities such as the strategic approach of attracting Mexican Graduate Students to study in Canada under the model of "triangular collaboration".

In 2010, a Letter of Intent for IDRC encompassing a research collaboration initiative entitled in "*Clean water and energy for healthy ecosystems in communities vulnerable to climate change: A Canada-Mexico Partnership*" was drawn. This proposed initiative is planned to be developed in collaboration with the Western Faculty of Sciences and Western Faculty of Health Sciences (Dr. Irena Creed and Dr.Jack Bend) and about ten other partner universities across Canada. This research is targeted to make both water purification and hydrogen production using solar energy, economically viable. This stored solar energy could be used very effectively both by aboriginal and remote communities in Mexico and Canada. It provides great flexibility for having readily available energy as well as photo-decontaminated drinking water during the winter season or during the night. This proposal encompasses joint research with Dr. B. Serrano, from the Universidad Autonoma de Zacatecas; Dr. H. de Lasa, Dr. A. Ray and Dr.M. Ray.

9) Governance

The IASCM counts with a Steering Advisory Board. The first meeting of the Steering Committee was held in Zacatecas on April 30,2010. The day began with the signing of the renewal Agreement of the IASCM-IEAMC between Universidad Autonoma de Zacatecas and the University of Western Ontario, for another 5 years. Following this, the presentations kicked off with an address by Dr. Hugo de Lasa, co-president of the IASCM-IEAMC. The subjects of the lecture were the objectives, the achievements, and the network of partner institutions of the IASCM-IEAMC as well as its 10 year action plan. Presentations and discussions involved Dan Sinai, Director of Research Development and Services at the University of Western Ontario; Francisco Dominguez Garay, Rector of the University Autonoma de Zacatecas; as well as a lecture by Roberto A. Contreras Zarate, Director of Evaluation and Innovation, Conacyt. There were also presentations by Aaron Gomez Ortiz, CFI Project Manager and by Jesus Moreira del Rio, who spoke about the benefits of the IASCM-IEAMC from the perspective of a graduate student. The day concluded with a Round Table discussion Coordinated by Prof. Benito Serrano. Overall, this event was very successful.

As a result of the deliberations of the Steering Committee, it was concluded that the IASCM-IEAMC should engage in asking for funds for its operation or overhead costs such as travelling expenses, paying staff, telephone calls, meetings, conferences and keeping the web site up-to-date. At the moment, the IASCM-IEAMC is looking for the sponsorship from:

i)University partners, ii) Industrial partners and if possible bi-national partners, iii) government federal and state agencies, iv) International agencies such as IDRC.

At the moment, the funding targets of the IASCM-IEAMC are as follows:

• *Target 1*: The funding for, at least, 10 scholarships per year from Conacyt or other institutions, for Mexican Students to study in Canadian Universities.

• *Target 2*: The funding for, at least, 10 scholarships per year from NSERC or other institutions, for short stays of Canadian students in Mexican Universities.

• *Target* 3: The funding for at least, 10 scholarships per year for Canadian and Mexican undergraduate students and PDFs to undertake studies in Canada and Mexico.

• *Target 4*: The funding to assist in the establishment of more PhD programs in Mexico.

In all cases, the arrangement of triangular collaboration would be implemented.

The IASCM-IEAMC goals for the next 10 years are as follows:

• To coordinate on a more extended basis Mexican and Canadian programs at the doctoral level, with the participation of all IASCM-IEAMC partners

• To receive the needed financial support from governmental organizations for: a) scholarships case of CONACYT-Mexico, b) special programs from NSERC-Canada such as CREATE program, c) special funds from partner universities.

10) Conclusions

a) The IASCM- IEAMC was established as a unique organization promoting and fostering collaboration between Canada and Mexico.

b) The IASCM- IEAMC provides to Canadian and Mexican students, PDFs and visiting professors unique opportunities to develop research activities in a framework of partnership called "triangular collaboration".

c) The IASCM- IEAMC focuses on research activities of great benefit for both Canada and Mexico. This includes economically viable and sustainable technologies such as water purification, hydrogen production and waste water treatment technologies.

d) The IASCM- IEAMC provides to participating researchers a great opportunity to be exposed to the cultures of Canada and Mexico.

Acknowledgments

The Institute of Advanced Studies Canada Mexico- is pleased to acknowledge the sponsorship of The University of Western Ontario, Canada http://www.uwo.ca/research/RDS/international/index.html, The Faculty of Engineering of The University of Western Ontario http://www.eng.uwo.ca/ as well as the Universidad Autónoma de Zacatecas, México http://www.uaz.edu.mx/and the Universidad Autónoma de San Luis Potosí http://www.uaslp.mx

(382c) A Systematic and Integrative Sequence Approach (SISA) for Computing Velocity Profiles

Rocio Tijaro, Chemical Engineering, Tennessee Technological University, Cookeville, TN; Ingenieria, Universidad Arturo Prat, Iquique, Chile and Pedro E. Arce, Chemical Engineering, Tennessee Technological University, Cookeville, TN

Viscous flows are a fundamental part of fluid mechanics and momentum transfer in any engineering degree; and most would agree that they bring a very challenging task to students. Textbooks commonly used to teach these subjects, do not offer students a logical and detailed approach to derive a plausible solution, either integrative connections with previous knowledge and actual situations simultaneously. Students frequently become frustrated when confronted with a challenging problem because conventional teaching methods incorporate confusing hidden assumptions to drive the calculation, instead of recalling familiar concepts from calculus, mass balance and momentum conservation. This situation is further complicated when students cannot, for example, identify a proper location for the coordinate system origin, or why one coordinate system is better suited than others.

Here, we describe a systematic, integrative and learning-based sequence for the calculation of fluid velocity profiles; this sequence clearly utilizes the students' knowledge and assists them in building an effective and efficient path to successfully obtain any velocity profile and its validation.

Initially, the student following our sequence sets up or distinguishes the geometry of the control domain. Following that step, a convenient coordinate system is chosen and logically anchored to simplify any existing symmetry within the control domain. Once that is complete, all assumptions of the control domain must be identified to then categorize the fluid and state all applicable properties of said fluid. The next step is to propose the kinematics of the system. The last thing that the student must identify is the type of flow or the driving force of the flow in the control domain, before recalling the Navier-Stokes equation. Up to this point of the analysis, the student has only tapped into the knowledge, comprehension and application categories of Bloom's taxonomy's cognitive domain, without any utilization of concepts related to conservation of total mass and momentum. Moving onto the analysis and synthesis categories, the student must recognize the boundary conditions present in the system and reduce the Navier-Stokes equation to only the applicable terms incorporates the information collected during the previous steps. After the differential model is specified and boundary conditions are selected, the final solution is obtained and verified. The authors will discuss details and illustration about the sequence as well as limitations and pedagogical reasons.

Future efforts will include the study of the pedagogical effects on student learning by comparing traditional approaches to the one described in this contribution.

(382d) A Distillation Experiment Linking Classroom with Industrial Processing

Tracy J. Benson, Chemical Engineering, Lamar University, Beaumont, TX, Peyton C. Richmond, Chemical Engineering Department, Lamar University, Beaumont, TX and Weldon Leblanc, Lamar Institute of Technology, Beaumont, TX

A pilot-scale distillation column was used as a unit operations experiment to provide a link between classroom lectures and real world applications. A 20-foot tall, 1-foot diameter packed distillation column, complete with reboiler, pumps, holding tanks, fin fan, control valves, and DCS control board, was used to separate a mixture of propylene glycol and water. The distillation unit is used as a teaching tool by a 2-year technology school (Lamar Institute of Technology) that is adjacent to Lamar University. Senior-level chemical engineering students were presented with an open-ended distillation experiment where they defined the experimental conditions to solve a generalized problem. Teamwork, chemical process safety, increased communication skills, and sensory awareness were the main objectives of this experiment.

Each of the four teams (each team had 4 -5 members) was tasked with a different testing parameter (feed flow rate, column vacuum pressure, reboiler temperature, and feed location) to help identify operating parameters for optimum separation and energy efficiency. The teams were further divided into sub-team groups with one sub-team at the DCS control board and the other "in the field" reporting instrument readings via 2-way radios. After performing the experiment, students used a combination of Excel and Aspen simulation to analyze and explain their results.

Each team was required to give a 15 – 20 minute poster presentation to disseminate their findings. As every student within the team was required to speak ~3 minutes, students had the opportunity to develop oral presentation skills. An assessment revealed that students were better able to comprehend the intertwined facets of typical chemical processes, developed increased communication skills, and were more aware of process safety than the typical unit operation experiment. Future plans for this experiment will include statistical design of experiments, other uses for old or existing columns, and multi-component distillation.

(382f) A Curriculum Review Process: A Top-Down Learning Outcome Approach to Revising the University of Dayton Chemical Engineering Curriculum

Donald Comfort[1], Michael Elsass[1], Amy R. Ciric[1], Elizabeth Hart[2] and Robert J. Wilkens[1], (1)Dept. of Chemical and Materials Engineering, University of Dayton, Dayton, OH, (2)School of Engineering, University of Dayton, Dayton, OH

The University of Dayton recently adopted a new general education standard, the Common Academic Program (CAP) to be implemented starting Fall 2013. To meet the new requirements for the program, the curriculum for the chemical engineering undergraduate degree at UD is in the process of being redesigned by identifying the desired attributes of the UD chemical engineer and the necessary course content required to achieve the associated learning outcomes. This top-down approach to the curriculum, while necessitate by CAP requirements, also has the desire to reduce the number of credit hours associated with the program while maintaining the quality of the education. Due to a lack of a chemical engineering body of knowledge, learning objectives were compiled from current classes, ABET requirements, FE exam topics, ASEE, other university programs, and advising board input to provide an initial list of potential course content. In addition, the revision of the Integrated Engineering Curriculum taken by all students in the UD School of Engineering needed to be accommodated. These learning outcomes are in the process of being evaluated collectively and the revised program curriculum will be proposed with primary emphasis on two student cohorts – those entering the workforce as process engineers and those headed to graduate school. This generated two different sets of important skills for these students. It was determined that all students should possess analytical thinking, problem solving, technical writing, leadership, and collaborative team work. Those headed to graduate school were additionally identified as requiring skills for experimental design and documentation, advanced mathematical skills, laboratory equipment operation and measurements, and the ability to read and summarize technical literature, whereas the process engineers were identified as requiring skills for process analysis and operation, understanding of process and personal safety, and increased understanding of unit operations and equipment. The ability to deliver course relevant content to these two cohorts will be incorporated into the final curriculum proposal presented in Fall 2012 and implemented with the incoming freshman in Fall 2013. The process of identifying and developing the learning outcomes, assignment of the learning outcomes to specific classes, and final curriculum will be discussed.

(382g) Integrating the Chemical Engineering Curriculum Into a Common Academic Program At the University of Dayton

Michael J. Elsass[1], Donald Comfort[1], Amy R. Ciric[1], Elizabeth Hart[2] and Robert J. Wilkens[3], (1)Dept. of Chemical and Materials Engineering, University of Dayton, Dayton, OH, (2)School of Engineering, University of Dayton, Dayton, OH, (3)Chemical and Materials Engineering, University of Dayton, Dayton, OH

Beginning with the incoming class of 2013, the University of Dayton will be implementing the Common Academic Program (CAP) to replace the current general education curriculum. CAP is designed to be integrated into all major curricula as well as co-curricular activities. The CAP is organized around the learning structure described in a 2006 report issued by the Marianist Education Working Group titled *The Habits of Inquiry and Reflection*. Seven learning outcomes lie at the core of the CAP: scholarship, faith traditions, diversity, community, practical wisdom, critical evaluation of our times, and vocation. In addition to these outcomes, CAP implementation is guided by a specific set of characteristics. CAP sequences courses over the entirety of a student's academic career to continually build upon the knowledge of each learning outcome as well as provide multiple opportunities to demonstrate this knowledge. This knowledge will be applied in the classroom through learning centered tasks. Additionally, the CAP will include professional and vocational learning opportunities by providing skills such as communication, decision making, ethics and critical thinking. Lastly, student learning will be assessed to guide continual course improvements.

In theory, the CAP curriculum is credit-neutral in that current programs will not be required to eliminate classes to accommodate the new general education curriculum. However due to class sequencing, it would be impossible to maintain the existing chemical engineering curriculum although it would be a simple task to rearrange classes to effectively recreate the same curriculum. Therefore, it was decided to fully re-examine the chemical engineering curriculum to better integrate with CAP.

The school of engineering's goal is to use the CAP as an opportunity to provide a more integrated and holistic education. Central to this is the idea of crossing boundaries classes. In the context of the chemical engineering curriculum, these crossing boundaries classes can be structured to allow for students outside of the school of engineering to gain exposure to the field of engineering. Since students begin chemical engineering classes during the sophomore year, the intention is to offer CAP based sophomore engineering classes that are structured in a way that helps to attract and retain a more diverse student body. More advanced CAP crossing boundaries classes will provide an opportunity to create and/or grow emerging high interest concentrations such as Renewable and Clean Energy and Bioengineering.

Work on CAP integration begun in January of 2012 and the curriculum reexamination and CAP integration tasks have been progressing in parallel. Following a review of the CAP program, the proposed implementation, and the potential impact on the chemical engineering curriculum, a top-down approach to both the chemical engineering curriculum review and CAP integration was started. The feasibility of merging a chemical engineering capstone experience with that of a CAP capstone experience was discussed and will be revisited once the chemical engineering curricula begins to take shape however, an initial possibility for the merger is a multidisciplinary design project. Next, a preliminary study on what learning outcomes can be directly implemented into the chemical engineering curriculum and which are better left to the general education portion of the curriculum was performed by comparing the CAP learning outcomes with those of ABET. For instance, the CAP learning outcome of scholarship is directly related to the following ABET learning outcomes: an ability to apply knowledge of math, science and engineering, an ability to conduct experiments, analyze and interpret data, an ability to identify, formulate and solve engineering problems. Some CAP outcomes, such as faith traditions, have no ABET or engineering counterpart and will be satisfied with general education classes. Overall, it was found that the CAP learning outcomes of Scholarship, Practical Wisdom and Community overlap with several ABET learning outcomes, the CAP outcomes of Critical Evaluation of our Times and Vocation each overlap with a single ABET outcome and the CAP outcomes of Faith Traditions and Diversity have no ABET overlap.

Once the university finalizes details related to CAP, such as credit requirements for each learning outcome, the specifics of the CAP integration into the chemical engineering curriculum can proceed. This work will center on which engineering courses and topics are needed to satisfy the CAP requirements. Of these classes and topics, it needs to be determined which can be offered by the chemical engineering department, which can be offered by the school of engineering, and which will be offered by other departments as part of the general education requirements. A key aspect is to create a progression of CAP topics and classes through all four years of a student's academic career. Work is ongoing, and the new curriculum must be submitted for review by the beginning of the Fall semester of 2012. Consequently, all ensuing work will be finalized by the time of this presentation.

(382h) Process Design Approaches for Seniors

Richard Long Jr., Chemical Engineering, NMSU, Las Cruces, NM

The problem for senior design is to deveop imagination and an independent mode of thinking as well as competence in simulation. Thus , there is a trade off between skill in simulation and design simulation. A recent design course showed that the independent mindset can be limited by challenges in detailed simulation. The result is that to some extent the students demonstrated that they can be confident but wrong. Indeed,sometimes professors at other universities can be equally confident but wrong as well.A group aasigned in a project derived the balance equation in the separation of a column and confused the transfer operation by a factor of 5 in the height of the column. In spite of the instruction of the correct height of the column to the group, the students simply ignored the advice and proceeded along with the incorrect design. They simply relied on the opinions of the group without following up or pursuing further questioning of the instructor. Several design instructors at other unversities confused the units and conversion factors in the contest problems. A suggestion to the students was that in the future, you should not ignore the engineering supervisor. Just because the simulations computes a result, it is not always correct. An caution is GiGo.

(382i) Critical Reflections of Their Freshman Engineering Design Project

Taryn M. Bayles, University of Maryland Baltimore County, Baltimore, MD

In a freshman level Introduction to Engineering Design course, students were required to design, construct, test and evaluate a system which would mimic the performance of a heart-lung machine. The goal of each design was to flow the 'blood' at an appropriate rate through their system, which was required to oxygenate and cool the 'blood'. Their design also had to minimize leaks and system volume. In addition to design, construction and testing, the teams were required to evaluate their system performance using theoretical calculations. This project provided an introduction to transport phenomena by combining momentum, heat and mass transfer. This freshman Introduction to Engineering Design course was composed of 200 Chemical, Computer and Mechanical Engineering students – and they worked in teams of 4-6 students to complete their designs.

Two and a half years later, approximately forty of these students have matriculated to the Chemical Engineering program and have completed junior level courses in Transport Phenomena I (Fluids) and II (Heat and Mass Transfer). As part of their Transport Phenomena II class, the students were given their original freshman engineering design reports and were required to provide an analysis of their design. The analysis assignment was two-fold: the students were required to submit a group report re-analyzing the technical merits of their heart-lung design project and the students were also required to submit an individual reflection paper. For the group report, the students worked in the original teams which included 1 to 3 members of their freshman engineering design teams. Students who were not in the freshman design class were paired with the students with one or two members. The group report analysis included: significance of design criteria, engineering design process and theoretical analysis of data collected. The idea behind the reflection assignment is to help the students clarify what they have studied and learned, integrate new knowledge with pervious knowledge, as well as to help them become an active and aware learner so that they can better understand ***how*** they learn. Their reflection topics included: the engineering design process, engineering/math/science connections and technical writing.

This presentation will feature a brief description of the design project challenge and descriptions of solutions. In addition, a summary of the group design analysis and the individual reflection assignments will be provided.

(416a) AIChE Leadership in Safety

Katherine S. Ziemer, Chemical Engineering, Northeastern University, Boston, MA

The Board of Directors believes that becoming the leader in Safety Education is an important way for the institute to serve its mission, its members, and our society. This is particularly true with undergraduate education where the knowledge gap has, unfortunately, been made very clear, and where the potential to truly impact society by preventing accidents in these students' university and professional careers is extremely high. We are working to leverage our already excellent services in the area of process safety to grow our impact and offerings to our members and to the outside world in a way that is self-sustaining in both costs and volunteer time. I have been working with SACHE, CCPS, CEOC and many dedicated individuals to develop effective strategies to enable AIChE's leadership in safety education. I would like to discuss with you our current plans and progress, gather feedback from you on these plans, and listen to your specific needs as department chairs. I appreciate your time and effort in this important area.

(416b) Best Practices In Teaching Process Safety

Said AbuBakr, Paper Engineering, Chemical Engineering and Imaging, Western Michigan University, Kalamazoo, MI, Daniel A. Crowl, Department of Chemical Engineering, Michigan Technological University, Houghton, MI, Thomas O. Spicer, Ralph E. Martin Department of Chemical Engineering, University of Arkansas, Fayetteville, AR, Vincent Van Brunt, University of South Carolina, Columbia, SC and Shashi B. Lalvani, Chemical and Paper Engineering, Miami University, Oxford, OH

The topic for this year panel is process safety and how it is taught and promoted in the chemical engineering curriculum. The panel members will share their experiences in this area and answer questions from the audience. In addition, the results of a survey monkey conducted last December will be discussed. 87 departments responded, 17 of them teach process safety in a dedicated course and 57 in several courses.

Said AbuBakr, Professor and Chair, Western Michigan University

Email: sabubakr@wmich.edu

Dan Crowl, Professor, Michigan Technological University

Email; crowl@mtu.edu

Tom Spicer, Professor and Head, University of Arkansas

Email: tos@uark.edu

Vincent Van Brunt, Professor, University of South Carolina

Email: vanbrunt@cec.sc.edu

Shashi Lalvani, Professor and Chair, University of Miami

Email address: lalvansb@muohio.edu

This talk is part of the invited session in the Department Heads Forum.

Description: A panel of department chairs will share their experiences on topics important to department chairs. This year's main topic is process safety in the chemical engineering curriculum. In addition, a panel representing Federal agencies will discuss funding opportunities, and informational and update sessions will cover ABET, ASEE Chemical Engineering Division summer school, the annual report on the chemical engineering faculty academic salary survey and CACHE.

(416c) ABET Update and Discussion

Jeffrey J. Siirola, Purdue University and Carnegie Mellon University, Kingsport, TN

This session will give an update from ABET regarding lessons learned from 2011-2012 visits and what to look for in 2013-2014. visits. The current criteria are:

The curriculum must provide a thorough grounding in the basic sciences including chemistry, physics, and biology, with some content at an advanced level, as appropriate to the objectives of the program. The curriculum must include the engineering application of these basic sciences to the design, analysis, and control of chemical, physical, and/or biological processes, including the hazards associated with these processes.

Jeffrey. J. Siirola,

Email: jjsiirola@gmail.com

This talk is part of the department heads forum.

Description: A panel of department chairs will share their experiences on topics important to department chairs. This year's main topic is process safety in the chemical engineering curriculum. In addition, a panel representing Federal agencies will discuss funding opportunities, and informational and update sessions will cover ABET, ASEE Chemical Engineering Division summer school, the annual report on the chemical engineering faculty academic salary survey and CACHE.

(416d) Funding Agencies Update

Thomas W. Peterson, National Science Foundation, Arlington, VA

The National Science Foundation promotes the progress of engineering in the United States in order to enable the Nation's capacity to perform. Its investments in engineering research and education aim to build and strengthen a national capacity for innovation that can lead over time to the creation of new shared wealth and a better quality of life. Most NSF programs in engineering are funded through the Directorate for Engineering, which also sponsors the NSF's Small Business Innovation Research (SBIR) program.

The Industrial Technologies Program (ITP) is the lead government program in the U.S. Department of Energy (DOE) that funds R&D to develop efficient technologies for industry, and works to increase the energy efficiency of U.S. industry—which accounts for about one-third of U.S. energy use. In partnership with industry, universities, and DOE national laboratories, ITP helps in research, development, and deployment of innovative technologies that companies can use to improve their energy productivity, reduce carbon emissions, and gain a competitive edge.

(416e) Summer School Report

Randy S. Lewis, Chemical Engineering, Brigham Young University, Provo, UT

The 2012 ASEE Chemical Engineering Division Summer School was held in July 2012 at the University of Maine. The one-week summer school included a one-day Teaching Effectiveness Workshop by Rich Felder and Rebecca Brent, the annual Chemstations Award Lecture, numerous educational workshops, and team/networking events. The summer school participants were primarily early-career faculty members and workshops were presented by nearly 50 faculty members. Workshop materials were provided to participants. Highlights of the summer school along with available resources will be presented.

This paper is part of the Department Heads Forum. A panel of department chairs will share their experiences on topics important to department chairs. This year's main topic is process safety in the chemical engineering curriculum. In addition, a panel representing Federal agencies will discuss funding opportunities, and informational and update sessions will cover ABET, ASEE Chemical Engineering Division summer school, the annual report on the chemical engineering faculty academic salary survey and CACHE.

Chemical Engineering Faculty Academic Salary Survey

Geoffrey L. Price
Department of Chemical Engineering
University of Tulsa
Tulsa, OK

Abstract

Each US Chemical Engineering department listed at www.che.utexas.edu/che-faculty/ has been contacted via e-mail through the chair/head/director of the department with a request to return a salary survey for their department. Responses will be averaged nationally and by region and presented at the Department Head's Forum. Results will also be compared to results from previous surveys.

(416g) CACHE Update

Joseph T. Golab, Nitriles R&D, INEOS Technologies, Naperville, IL, David A. Kofke, Chemical and Biological Engineering, University of Buffalo, Buffalo, NY and Thomas F. Edgar, Department of Chemical Engineering, The University of Texas at Austin, Austin, TX

CACHE (Computer Aids for Chemical Engineering) is a not-for-profit organization founded in 1969 to promote cooperation among universities, industry, and government in the development and distribution of computer-related educational aids for the chemical engineering profession. To this end CACHE is engaged in numerous and varied activities including: development of software; organization of conferences; commissioning of educational materials; dissemination of engineering and educational information; maintenance of an academic jobs board; sponsorship of awards; and leadership of new initiatives in engineering education. CACHE is staffed by a volunteer board of 28 trustees from academia and industry and its activities are supported through membership of more than 150 chemical engineering departments worldwide. In this presentation, we will review the activities fostered by the CACHE organization, update the latest developments, and suggest how others can become involved.

This paper is part of the Department Heads Forum. A panel of department chairs will share their experiences on topics important to department chairs. This year's main topic is process safety in the chemical engineering curriculum. In addition, a panel representing Federal agencies will discuss funding opportunities, and informational and update sessions will cover ABET, ASEE Chemical Engineering Division summer school, the annual report on the chemical engineering faculty academic salary survey and CACHE.

(421a) Building a Service-Learning Program Through Freshman Engineering and LEGO Nxttm Robotics

Bill B. Elmore, Dave C. Swalm School of Chemical Engineering, Mississippi State University, Mississippi State, MS

Chemical Engineering freshmen in the Swalm School of Chemical Engineering at Mississippi State University are immersed in chemical engineering fundamentals using LEGO NXT™ robotics, Vernier® sensors and small-scale pumps, pipes, tanks, to construct simple, automatically-controlled process unit operations. From their studio experience, we are working to translate these activities into K-12 outreach. Now a regular feature of our K-12 outreach, these activities are enthusiastically embraced by students—elementary through high school, and are proving to be a relatively-inexpensive means of providing a significant number of students (e.g. 50-60) an engineering design experience in small groups of three to four each. Such hands-on activities in a high-energy problem-based environment lead to numerous learning opportunities among our chemical engineering and K-12 students alike. In this presentation, we report on several features of this program including: 1) improved techniques for recruiting students to engineering studies, 2) stronger "visualization" of engineering concepts among our engineering freshmen and 3) a stronger sense of the need for life-long learning and community service among our engineering undergraduates.

(421b) Nanoexposed! – A Freshman Introduction to Nanotechnology

Priscilla J. Hill, Dave C. Swalm School of Chemical Engineering, Mississippi State University, MS State, MS, Oliver Myers, Mechanical Engineering, Mississippi State University, Mississippi State, MS, Yaroslav Koshka, Electrical & Computer Engineering, Mississippi State University, Mississippi State, MS, Giselle Thibaudeau, Biological Sciences, Mississippi State University, Mississippi State, MS and Carlen D. Henington, Counseling and Educational Psychology, Mississippi State University, MS State, MS

NanoExposed! is a 1 hour weekly seminar course that aims to familiarize freshmen with basic nanotechnology concepts in an exciting way. The course is taught by a multidisciplinary team of faculty from chemical, electrical, and mechanical engineering, and biological sciences to emphasize the interdisciplinary nature of nanotechnology. This presentation includes an overview of the course structure and topics covered. The topics included synthesis of nanoparticles and nanostructures, characterization of nanomaterials, current applications, and possible future applications. The focus in this course is on property and functional differences between the nanoscale and the bulk scale. Examples given in this presentation will predominantly come from the chemical engineering section of the course, but the overlap with other disciplines will also be discussed.

Portions of this presentation, but not the entire presentation, have been presented previously as a poster[1] and as an oral presentation[2]. This material is based upon work supported by the National Science Foundation under Grant No. 1042114.

1. Hill, P. J., Koshka, Y., Myers, O. J., Thibaudeau, G., and Henington, C. D., NanoExposed! – An Introduction to Nanotechnology. *2012 ASEE Annual Conference & Exposition*.
2. Hill, P. J., NanoExposed! – Chemical Applications in Nanotechnology, ASEE Southeastern Section Annual Conference, Mississippi State, MS, April 2012.

(421c) Introducing Freshman Students to the Multi-Faceted World of Engineering and Sustainability Through Biofuels Synthesis From Waste Cooking Oil

Justinus A. Satrio, Chemical Engineering Department, Villanova University, Villanova, PA and Laura-Ann Chin, Chemical Engineering, Villanova University, Villanova, PA

Well-designed first year experience courses are mainstays in the curriculum for freshman engineering students, as well as for freshman in other fields. First year courses are building blocks in helping a freshman navigate through and discover a new major. It is imperative for a freshman to be exposed to different engineering disciplines to experience first-hand the expectations and variations among them.

VU Engineering has strategized this teaching opportunity through a series of freshman mini projects designed by different engineering majors. The curriculum begins with a seven-week core course that incorporates engineering fundamentals with hands-on group micro-projects that brings classroom lessons to life. Following the core course, the students are presented to select two of six interdisciplinary, hands-on mini projects that expose students to a minimum of two major disciplines each throughout the second half of the fall and the first half of the spring semester. Mini projects offered include:

· Application of Acoustic Technologies for Predicting Structural Failure

· Biofuels Process and Sustainability: Biodiesel Synthesis from Waste Cooking Oil

· Electric Car Design

· Robotics and MATLAB Programming

· The Load/Deflection Character of a SMARTBEAM

· Adsorption – Drinking Water Treatment Process

By mid-second semester, students select their intended major discipline and spend the remaining seven-weeks in the chosen disciplinary field.

This paper will be focused on the Biofuels Process and Sustainability mini project, which started to be offered in Fall 2011 semester. This project was developed with the idea of making our young students to be aware of the changes in our society. In responding to the need of our reliance to non-renewable fossil fuels due their depletion and primarily due their negative impacts to our environment, our society is slowly moving into an era of so-called bioeconomy, where we attain vital sources of carbon and energy from biorenewable materials, such as biomass. It is very crucial that future generations of our students embrace this transformation. For this to happen, our students first need to learn about various aspects involved in bioeconomy. This mini project exposed Villanova freshman engineering students to two very important aspects in bioeconomy, i.e. how transportation liquid fuels can be produced from biorenewable materials and the sustainability issues related to how biofuels are produced and utilized. Specifically, the goal of this project is for students to use basic engineering and chemistry principles to synthesize and characterize biodiesel from a renewable resource, i.e. waste cooking oil from VU dining centers, use the experimental data to design a biodiesel processing plant and finally assess the sustainability of the process.

The 7-week freshman engineering project was presented as a combination of lectures and in class group exercises on various aspects on biofuels production and sustainability. Hands-on laboratory experiments on biodiesel synthesis and characterization, analysis on energy usage and heat transfer of the synthesis process, and the synthesis of soap from glycerin by-product were performed in the weeks following the lectures. Students go through the elementary transesterification of WCO into biodiesel while producing a by-product of glycerin. From the experiments, students prepared laboratory reports, one on the chemistry and mass balance aspects of biofuels synthesis and the other the energy and heat transfer aspects of the biofuels synthesis. The report preparation involves students to conduct technical literature searches, perform data collection, and use spreadsheets to analyze data and perform numerical analysis In addition to the chemistry and heat transfer aspects of biodiesel production; the element of entrepreneurship is also incorporated in the soap production via glycerol byproduct. Students are also exposed to scale-up of lab synthesis to a commercial level of a biodiesel production facility. The schedule of learning activities and gained skills from each activity are described in the Table below.

Week	**Learning Activities**	**Learned Skills**
1	- Introduction to processes of biofuels production and concept of sustainability. - Sustainability and carbon foot prints: how much CO2 does my household emit per year? Develop carbon footprint calculator.	Concepts on biofuels production and sustainability, carbon foot print, programming using Matlab , ethics, basic organic chemistry
2	Introduction to mass & energy balances and heat transfer phenomena	Concepts of mass & energy balances and heat transfer phenomena, math
3	Going to the lab: making biofuels and collecting experimental data	Laboratory skills, collecting data, team work, time management
4	Going to the lab: characterizing biofuels and collecting experimental data	Laboratory skills, collecting data, team work, time management
5	- Analyzing data and using data to prepare a report on energy and heat transfer. - Utilizing byproduct: soap making	Data analysis, math, team work, computation, time management, laboratory skills, report writing.
6	- Analyzing data and using data to prepare report on chemistry and mass balance. - Lecture and in class activity: Concept of entrepreneurship.	Data analysis, math, team work, computation, time management, report writing. Concept of entrepreneurship, brain storming, oral presentation.
7	Final presentations: poster and final report	Oral presentation, team work, time management, poster preparation, executive summary writing.

The freshman engineering project on biofuels was also developed in conjunction with the on-going student-run Villanova Biodiesel program in the department of chemical engineering. In the 2-year old biodiesel production program, students operate a biodiesel production facility located in the department of chemical engineering's unit operation lab to convert waste cooking oil to into biodiesel. The production facility has the capacity to process all of the approximately 7000 gallons waste cooking oil that is produced from the dining service yearly. Currently, only about 10-15% of the waste cooking oil is processed for biodiesel production. In order to increase the capability of the facility to produce more biodiesel, more student participations are needed. The proposed mini project can serve as a recruiting tool by introducing new engineering freshman students to various aspects involved in the production of biofuels, which will make the students better prepared should they decide to participate in the Villanova Biodiesel program.

Overall, the freshman mini project is geared to provide engineering students with motivation for the consideration and recognition of engineering design and development examples in daily life. Apart from introducing students to a variety of engineering design and development tools, the mini projects have been a seed in helping students decide early on in their journey for a successful undergraduate career.

****

(421d) Design & Assembly of a Freshman Design Laboratory

Anthony Butterfield, Chemical Engineering, University of Utah, Salt Lake City, UT

Active and collaborative learning environments have been found to increase student conceptual learning, satisfaction with their department, and student retention, particularly of underrepresented groups. However, the majority of chemical engineering departments use such teaching methods minimally until their capstone courses, leaving much of their effectiveness underutilized. To address this problem in our department, we have developed and tested many of the components needed to create an active-learning freshman design laboratory experience.

Students in this laboratory are to collaborate in teams, and, through guided inquiry, achieve a variety of design goals related to basic chemical engineering topics. To build this course, we have developed and tested several teaching modules geared towards a freshman's capabilities and needs. Each module in this introductory lab is meant to act as an anchor point to which students may affix important chemical engineering concepts, and through which difficult theory may be revisited later in their academic career.

Execution of the design projects are based upon the open sourced Arduino DAQ boards, due to their ubiquity and high degree of capability relative to their costs. Through this data aquisition platform and with minimal instruction on electrical devices and circuitry, freshmen may design and control various systems and experiments, and their data may be collected, visualized, and analyzed within the environment of their student edition of MATLAB, without the need for additional toolboxes.

Among the modules that have been developed to date are a flow cell spectrophotometer module, meant to ease students into simple design concepts, data acquisition and data analysis, and introduce concepts of reaction kinetics and fluid dynamics. Principles of heat transfer and cost considerations in design may be introduced with a heat exchanger design module. Product design may be explored through the dissection of a transdermal drug delivery patch and a related diffusion cell module, to introduce the importance of mass transfer considerations to the chemical engineer. A total of nine design projects have been created to be used through this semester-long course. A final project is to be conducted through the last three weeks of the course which may involve several options, including the design of a process using multiple student-built apparatuses from previous modules as building blocks.

This freshman course is also meant to act as a launching platform for effective, evidence-based pedagogy throughout the remainder of our curriculum. As such, some of the teaching modules have been brought into upper level courses to create active learning projects and address relevant theory with newly-gained sophistication. These teaching modules have also found use and have been validated within our K-12 outreach programs.

(421e) Investigating Cooperative Learning Grouping Strategies in an Introductory Engineering Course

Edna Margarita Prieto and Elizabeth J. Adolph, Chemical and Biomolecular Engineering, Vanderbilt University, Nashville, TN

ES140 is a required freshman course designed to teach students about engineering analysis and design and to expose them to problem solving in discipline-specific contexts. Engineering freshmen choose 3 modules to take during the semester to learn about 3 different engineering fields. Previous instructors of the ES140 chemical engineering module have noted that 25-50% of students, particularly non-majors, are not interested in the module and are not engaged during class. The aim of this teaching as research project was to apply cooperative learning to improve the academic achievement and interest level of students in the ES140 chemical engineering module, particularly of those who are non-majors. Cooperative learning involves the use of small groups of students working together to learn the material. This method has been shown to improve academic achievement, higher-order thinking skills, communication, and social skills of students compared to individualistic and competitive learning styles (Totten, et al., 1991). In this Teaching As Research project, we hypothesized that by having Chemical Engineering students (majors) and non-Chemical Engineering students (non-majors) collaborating in teams, the academic achievement and interest level of the students would increase due to the exposure of the non-major students to the views and perspectives of the major students. We tested three different grouping strategies: (a) Self-selected working teams, (b) Homogeneous working teams (students were grouped by majors), and (c) Heterogeneous working teams (majors and non-majors were grouped together). 7 out of the 10 groups in the self-selected module ended up being homogeneous groups.

The module comprised 5 classroom sessions of 50 minutes each. The main topic of the module was Biomaterials, and the topics covered offered an introduction to tissue engineering, drug release mechanisms, and mechanical properties. Each session consisted of a combination of lectures and hands-on group activities. These activities included the preparation of a polyurethane scaffold, the observation of the release of a dye from different biomaterials, fitting release profiles in Matlab, and a jigsaw activity in which students became "experts" in two mechanical properties and taught their group members about them. Academic achievement and interest level were measured with a quiz and a survey given both at the beginning and at the end of each module. Although there was a significant increase in students' knowledge of the material after taking the module, there were no significant differences in academic achievement or interest level between the grouping strategies. In general, the students liked the activities more than the lectures. It is possible that the short time that we had with the students was not enough to generate changes in interest level by working in teams. We would like to repeat the experiment in the future by focusing on developing group activities in which students need to rely more on each other to achieve learning. As instructors, we were able to develop teaching skills related to course design, creating assessments, and lecturing. We also identified opportunities for further improvement that we look forward to implementing in the future.

(421f) Attrition in Engineering Education

Lizzie Santiago, West Virginia University, Morgantown, WV

Engineering attrition is a concern for first year engineering programs and engineering colleges. The stress related to making the transition from high school to college has been suggested as one reason for the high attrition rate. Not only is there a disruption to student-family relationships, but students need to learn how to manage their time and resources, as well as to meet deadlines without the guidance and close supervision of parents and relatives. Many first year engineering programs provide extensive academic and social support to help students make the transition and succeed academically. While necessary, are these programs sufficient to keep students in an engineering program? Are students who leave engineering academically successful in their non-engineering field of study? This study was designed not only to address why students transfer out of engineering, but to determine if those students who leave engineering are able to succeed in their new discipline and graduate from the university.

All "engineering" students at this large land grant university in the mid-Atlantic region, both calculus-ready and not calculus-ready," must complete a common "first year experience" before moving to a discipline major. Students who are not calculus-ready at entry usually take 1.5 to 2 years to complete the required courses, depending on their initial math placement. The authors studied 527 students who transferred out of engineering during their first or second year of that general engineering program. The students were mostly men who changed majors between January 2007 and December 2010. An exit questionnaire administered at the time of the transfer was utilized to determine their exit grade point average (GPA) and the reason for the switch. Furthermore, university databases were utilized to determine if those students were able to graduate from, or are still pursuing a degree at, the university. The number of students who withdrew from the university, were suspended, or never returned to the university was also assessed, as was the percent of students who left engineering, but were later readmitted into the program.

Analysis of exit surveys provided insight into the academic characteristics of those first year students who transferred out of engineering, reasons why they left, and the degree to which these students persisted to degree completion in another major at the university. Results indicate that factors different from academic difficulty are leading to the change of discipline among general engineering students. Students who are in good standing academically are leaving engineering because they lack interest in the subject. Additional explanations are considered and presented, as well as the implications for potential intervention programs to address increasing student interest as well as academic success in engineering.

The percentage of students who leave engineering and who also leave the university is a source of concern for both engineering and university administrators. Influencing factors for leaving both engineering and the university are explored and presented.

(421g) Preparing Students for 2020

Joseph J. Biernacki and Pedro E. Arce, Chemical Engineering, Tennessee Technological University, Cookeville, TN

The freshman student at large is unprepared to think critically, find creative solutions to real-world problems and to work efficiently on teams. Ultimately, it is our job, the faculty, to train students in the necessary skills to enter the 21st Century workforce. The days of lecturing, assuming that students will perform well on teams later and hoping that they might discover how to think critically on their own are long gone. We can no longer assume that our lectures and projects will help students to become problem solvers, critical thinkers or team players or that someone else will teach them those skills. So, what can be done and how can we integrate these skills into our courses and still have time for all those equations? What do we know about teaching teamwork, problem solving and critical thinking anyway? Examples from a curriculum involving a number of new paradigms will be presented. Continuous contact starting from the freshman year, introducing teamwork immediately and pervasively across the curriculum, deliberately teaching teamwork and critical thinking skills as well as focusing on developing proficiency in the application of unifying formalisms rather than compartmentalization and formula driven engineering practice are among some strategies for tackling both the technical and skill-based content necessary to produce the Engineer for 2020.

(502a) Mentoring Through the AIChE Student Chapter

Bill B. Elmore, Dave C. Swalm School of Chemical Engineering, Mississippi State University, Mississippi State, MS

Mentoring through the AIChE Student Chapter

ABSTRACT

As faculty advisor for an AIChE student chapter each of my 22 years in chemical engineering education, I have enjoyed the privilege of mentoring a number of students through a variety of outreach and service activities. Working side-by-side with students on projects ranging from our Boy or Girl Scout Merit Badge Days to Habitat for Humanity to LEGO NXT robotics outreach in middle schools, I get to observe and encourage our students as they develop their own leadership skills and grow in their vision for giving back to both the profession and society. Service learning activities offer an abundance of opportunities to build, among many positive characteristics, students'

- "understanding of professional and ethical responsibility" to society (ABET Student Outcome "f")
- "recognition of the need for, and an ability to engage in life-long learning (ABET Student Outcome "i") and
- "knowledge of contemporary issues" (ABET Student Outcome "j")

In this talk, I will report on the considerable number and variety of activities our student chapter engages in each academic year comprising a strong service-learning component. I will also share anecdotes about the mentor/protégé relationship that I have experienced with students and some of the contributions this effort makes to our chemical engineering program assessment of Student Outcomes.

(502b) Incremental Development of Student Success Through Leveraged Resume Development in Class and Advising Interactions with Students

Paul Blowers, Chemical and Environmental Engineering, The University of Arizona, Tucson, AZ

The University of Arizona has built a strong mentoring and advising system that fully integrates activities from freshman through senior years and has been fairly robust even as new faculty members rotate into courses and take on new responsibilities. Activities begin in the freshmen year with an elective course for where how to develop a resume is covered along with topics like interviewing skills, time management, study strategies, test taking ideas, and other useful topics for those entering a rigorous academic environment as they transition from high school. Approximately 10-20% of chemical engineering freshmen take this course.

The sophomore year integrates a discussion of career possibilities and what the different types of jobs are in Material and Energy Balances. This course also has a companion computer laboratory course where the students will use Visual Basic in Excel. The first meeting is dedicated to the elements of a strong resume and students are then required to submit one that gets critical review. The sophomore year is where most students will begin meeting with advisors more regularly. The core faculty who do the advising continually stress the places alumni have gone and how they achieved their successes. The alumni are also referred to directly connect with and mentor current students and they do not hesitate to give advice as the same mechanisms benefitted them as students. This culture of past students helping current students has taken 15 years to build and is incredibly valuable. It is noted that this network was built by one to two faculty through their deep engagement with students as they transition between school and employment.

The junior year involves increased advising interactions as students select electives. Students are continually reminded of the need for improving their resumes through leadership in clubs or other organizations, by being involved in research, by building a strong GPA, and by finding technical employment. Students are shown resumes of past and current seniors and discussions are held that help them find out ways to build from their base. Resumes are required to be submitted in at least one junior and one senior course for evaluation, with the senior ones being used for ABET assessment of the ability to meet both lifelong learning needs and the ability to solve engineering problems through internships and research.

Senior year involves practice interview questions in the design course where students are trained how to answer questions with specific stories, relying on their resume development over the years and their experience of quantifying, selling skills, being positive, and being specific, which are the same core ideas stressed in building strong resume bullets across the years.

As a result of the mentoring and advising both inside and outside of courses, retention is high compared to other anecdotal information from other programs. Eighty to ninety percent of students who enter Material and Energy Balances with chemical engineering as a major continue on to the second sophomore semester. Approximately 70% of the total students who entered Material and Energy Balances continue on into the junior year, and only 5% or so of students are lost after that point. Due to the intensive resume focus, student placement is very high, with only 10% of students having no signed offer upon the day of graduation, typically because they did not work on their resumes or because they are foreign students with visa issues.

(502c) Gotta Keep It Separated: Engineering a Centrifuge From a Simple Salad Spinner

Carolyn A. Nichol, Center for Biological and Environmental Nanotechnology, Rice University, Houston, TX

The isolation of compounds of interest from reaction media or source materials presents challenges to the field of chemical engineering that often require creativity and ingenuity to solve. These problems provide high school chemistry educators a unique opportunity to introduce engineering concepts into their classrooms as various separation techniques are also included in many states' chemistry standards. In this session, an example of a unique engineering challenge posed to Rice University undergraduates will be presented, as well as the novel solution two of those undergraduates devised. In addition, the presentation of this case study to high school chemistry teachers as part of a K-12 outreach program will be discussed, including specific strategies for incorporating both the case study and the engineered product itself into the high school introductory chemistry curriculum. The specific engineering challenge discussed involved finding a way to diagnose anemia without access to electricity that would also be cost-effective and portable. The solution involved the re-engineering of a simple salad spinner into a cheap centrifuge capable of separating blood samples, thus allowing for the measurement of the ratio of red blood cells to the total volume that is required for the diagnosis of anemia. While the development of this affordable separation technique has obvious benefits for the global health community, it has also had unforeseen benefits for the Houston area high school chemistry community. As it turns out, the same device that can separate blood can also separate precipitates from aqueous solution and is therefore a much more cost-effective means of introducing this separation technique to introductory chemistry students. Presenting centrifugation to students in the context of the engineering challenge for which the salad spinner centrifuge was designed provides students with both a real-world application of this chemistry concept and an example of engineering in practice. Instructions on constructing salad spinner centrifuges will also be provided to session attendees.

(502d) Chemical Process Videos: A Long-Distance Partnership for Outreach and Communication Skills Development

Shannon Ciston, Chemical and Biomolecular Engineering, University of California, Berkeley, Berkeley, CA and Lindsey Own, The Evergreen School, Shoreline, WA

Modern communications technologies create new opportunities for learning and collaboration while demanding new skills of engineering students of all ages. This project is a partnership between seventh graders at The Evergreen School and chemical engineering undergraduates from the University of California, Berkeley that leverages these opportunities for mutual learning and exchange.

In this video project, teams of junior-level chemical engineering students were challenged to create an educational video to teach a chemical process or science concept to a middle-school-aged audience. The project was set in the context of a Technical Communications for Chemical Engineers course, in partnership with seventh grade science students at The Evergreen School. The seventh graders acted as consultants for the undergraduates, drawing on their previous experience watching educational videos in school and their life experience and interests as middle school students.

There were three learning objectives for the undergraduate students:

1) Tailor technical content to the interests and needs of a non-technical audience

2) Use verbal and visual channels to effectively communicate ideas through modern communication tools

3) Contribute to the community by teaching science and engineering concepts to children

This project was an interactive process in which the students worked together to guide the video creation process, from idea generation to final production:

1) Undergraduates suggested ideas

2) Seventh graders chose ideas based on interest and appeal

3) Undergraduates Received training in audience analysis of middle school students and fair use and copyright of videos and audio materials

4) Undergraduates developed story boards

5) Seventh graders rated story boards and provided feedback during a Skype video conference with the undergraduates

6) Undergraduate teams developed videos

7) Seventh graders watched and critiqued videos

8) Undergraduates reflected on the experience

Fifteen teams total created video projects in the Spring 2012 semester, including video topics such as “Specific Heat Capacity,” “Biofuels,” and “Schrödinger’s Cat.” When asked to reflect on the project’s effectiveness for attaining each learning objective, the undergraduates rated the effectiveness at averages of 4.2, 4.4, and 4.4 out of 5, respectively. During the process, undergraduates learned to adjust the level of technical detail and terminology in response to seventh-grader feedback, and to respond to specific audience questions and requests. They gained a functional working knowledge with using Skype for video chat, and effectively combined multiple media and story-telling elements to create effective informational videos, but reported a need for more support in the technical details of video editing. The undergraduates were pleasantly surprised at the responsiveness and interest of the middle school students, and were impressed that outreach is easy; they are enthusiastic about doing more science and engineering outreach to pre-experts. The seventh grade students engaged fully in their roles as consultants and advisors, and enjoyed the interaction with the chemical engineering students. They learn interesting new science content, and felt accomplishment and pride in having supported older students in furthering their communication skills.

In the course of two semesters of this video project, the instructors learned several valuable lessons for successful implementation of a project like this.

1) Start the process early

2) Find interested and engaged partners with the curricular flexibility to incorporate the project

3) Make resources including video cameras, video chat-ready laptops, video editing software and resources available for student use

4) Facilitate student interactions with focused questions

(502e) Introducing K-12 Students to the Field of Pharmaceutical Engineering

Daniel Lepek, Charmian Wu and Ryan Poling-Skutvik, Department of Chemical Engineering, The Cooper Union, New York, NY

The design, development, and engineering of drugs provide chemical engineers with many opportunities and challenges in the pharmaceutical industry. In an effort to engage the surrounding communities, New York City public high school students were introduced to the field of pharmaceutical engineering over the course of six weeks. Through the use of lectures, teamwork activities, and laboratory experiments, students learned about the fundamentals of oral solid dosage forms, drug dissolution, and experimental design. Examples of experiments performed include building their own "in-house" drug dissolution devices, studying the effect of impeller geometry and velocity on dissolution rates, and obtaining drug dissolution profiles for various oral solid dosage forms containing Ibuprofen using UV-Vis spectroscopy. Students were also trained in communication skills, such as writing a technical report and giving an oral presentation. An overview of the program will be provided, as well as suggested laboratory exercises for those who might consider developing a similar K-12 outreach program centered on pharmaceutical engineering.

(502f) Motivating K-12 Students to Study Pharmaceutical Engineering with Hands-On Modules, Guided Visits and a One-Week Summer Camp

David A. Mota-Aguilar, **Daniel Mateo**, Miguel Florian, Sonia L. Aviles-Barreto, Rafael Mendez, Carlos Velázquez and Nelson Cardona-Martínez, Chemical Engineering Department, University of Puerto Rico - Mayaguez Campus, Mayaguez, PR

The pharmaceutical industry is important for the economy of Puerto Rico offering approximately 22% of the total manufacturing employment in the island. Since qualified human resources are fundamental, it is important to develop and support initiatives to help students increase their awareness of this area in the early stages of their academic careers.

The Pharmaceutical Engineering Education and Outreach Program of the University of Puerto Rico at Mayagüez (UPRM), as a member of the Engineering Research Center for Structured Organic Particulate Systems, organizes periodic K-12 activities to improve student recruitment and help in the training of current students who will be able to apply their knowledge across disciplines and advance technology.

Here, we report the results of two educational initiatives: monthly guided visits and a yearly one-week summer camp. In both of them, a team of graduate and undergraduate students of the Chemical Engineering Department of UPRM guide K-12 students through a series of demonstrations and hands-on modules in a pharmaceutical operations laboratory. These initiatives expose the participants to interactive experiences where they can learn and apply pharmaceutical engineering concepts, such as: laboratory safety (combustible dusts), powder characterization, solid mixing, granulation, tablet compression, coating, tablet dissolution, friability, etc.

The aim of these activities is to introduce and attract middle and high school students to basic concepts of Pharmaceutical Engineering in interactive ways. To date, the program has impacted 375 students and 18 K-12 teachers in addition to the direct impact to 20 undergraduate students and 6 graduate students of UPRM.

A comparison of the preliminary and final tests demonstrated that the students are learning new pharmaceutical concepts and the results of the satisfaction surveys indicate that almost 89% of the participants rated the activities as excellent.

(502g) Student Recruitment and Community Outreach Through an Undergraduate Mentoring Program

Ahmed Elmadhoun, Colin Young and **Anthony Butterfield**, Chemical Engineering, University of Utah, Salt Lake City, UT

Community outreach programs are vital promoters of growth in any chemical engineering department. Students in K-12 classrooms often have little to no experience with engineering and are unaware of the various types of career paths available to them through higher education, particularly the path of chemical engineering. Community outreach is also an important portion of the societal mission of a university and it contributes greatly to the general public's understanding of the role of various professions in society. We have developed an efficient and effective model for a student-managed outreach program. Through this program we have been able to connect with many of our local high schools and spark interest in engineering for a great number of students, setting them on rewarding career paths of which they had been previously unaware.

Our department employs a team of undergraduate Chemical Engineering students. These students travel in groups of three to high schools around the state and engage K-12 classrooms in an active learning environment. To create community connections we use a range of contact methods, and offer a simple internet form to allow K-12 teachers to request visits, clarify their curricular needs, and request specific demonstrations and activities.

At class visits our teams cover an introduction to engineering and then, more specifically, chemical engineering. Included with the introduction, the team members discuss the wide variety of possible career paths open to chemical engineers, along with the notable benefits of becoming an engineer. Afterwards, we conduct interactive demonstrations, using developed teaching modules and small experimental apparatuses. In an effort to make our outreach program more effective, our team has prepared and documented over 30 teaching modules that illustrate fundamental principles of core chemical engineering courses such as fluid dynamics, heat transfer, separations, thermodynamics, and process control.

This program has developed course segments of various lengths, which are directly related to a teacher's lesson plan and other pedagogical needs. One example of an in depth outreach activity, is a two week introduction to chemical engineering segment taught in a local high school's nascent Introduction to Engineering course. In this course students were presented with a wide range of chemical engineering concepts though the design and assembly of their own photobioreactors. During the periods in which students were waiting for their algae to grow, they created biodiesel from vegetable oil via trans-esterification, and they were able to compare the energy released by the regular oil and their diesel. These in-depth outreach experiences gave students a grounded example of process design, as well providing them an example of product engineering, and resulted in significant interest in our department.

Over this college-wide program's four years of existence we have gone from 389 high-school students visited in the 2007-2008 academic year to 3381 students in 2010-2011, and we are poised to significantly surpass that level by 2012. Through this program, we have also conducted a camp each summer, accommodating approximately 20 high-school students through a weeklong experience with a variety of engineering fields.

It is difficult to quantify the exact effect of such a program. However, we have seen a dramatic influx of students into our department; enrollment is up nearly 50% over the outreach program's existence. Furthermore, the attrition rate of undergraduate outreach students participating in the program is zero, compared to 40% for our typical undergraduate, indicating that involvement in such an outreach program may aid in student retention. An increase in high school student interns to our program has also been observed as a direct result of our outreach visits. Currently we have four high school interns working with our outreach teams and laboratories in our department, and the quality of students directly attracted through this program has been abnormally high; our first intern has recently received full scholarships to several universities. Overall, this community outreach program and its structure seem to be an excellent catalyst for attracting quality students to chemical engineering and creating a healthy and robust connection to our surrounding community.

(502h) Hands-On Chemical Engineering Demonstrations for Effective K-12 Outreach

Colin Young, Ahmed Elmadhoun and **Anthony Butterfield**, Chemical Engineering, University of Utah, Salt Lake City, UT

Figure 1: Screenshot of Our Outreach Teaching Module Database.

Hands-On Chemical Engineering Demonstrations for Effective K-12 Outreach

Hands on experiments and real-world demonstrations are among the most memorable experiences that science and engineering students will encounter in their education. An effective demonstration can help cement fundamental principles in science and engineering courses and aid students, including those with a variety of learning styles, in securing foundational insights into concepts they might not otherwise find from a traditional lecture. While such experiences are demonstrably important to student learning, in the field of chemical engineering there remains a need for additional resources and consolidation of teaching modules.

In order to address this need, our department has created an online, searchable database of over 30 hands-on teaching modules (Figure 1). Each module has a detailed list of materials required, a procedure, a description of the module's career connections, intended learning outcomes, and the theory governing the experiment. Additionally, assessment questions are provided to help instructors develop a lesson plan, and test the students' retention and understanding of the concepts presented.

Figure 1: Screenshot of Our Outreach Teaching Module Database.

Our modules highlight a wide variety of core chemical engineering concepts, and most are able to be addressed through the lenses of multiple chemical engineering courses. For example, we use an exploding can demonstration to illustrate concepts of flow through an orifice, buoyancy, flame speed, and combustion. A ping-pong ball positioning demonstration is used to illustrate both drag forces, and process control of an integrating system. Kinetics, thermodynamics and nanotechnology are demonstrated through our magnetic ferrofluid module. Furthermore, many of the developed modules have been designed to build upon one another. For example, biochemical engineering, data acquisition, and process and product design may be illustrated through the use of our algae photobioreactor module, in which a homemade spectrophotometer module monitors cellular growth, culminating in a biodiesel module to produce a useful product.

This resource of teaching modules and our method of collecting, presenting, and assessing them have been successfully used for over two years in both our extensive K-12 outreach program and in various undergraduate courses. Student and instructor feedback has indicated that this online collection is an effective and engaging tool in meeting the educational mission of our chemical engineering department as well as community high schools. It is our goal that this module database can be used by other departments and expanded upon, providing benefits to students outside our direct sphere of influence. By creating such an easily accessible database, we hope to add to the body of online chemical engineering teaching material and provide educators from across the country with teaching modules which may be used to supplement their lectures and conduct outreach within their communities.

(523a) Teaching Product with Process Design

Warren D. Seider, Chemical and Biomolecular Engineering, The University of Pennsylvania, Philadelphia, PA

This presentation addresses the challenges in teaching product design with process design. For the simplest chemical products, i.e., basic chemicals, which are often manufactured as commodities, the design focus is usually on manufacturing efficiency. Other chemicals involve special properties to meet customer needs. For these chemical products, the focus is on matching the latest technologies with the *voice of the customer*; i.e., on product design.

As the field of chemical engineering diversifies, college graduates increasingly become engaged in product design. Classical process design, which meshes nicely with most core chemical engineering courses, normally provides students with few, if any, product design experiences and perspectives. For product design, it's often important to find chemicals and chemical mixtures having desired properties and performance. Often small quantities of specialty chemicals are involved; e.g., pharmaceuticals, requiring the design of multipurpose batch chemical processes. Often to satisfy customer needs, product design focuses on the 3-dimensional product structure; e.g., LCDs, labs-on-a-chip, colloidal structure, ..., and their more complex process designs.

In this presentation, approaches are suggested for introducing product design topics while teaching a classical process design course. Subjects added include the product- and technology-development framework, the innovation map (linking new technologies with customer needs), the Stage-Gate™ Product-Development process (Cooper, 2002, 2004), and molecular-structure design; i.e., the identification of molecules that satisfy property specifications for new chemical products. These and related subjects are interspersed while our students solve process design homework exercises. Included are an example of configured-consumer product design; e.g., a lab-on-a-chip for high-throughput screening of kinase inhibitors, and an industrial product; e.g., a thin-glass substrate for liquid-crystal displays (LCDs).

While some groups of 3-4 students solve classical process design projects, others solve product design projects; e.g., to produce encapsulated drugs with controlled release profiles, and to model physical vapor deposition of thin films on integrated circuit substrates (using kinetic Monte-Carlo methods). While our entire senior class is not trained to be expert in each technology platform, each student group learns enough basics from its faculty advisor to work effectively on its product design.

(523b) The Teaching of Capstone Design

David L. Silverstein, University of Kentucky, Paducah, KY, Lisa G. Bullard, Chemical and Biomolecular Engineering, North Carolina State University, Raleigh, NC, Warren D. Seider, Chemical and Biomolecular Engineering, The University of Pennsylvania, Philadelphia, PA and Margot A.-S. Vigeant, Chemical Engineering, Bucknell University, Lewisburg, PA

The authors will present the results of the fourth survey in the resumption of the Chemical Engineering Education Special Projects Committee series of annual surveys on how chemical engineering courses are taught. The surveys are now conducted under the auspices of the Education Division of the Institute. This year's survey covers capstone design, and seeks to determine what concepts are being taught, who is teaching the course(s), and what tools are being used. The role of industry in preparing instructors and in developing projects and case studies is also studied. Teaching methods and novel approaches are, as always, focus elements of this year's survey. Some comparison with previous studies on the subject will be included.

Following the presentation of the survey results, a panel consisting of selected participants in the survey will lead a discussion with the audience covering common issues in teaching capstone design. Ideas and approaches discussed during the session will be incorporated into the final survey report.

(543a) Technology Forcing with Clean Air Act Best Available Control Technology

Mary Ellen Ternes, McAfee& Taft, Oklahoma City, OK

The Clean Air Act's Prevention of Significant Deterioration program (CAA PSD) was enacted to ensure that ambient air quality continues to meet ambient air quality standards through applicable of the Best Available Control Technology (BACT) to new major stationary sources of air pollutant emissions, or major modifications of a major stationary source. With the CAA authority being extended to greenhouse gas emissions as a result of the 2007 case Massachusetts v. EPA, and following EPA CAA greenhouse gas rulemakings, BACT as applied through PSD to major stationary sources of greenhouse gas emissions will result in driving technology and research toward methods to achieve greenhouse gas emission reductions. This presentation will discuss the existing PSD BACT program and how it has been applied recently to major stationary sources of greenhouse gas emissions.

Avoiding Backdoor Licensing and Other Tech Agreement Traps

By: Anthony P. Venturino, Esq., Partner of Novak, Druce + Quigg, LLP Washington, DC

AIChE Fall Meeting - October 31, 2012 Pittsburgh, TX

The views presented here are those of the author and not attributable to his firm or its clients.

I. Engineering is Full of Technology Agreements

Engineers may encounter agreements having technology or other forms of intellectual property in the course of business, for example organizing a startup company. Some typical agreements are listed as follows:

- Confidentiality Agreements
- Vendor Supplier Agreements
- Joint Development/Commercialization Agreements
- Exclusive Licenses / Non-exclusive Licenses
- Co-promotion Agreements
- Employee Agreements
- Settlement Agreements
- Joint Venture Agreements
- Contract R&D Agreements
- Consulting Agreements

II. Things Go Wrong

A few examples of where things can go wrong:

- Evaluation of the technology
- Selection of partners
- Conduct of the negotiations
- The agreement itself
- Implementing and servicing the agreement

Classic Licensee mistakes:

- Payments made under a trademark license when the trademark is not used
- patent royalties are still being paid even though the patents covered in the license expired or no longer cover the products
- Royalties paid at the wrong rate, i.e. one higher or lower than provided

Source: *Licensing – A Strategy for Profits*, Edward P. White (1990)

Classic Licensor mistakes:

- Failure to take action if royalty payments cease when they should not, e.g., prior to expiration of the patents.
- Failure to monitor royalty payments, e.g. sudden reductions or failure to meet minimum royalties that could trigger a change from exclusive to nonexclusive or termination of the licenses.
- Failure to stay in technical contact with the licensee, with the result of licensee not interested in paying royalties, in maintaining quality, or keeping good records.

Licensing – A Strategy for Profits, Edward P. White (1990)

III. Before you do the deal.

- Know your business goals, strengths and weaknesses
- Know the available protection for the technology
- Have a realistic idea of the value of the technology
- Be open to creative technology agreements
- Admit when you are in over your head and get help
- Do not let paranoia get in the way of a good deal
- Watch out for foreseeable legal and technical problems
- Admit when you are in over your head and get a lawyer or other licensing professional

IV. Avoiding Common Traps

A. Controlling Backdoor Licensing by University Researchers

Backdoor licensing is tech transfer traveling outside the normal or accepted route. It can be communications between a consultant and its client during the consulting. It can be an improper disclosure of information depending on contractual obligations on the discloser when the disclosure was made. For example, an employee might transfer his employer's technology to a customer in violation of his employment contract.

This can be a problem for universities. University technology transfer offices (TTOs) collect and organize invention disclosures as well as help to protect and commercialize the inventions. However, not all universities have a tech transfer office. Also, even if there is a tech transfer office, sometimes a university researcher bypasses the tech transfer office.

Taylor Aldridge and David B. Audretsch developed a database of the largest 20% of National Cancer Institute grant awardees between 1998 and 2002. Of this group 392 scientists were identified that earned patents and the authors examined how the research was commercialized. Taylor Aldridge and David B. Audretsch, "Does Policy Bias the Commercialization Route? Evidence from National Institutes of Health Funded Scientists," *Research Policy*, June 2010. The study reported 30 percent of the researchers did not assign at least one of their patentable inventions to their university employers.

Some points raised by the study are as follows:

1. "Of the 392 patenting scientists, 71.20% selected the TTO commercialization route, in that they assigned all of their patents to the university. However, 29.80% of the patenting scientists resorted to the backdoor for at least one of their patents."

2. " Scientists choosing the backdoor route for commercialization, by not assigning patents to their university to commercialize research, tend to rely on the commercialization mode of starting a new firm. By contrast, scientists who select the TTO route by assigning their patents to the university tend to rely on the commercialization mode of licensing. "

3. "The likelihood of a scientist commercializing through the backdoor may be influenced by characteristics of the TTO, the university, the region, the scientist, and the technological nature of the invention itself."

4. "[T]here is at least some evidence suggesting that the commercialization path is not neutral with respect to the significance of the potential innovation. Rather, those innovations with a potentially greater value are less likely to be commercialized through the technology transfer office."

Another type of backdoor licensing is informal technology transfer between university scientists and industry personnel. For example, a company hires a university researcher as a consultant and inventions are made on company time with company resources.

How can a University respond?

One approach is to give researchers incentives to go through the TTO. For example, universities could make patenting one criterion for promotion and tenure. Christoph Grimpe and Heide Fier, Informal University Technology Transfer: A Comparison Between the United States and Germany, Discussion Paper No. 09-033, ZEW Zentrum für Europäische Wirtschaftsforschung GmbH, Centre for Economic Research (2009); URL<ftp://ftp.zew.de/pub/zew-docs/dp/dp09033.pdf>.

Another approach is to have a patent policy and a culture encouraging disclosure. Taylor Aldridge and David B. Audretsch, indicated "There may also be systematic differences between public and private universities. Contributing to state economic development, growth, employment creation and technology transfer is often an explicit mandate at publicly funded state universities. This is rarely the case at private universities. Thus, public universities may be more sensitive to controlling, monitoring and capturing the flow of intellectual property going from university scientists to commercialization."

B. <u>Define how to handle joint invention of improvements</u>

By US statute joint inventions are jointly owned. It is better to have one party own the invention and, if necessary, grant the other party an exclusive field of use license. Thus, a clause such as the following should be included if a jointly developed invention may occur during performance of the agreement:

> In the event of a joint development my company owns the invention and controls patent prosecution and your company gets an exclusive use license in the field of _______________.

C. Keep good records of what the agreement covers or when it expires

The company, its legal department or a law firm should docket the agreement in a database. The docketing record should include: the type of agreement, the parties, the technology covered by the agreement, the deadlines, the expiration date, the contact person and backup contact person, and have a miscellaneous portion for anything unusual.

Preferably an institutional person, for example, a docket clerk or contracts administrator, should be responsible for tracking notices in the docketing system and then relay the information to the business people. The more automatic you can make the timely conveying of such information, the better.

D. Do not let an agreement to agree accidentally become an agreement

The first time you negotiate with a party, especially if the other party sees this agreement as a one shot deal, be careful not to make records during the negotiations which unexpectedly bind you. Do not accidentally create an agreement.

Summaries of the negotiations signed by both parties can lead to binding agreements even when a clause that might be important to you is missing. Let's take a hypothetical scenario where two parties bring their business people and lawyers to a meeting to negotiate a potential IP license. Mid-afternoon the attorney for the IP owner (potential licensor) had to leave to catch a plane. The potential licensor's business person (let's call him "Bill") had a later plane. The potential licensee kindly offered to take Bill to the airport later so he would not have to waste time in a boring airport. While Bill was by himself with the potential licensee's attorney and business people one of the business people suggested they draft and sign a summary of the day's discussions. They did this and then Bill caught his plane. The next day Bill's company was asked to sign an Assignment document to reflect the assignment they agreed to yesterday. Bill's response was "What assignment?" After a lawsuit the parties settled. The moral of the story is train the business people not to sign anything.

A deal can also inadvertently arise if the parties act like there is an agreement can create an agreement. This sometimes arises where the parties are in a rush to do joint development. While the deal is only a draft they start working together. Sometimes the unsigned draft becomes the *de facto* agreement.

The opposite can also happen. Sometimes you think you have an agreement but in reality you do not, or at least not all the clauses apply. Take the case of *DataTreasury v. Wells Fargo,* 522 F.3d 1368 (Fed. Cir. 2008). In 2003 WMR e-Pin licensed a patent to Wells

Fargo Services Corp. a subsidiary of Wells Fargo & Co. The license included an arbitration clause. In 2006 WMR e-Pin assigned 4 patents to DataTreasury. At least one of the 4 patents was a patent of the license. Afterwards DataTreasury sued Wells Fargo & Co. for patent infringement in court. Wells Fargo & Co. moved to dismiss or stay pending arbitration. Under 5th Circuit law, as applied by the Federal Circuit, "to be enforceable, an arbitration clause must be in writing and signed by the party invoking it. *DataTreasury* at 1373, quoting *Westmorland v. Sadoux*, 299 F.3d 462, 465 (5th Cir. 2002). Neither Wells Fargo & Co. nor DataTreasury signed the original patent license. Thus, the court held these parties were not bound by the arbitration clause.

E. Remember important clauses

Clause checklists are important tools when reviewing a license to see if all the necessary clauses are present. However, use the right checklist for the type of agreement and the type of IP. Do not assume the clauses in your standard patent license belong in your new trademark license and *vice versa*. Unlike patents which protect the patent holder, trademarks protect the public by assuring the public a good labeled with a particular trademark will have the characteristics they expect. For example, a PIRELLI brand tire has certain handling properties. Trademark licenses without quality safeguards are termed naked licenses and invalidate the trademark rights.

A short checklist is provided in TABLE 1:

TABLE 1
Introductory Clauses
Definitions
Grant of Rights -patents-trade secrets
Consideration/Compensation
Diligence
Product Quality Control
Advertising
3rd Party Infringement Suits
3rd Party Product Liability Suit
Licensor/ Licensee Suits
Termination
Continuing Obligations

A detailed set of clause checklists for a licensing agreement is available upon request from the author.

F. Remember to protect the Trade Secrets and Know-how

- Define trade secrets
- Set up a manageable way to transfer the information

G. Keep control of the technology after making the deal

Proprietary technology is a competitive advantage. If the licensor loses control of the technology he loses control of the competitive advantage. Thus, the licensor should determine the level of control he is willing to lose. Maybe the licensor thinks he can permit the licensee to sublicense because he trusts the judgment of licensee's president. What if the president leaves? Thus, the agreement preferably precludes sublicensing or assigning to competitors.

Another common way to lose control is for the licensee to be bought by the licensor's competitor. The Agreement should specify it may be assigned by ONE of the following:

1. Anyone;
2. Goes with the business;
3. Goes with the company;
4. Licensor maintains right of first refusal;
5. Licensor maintains right of first refusal which will not be unreasonably withheld;
6. No one.

H. Promote good communication after making the deal

Typically problems arise from misunderstandings, lack of thoroughness, carelessness, or failure to administer the agreement at all, i.e., no one in authority is responsible for administration (LES 1983-84 study).

The lawyer can structure the agreement to include mechanisms for communication between the licensor and licensee so little problems and misunderstandings do not become big ones. You cannot force the business people to communicate but the mechanisms provide fertile soil for communication. For example, the agreement should include a contact person and back-up contact person. The agreement can also mandate times for periodic meetings.

I. Include Clean Termination Provisions

Relationships can end badly. Thus, you need the equivalent of a pre-nuptial agreement. Include termination provisions in the agreement to give each party a reasonable way to terminate the agreement. This includes planning for unexpected events, for example, bankruptcy or purchase of the other party. This also includes providing ways to terminate or renegotiate if the economics of the deal no longer make the current agreement viable.

V. Organize A Licensing Department – Overview

Some companies only occasionally do a license or may be at the beginning of setting up a license project group. In these cases, the company might ask his or her lawyer for advice on how to organize the licensing project group.

The license project group will need the support of a top executive. Without this the license project group will generally not get the resources it needs.

Recognize the different functions to be performed. Then, divide or share these functions depending on available resources, business people, consultants or lawyers:

- Profit Center Group
    - optimizes the licensing in and licensing out to make money for the company
- Support Group
    - coordinates technical, legal and accounting aspects of tech transfer
- Administrative Group
    - Handles accounting, record keeping and day to day details of maintaining the license.

VI. Conclusion

University researchers and their Technology Transfer offices can work together regardless of whether the university researcher wants to license the invention or commercialize it herself. However, Universities and their Technology Transfer Offices should strive to have policies, procedures and culture to be a bridge to encourage partnerships with researchers and the outside industrial community.

Agreements are necessary to manage intellectual property. This applies regardless of whether the agreements involve the university researcher, the industrial entrepreneur or the Fortune 500 Company If you know your goals and avoid the traps your agreements can help turn technology into profit.

(543c) Quantifying the Microeconomic Impacts of State-Level Incentive Programs On Biorefineries

Tristan Brown[1], Rajeeva Thilakaratne[1], Guiping Hu[2] and Robert C. Brown[3], (1)Bioeconomy Institute, Iowa State University, Ames, IA, (2)Iowa State University, Ames, IA, (3)Mechanical Engineering, Iowa State University, Ames, IA

A growing number of U.S. states are implementing laws and policies that incentive the construction of biorefineries within their boundaries. These incentives range from wide-ranging Renewable Portfolio Standards (RPSs) to narrowly-drawn interest-free loans, such as the $75 million loan recently granted to catalytic pyrolysis company KiOR by the state of Georgia for the construction of a biorefinery in the state. Corporate income tax rates are another factor affecting the economic feasibility of biorefineries that vary widely across states.

We compile and assess the major state-level incentive programs currently being employed by different states to incentivize biorefinery construction. A techno-economic Chemcad model of a biorefinery employing the fast pyrolysis and hydroprocessing pathway for the production of biofuels and bioproducts is built alongside a discounted cash flow rate of return (DCFROR) spreadsheet. The effects of these incentive programs are simulated within the model and their microeconomic impacts are quantified and compared. The result ranks the major incentive programs according to both their direct costs and their net benefit (or harm) to the biorefinery.

(560a) Reorganizing the Junior and Senior Years At UConn

Daniel D. Burkey[1], Aravind Suresh[2] and Daniel Anastasio[1], (1)Chemical, Materials, and Biomolecular Engineering, University of Connecticut, Storrs, CT, (2)Chemical, Materials and Biomolecular Engineering Department, University of Connecticut, Storrs, CT

In this presentation we discuss the reorganization of the junior and senior year curriculum at the University of Connecticut to meet a number of educational goals. First off, we discuss the reorganization of the laboratory classes from a monolithic senior year course to two junior-level labs tied closely to core courses, and a senior laboratory course focused on multiple larger experiments that tie in with a theme in chemical engineering (for example, energy, bio, water, etc.). Next, we discuss the reorganization of the transport sequence from a two-semester sequence to a three semester sequence, with the first semester in the senior year being a Unit Operations and Introduction to Design course. This allows us to have a full year for senior design, allowing expanded projects, more discussion of process safety in design, and a full semester of instruction in process simulation using modern chemical engineerng tools. Lastly, we discuss the addition of a chemical engineering analysis course to the core curriculum, bolstering students' mathematical preparation and giving them the chemical engineering context for many of the techniques that they have started to learn in their other math courses. All in all, through modification of existing courses, introduction of new ones, and elimination of extraneous ones, we have been able to shift the chemical engineering towards a more experiential, design focused track without increasing the total number of credit hours.

(560b) Bringing Industry to Students: Btec's Design of Courses in Downstream Bioprocessing

Baley Reeves and Gary Gilleskie, Biomanufacturing Training and Education Center, North Carolina State University, Raleigh, NC

Among chemical engineers graduating from NCSU with a bachelor's degree, 80% begin their careers in industry. This fact compels educators to design courses that are industry relevant to ensure students get off to a successful start as they begin their careers.

The Biomanufacturing Training and Education Center (BTEC) at NC State University has developed a series of industry-relevant, hands-on courses in downstream bioprocessing. These courses are designed to take students through the steps required to put a biopharmaceutical manufacturing process in place - from process development through process transfer to execution of the manufacturing process. Along the way, students are required to demonstrate knowledge of the principles underlying relevant unit operations and knowledge of basic engineering principles (such as mass balances, fluid dynamics, and mass transfer) in the context of these unit operations, just as they would in a traditional unit operations course. The unit operations covered include centrifugation, homogenization, chromatography, and filtration.

Throughout the semester, students are tasked with designing, scaling up, and executing at manufacturing scale a process to recover and purify recombinant green fluorescent protein (GFP) produced in *E. coli*. Students determine "ranges" for key process parameters for each unit operation in BTEC's intermediate-scale development labs. Students scale up operations to process in BTEC's 300-L (bioreactor scale) "large-scale" labs. Lastly, students execute the process they have designed under simulated cGMP conditions to mimic a biopharmaceutical production environment. The laboratories house equipment that is commonly used in industry. Throughout the courses, concepts are presented in light of regulatory requirements that apply to biopharmaceutical manufacture. These include ICH Q8(R2), Guidance for Industry: Pharmaceutical Development and 21 CFR Parts 210 & 211, Current Good Manufacturing Practice for the Manufacture, Processing, Packing, or Holding of Drugs.

This presentation will provide details on the content of these courses, lessons learned, and results from follow-up surveys with students showing how well knowledge was transferred to the workplace.

(560c) Developing an Operator Training Simulator As a Class Project in a Simulation Course

Richard Turton, Department of Chemical Engg., West Virginia University, Morgantown, WV

An operator training simulator (OTS) is to the chemical engineer what a flight simulator is to the aerospace engineer. The basis of an OTS is a dynamic model of a chemical process that allows an engineer to simulate the start-up, shut-down, and normal operation of a chemical process. It can also be used to test the skill of an engineer or operator to respond and control unforeseen situations through the use of programmed malfunctions.

The purpose of this presentation is to discuss the author's experience in teaching a course in chemical process simulation for the first time. A brief summary of the course content, which includes basic information and examples of steady state and dynamic simulations, will be given. The guided development of a steady state and dynamic simulation for a depropanizer column, used during the first 8-10 weeks of the course, was a key element in the pedagogy used in teaching the course and a discussion of student progress made through structured problem sets will be discussed.

In order to gain experience with a state-of-the-art OTS, students spent approximately 6 hours of training on the Integrated Gasification Combined Cycle (IGCC) power plant OTS that is the cornerstone of the AVESTAR (**A**dvanced **V**irtual **E**nergy **S**imulation **T**raining **A**nd **R**esearch) Center with sister facilities at DOE/NETL, Morgantown and WVU. During this training, students followed a set of highly detailed start-up procedures that were required to start-up the Claus sulfur capture process within the IGCC plant.

The final projects in the course required student groups (one group of undergraduates and one group of graduates) to develop dynamic models of two chemical processes (a dimethyl ether plant for the group of undergraduates and a toluene hydrodealkylation plant for the group of graduate students) and to write a detailed procedure for the start-up of each process. The scope and success of this final project and a summary of students' impressions of the course and suggested improvements will be given.

(560d) Development, Implementation, Observations and Abandonment of a Comprehensive Concept Inventory in Chemical Engineering

Timothy Raymond, Margot Vigeant and Michael Prince, Chemical Engineering, Bucknell University, Lewisburg, PA

In an attempt to better assess our entire chemical engineering curriculum for both self-analysis and as a form of direct assessment for ABET, the Chemical Engineering Department at Bucknell University has development and implemented a Comprehensive Concept Inventory (CCI). Over the past 5 years, we have worked with every member of our department to create and implement the CCI. Faculty members identified a list of key concepts from each course and worked in groups of two or three to develop multiple choice concept questions. As a department, we narrowed the concepts and questions down to a total of 30 that covered the key curriculum and represented concepts we believe every chemical engineer should understand by graduation. The authors hosted a workshop on CCI development at ASEE in Pittsburgh in 2008. The CCI was implemented first to all faculty in the department and subsequently to the last four senior classes within one month of graduation. After careful analysis of the efforts and results of this implementation, we have decided to abandon the use of the CCI instrument at this time.

This paper will present the CCI concepts, questions and results from the past four years. We will explain what we believe the CCI results tell us, what they do not, and why we are abandoning the use of the CCI at this time. We hope these results will initiate a lively discussion of the potential future uses of instruments such as the CCI for both ABET and self-assessment purposes.

(560e) Enhancing the Design Experience in the Junior and Senior Year At Villanova

Dorothy W. Skaf and Jeremy Kuhn, Chemical Engineering, Villanova University, Villanova, PA

In this presentation, efforts to improve the design component in the junior and senior year at Villanova University will be discussed. The current curriculum includes a single-semester design class, and prior to this project, students were not exposed to process simulation tools until the design course. We will discuss how experiences with process simulation software and equipment design were incorporated into junior year courses, including thermodynamics, heat transfer, reaction engineering and mass transfer. Tools such as self-paced assignments and student guides were used, to avoid detracting from the existing technical content of the classes. Problems were selected to complement course content or to emphasize the practical importance of course content. The impact of these experiences on student preparedness in the fall semester design course and their ability to address more complex design issues will be discussed. Student impressions regarding their experiences and improvements to incorporate in the next class cycle will also be presented.

(560f) Project-Based Learning in Chemical Engineering Core Courses

Kevin Dahm[1], Stephanie Farrell[1] and Concetta La Marca[2], (1)Chemical Engineering, Rowan University, Glassboro, NJ, (2)DuPont Engineering Research & Technology, Wilmington, DE

Project-based learning is a learner-centered, constructivist approach to teaching which offers several advantages over traditional classroom methods. Problem-based learning (PBL)has been shown to result in better content knowledge, problem solving ability, metacognitive skills and attitude toward learning [[1]]. In addition, PBL results in enhanced communication and teamwork skills, understanding of professional practice, and ability to apply learning to solve problems [[2]]. PBL is currently the most popular pedagogy for teaching design and has also been widely used for first-year cornerstone courses [[3]]. Descriptions of PBL in Chemical Engineering core courses are relatively scarce in the literature. The presentation will include an overview of published resources (primarily in the journals Chemical Engineering Education and Education for Chemical Engineers) available to faculty members who are interested in implementing PBL. Practically speaking, making the transition to PBL from more traditional classroom approaches is daunting and unlikely to be accomplished in a single step. The authors will present their personal experience teaching a Chemical Reaction Engineering course. A large design project was integrated into the course, and its specific role in the course evolved over several years, moving towards the PBL approach in a stepwise manner. This stepwise implementation can be mirrored with other projects in other courses.

---

[1]. Thomas, J.W., "A review of research on Project-based Learning," San Rafael, CA: Autodesk Foundation, 2000.

[2]. Mills, J.E., and Treagust, D.F., "Engineering Education—Is Problem-Based or Project-Based Learning the Answer?" Australasian Journal of Engineering Education, http://www.aaee.com.au/journal/2003/mills_treagust03.pdf.

[3]. Dym, C., Agogino, A., Eris, O., Frey, D., and Leifer, L., "Engineering Design Thinking, Teaching, and Learning," Journal of Engineering Education, Vol. 94, No. 1, 2005, pp. 103–120.

(560g) The AIChE Concept Warehouse Project: Promotion of Active, Concept-Based Learning Pedagogies

Debra Gilbuena[1], Bill J. Brooks[1], David L. Silverstein[2], John L. Falconer[3], Ronald L. Miller[4], Marina Miletic[5] and **Milo D. Koretsky**[1], (1)Chemical, Biological, and Environmental Engineering, Oregon State University, Corvallis, OR, (2)University of Kentucky, Paducah, KY, (3)Chemical & Biological Engineering, University of Colorado at Boulder, Boulder, CO, (4)Colorado School of Mines, Golden, CO, (5)Department of Chemical and Biomolecular Engineering, University of Illinois Urbana-Champaign, Urbana, IL

The AIChE Concept Warehouse was initiated with the goal of creating a community of learning within the discipline of chemical engineering (ChE) focused on concept-based instruction. We have been developing and promoting the use of a cyber-enabled infrastructure for conceptual questions, *the AIChE Concept Warehouse*, which can used throughout the core ChE curriculum (Material and Energy Balances, Thermodynamics, Transport Phenomena, Kinetics and Reactor Design, and Materials Science). Conceptual questions, both as Concept Inventories and ConcepTests, are available through an interactive website maintained through the Education Division of the American Institute of Chemical Engineers (AIChE), the discipline's major professional society. The overall objective is to lower the activation barrier for using conceptual instruction and assessment so that many more chemical engineering faculty incorporate concept-based learning into their classes.

The innovative software application allows interactive electronic use, as well as PowerPoint, Word, and pdf formats to be automatically generated so that conceptual learning and evaluation can be incorporated into instruction in various forms: in-class ConcepTests with student response (clickers, laptops, cell phones), concept inventories to evaluate student learning (or student preparation for a course), exam and homework problems. A key element to this approach is to provide a context for conceptual learning that reflects the richly interconnected and interrelated knowledge structures of experts. The pedagogy includes cross development and testing between items on a Concept Inventory and in ConcepTests.

In this paper we report on the initial development, promotion, limited deployment, and future plans for the AIChE Concept Warehouse. Activities intended to build community, including conference special sessions and workshops will be described. Data collected includes website and database statistics, pre- and post-use semi-structured interviews with beta testers as well as development documentation. Semi-structured interviews were analyzed in order to identify common likes and dislikes with the conceptual resources and user interface for the AIChE Concept Warehouse. Interviews were also analyzed to identify common issues with the implementation process both from a software perspective as well as a curricular adoption perspective. Results from interviews are being used to iterate on the tool development and improve the overall design.

(560h) Role of LES On Fostering Innovation in the "Composer-Style Engineer"

Pedro E. Arce, Chemical Engineering, Tennessee Technological University, Cookeville, TN and **Lacy Loggins**, C&I-ChE, TTU, Cookeville, TN

This preesentation will describe the collaboration betweeen Chemical Engineering and the Dpt of Curriculum Instruction at the Tennessee Technological Unversity, Cookeville, TN on the implementation and role of the Linear Engineering Sequence (LES) on the develoment of the "Composer-Style Engineering" as one of the most effetive appoeaches for the NAE 2020 MNodel. The discussion will share the design, implemnetation and assesment of LES for the Junior Level Fluid Mechanic of the Chemical Engineering Curriculum. Students have been introduced to various approaches for fostering innovation in engineering education and "coached" to identify "needs" that they can connect with the fundamental aspects of the course in order to produce an innovation. This needs to address several items including planning, design, testing and bussiness plans for possible commerciallization of the innovation. Students are not given any need or help to select a possible solution for such a need. The students are also exposed to various model of cirtical thinking such as the Bloom Taxonom and require to make a connection with them durign the process of innovation. A pannel of judges assess the final innovation in a poster format, learning session of three hours. The panel is composed of the same number of judges as teams so that during the 'learning-assesment" session a constant flow of assessors are with the teams.

AUTHOR INDEX

AUTHOR INDEX

AUTHOR INDEX

AUTHOR INDEX